Plant Science
An Introduction to Botany

Catherine Kleier, Ph.D.

PUBLISHED BY:

THE GREAT COURSES
Corporate Headquarters
4840 Westfields Boulevard, Suite 500
Chantilly, Virginia 20151-2299
Phone: 1-800-832-2412
Fax: 703-378-3819
www.thegreatcourses.com

Copyright © The Teaching Company, 2017

Printed in the United States of America

This book is in copyright. All rights reserved.

Without limiting the rights under copyright reserved above,
no part of this publication may be reproduced, stored in
or introduced into a retrieval system, or transmitted,
in any form, or by any means
(electronic, mechanical, photocopying, recording, or otherwise),
without the prior written permission of
The Teaching Company.

CATHERINE KLEIER, PH.D.
PROFESSOR OF BIOLOGY
REGIS UNIVERSITY

Catherine Kleier is a Professor of Biology and former chair of the Department of Biology at Regis University in Denver, Colorado. She has also served as the associate director of the Regis College Honors Program. She teaches courses in plant physiology, ecology, and environmental science.

Professor Kleier holds a Ph.D. in Organismic Biology, Ecology, and Evolution from the University of California, Los Angeles (UCLA). While at UCLA, she received a Mildred E. Mathias Graduate Student Research Grant, a Dissertation Year Fellowship, and the Lasiewski Award for outstanding graduate research.

Professor Kleier also holds a B.A. in Ecology, Population, and Organismic Biology from the University of Colorado Boulder and an M.S. in General Science with an emphasis in Botany and Plant Pathology from Oregon State University.

Professor Kleier's current research interests include long-term restoration ecology on trails in the Colorado Rocky Mountains, plant impacts from skiing, and urban ecology. Her laboratory work has focused on the effect of magnesium chloride on the growth and reproduction of plants.

Professor Kleier was awarded a National Geographic Society/Waitt Grant in 2011 to travel to northern Chile to explore populations of a rare, giant alpine cushion plant, *Azorella compacta*. She has presented many conference papers and posters at numerous professional societies, including the Botanical Society of America, the Ecological Society of America, the Society for Conservation Biology, and the Association for Environmental Studies and Sciences.

In 2010, Professor Kleier was selected as a scholar for the Aspen Institute's Aspen Environment Forum. In 2013, she was a Visiting Fulbright Scholar in the Department of Botany at the University of Otago in New Zealand, where she investigated facilitation in the alpine cushion plant genus *Raoulia*.

In 2014, Professor Kleier was elected Faculty Lecturer of the Year at Regis University, and in 2015, she was named the Colorado Professor of the Year, sponsored by the Council for Advancement and Support of Education and the Carnegie Foundation for the Advancement of Teaching. She is on the board of the Colorado Native Plant Society, and she often leads field trips to the beautiful Colorado Rocky Mountain alpine zone. She is also the creator and director of the Certificate in Applied Craft Brewing at Regis University. ∎

TABLE OF CONTENTS

INTRODUCTION

Professor Biography i
Course Scope 1

LECTURE GUIDES

1   The Joy of Botany 4

2   Plants Are like People 23

3   Moss Sex and Peat's Engineered Habitat 42

4   Fern Spores and the Vascular Conquest of Land 63

5   Roots and Symbiosis with Non-Plants 82

6   Stems Are More Than Just the In-Between 102

7   The Leaf as a Biochemical Factory 121

8   Photosynthesis Everyone Should Understand 141

9   Days and Years in the Lives of Plants 161

10   Advent of Seeds: Cycads and Ginkgoes 179

11   Why Conifers Are Holiday Plants 198

12   Secrets of Flower Power 218

13	The Coevolution of Who Pollinates Whom	242
14	The Many Forms of Fruit: Tomatoes to Peanuts	262
15	Plant Seeds Get Around	281
16	Water Plants Came from Land	299
17	Why the Tropics Have So Many Plant Species	318
18	The Complexity of Grasses and Grasslands	337
19	Shrublands of Roses and Wine	357
20	The Desert Bonanza of Plant Shapes	375
21	How Temperate Trees Change Color and Grow	394
22	Alpine Cold Makes Plants Do Funny Things	414
23	Bad Plants Aren't So Bad	433
24	Modifying the Genes of Plants	453

SUPPLEMENTARY MATERIAL

Bibliography 473
Image Credits 478

PLANT SCIENCE
AN INTRODUCTION TO BOTANY

Although the world is full of plants and we see them every day, we are often blind to their presence. More than that, our knowledge of these plants may not extend much beyond the idea that they use sunlight and carbon dioxide to produce oxygen and sugars, the very things that sustain our life on Earth. This course will help you learn to see plants, begin a relationship with them, learn their stories, and appreciate their amazing diversity.

The course will take a chronological approach in terms of when certain plants and plant traits appeared on Earth. You will begin by considering the differences between plants and animals. Plants are certainly different, but there are some similarities, too, in the way that plants and animals sense the world. Because the first true plants to colonize the land were mosses, that's the first plant group you will consider in depth. What makes a moss a moss, and what's important about mosses? Next, you will consider the ferns, and you will determine why ferns were an important step in the evolution of seed plants, which will come next.

Because most plants have a basic anatomy of roots, stems, and leaves, you will consider each part in detail. These structures are intimately linked to function. Many roots have a symbiotic relationship with fungi that enable absorption from more of the soil. Stems come in many different sizes and shapes—as do leaves, from small needles to leaves that are more than a square meter. You will discover why this is the case.

Because all plants undergo the life-support process of photosynthesis, you will next take a detailed look at the biochemistry of photosynthesis, beginning with carbon dioxide entering the leaf. This process is a cycle that depends on 2 reactions, one of which captures light energy and the other of which uses that energy to turn carbon dioxide into sugars.

Because they are light-sensing organisms, plants have a well-developed circadian rhythm that determines when they flower and fruit, and you will examine these mechanisms.

The seed was a remarkable adaptation that allowed plants to colonize more types of environments. The conifers and related plants were the first to have seeds, and you will learn that even though these plants lack flowers, they are uniquely adapted to the environments in which they live. When and how the first flower appeared is still a bit of a mystery, but you will consider the latest evidence about the evolution of flowering plants. You will learn about some important adaptations of flowers and discover the intricate relationship between flowers and their pollinators.

The exchange of pollen is necessary for the development of the fruit, and there are many types of fruits, each with a unique dispersal mechanism to spread the seeds of the next generation. Fruits have interesting botanical names, and you will learn a few of these. For example, a strawberry is not a true berry, but a tomato is, and a peanut is not a nut, but a hazelnut is. Botanists categorize flowers and fruits to put plants in taxonomic groups, and you will explore some of the major plant groups.

At this point, the course will explore plants in various habitats and examine the science behind adaptations to different environments. This tour will begin with underwater plants. You will discover that there are not just seaweeds that live in the sea, and water lilies and other plants have unique tissues that can keep them afloat.

Then, you will uncover some of the amazing diversity of the tropical rain forests. You will also examine evidence that suggests why these areas are the most diverse habitats on land. You will explore unique life-forms and discuss why these types of plants were so successful in the tropics.

The next habitat is unique because the grasslands are not only a community type but a dominant group of plants. They have a different structure than non-grasses and a different way of growing that enables them to thrive under pressure from grazing animals. Grasses are also the most economically dominant group of plant, because all of our major cereal crops are grasses, such as corn, rice, and wheat.

An often-overlooked community type is that of the shrubland, but shrubs are some of the oldest and hardiest woody plants on the planet. Shrubs also include important ornamentals, such as roses.

Shrublands generally give way to true deserts. Desert plants undergo a different type of photosynthesis that allows them to conserve water. Because water is the most limiting factor for most plants, deserts have a myriad of unique plant forms to cope with this stress.

Deciduous trees may look alike, but trees are different in their reproductive capacities and susceptibility to pathogens. You will encounter a number of interesting trees in the deciduous forest.

Next, you will discover the wonderful world of alpine plants. You will learn how these plants survive in such extreme environments.

Although plants make life on Earth for us possible, there are some plants that strike fear in our hearts, such as stinging nettle and poison ivy. Some plants also get introduced to a new area and become invasive, which is also unpleasant. You will consider the why and how of these plants, as well as the strange carnivorous plants.

Finally, you will explore genetic modification of plants and the future of this new technology and botany.

LECTURE 1

THE JOY OF BOTANY

Botany is the study of plants—everything from biochemical reactions to whole forests. At first glance, plants might seem boring; they may not ignite our compassion like animals do. But plants have a quieter way of igniting our passion, because we're dependent on them. The goal of this course is for you to understand the fascinating differences, and similarities, between plants and animals; to get better acquainted with the various parts of the plant and what they do; and to open your eyes to the science of botany that is all around you.

PLANT DIVERSITY

- There is incredible diversity within the more than 300,000 species of plants.

- The dodder plant is a parasitic plant, which means it gets all of its nutrition from other plants. Dodder is a common name, and it refers to an entire genus of plants called *Cuscuta*.

GREATER DODDER
Cuscuta europaea

- Because this plant is parasitic, it will often have no chlorophyll of its own, leaves so tiny they're easily overlooked, and a bright orange color. It's found all over the United States. Research with the dodder plant shows that it doesn't attack a host randomly; it actually searches out a host it prefers.

- Research into chemical signaling in plants and animals by Dr. Consuelo De Moraes showed that dodder vines grow toward plants and don't grow toward empty pots. What's more, if the dodder plant is given a choice between a tomato plant or a wheat plant, it will always choose the tomato plant.

- Dr. De Moraes wanted to see just how strong the dodder's responses were to tomato plants. In fact, according to a 2012 article in *Scientific American*, she wanted to test the hypothesis that the dodder plants could smell the tomato. What if the definitive marker for a sense of smell was just a response to an olfactory cue? If that's the case, could the dodder plant respond to the smell, or chemical signal, from the tomato?

- To test this hypothesis, Dr. De Moraes made a tomato perfume concocted from the chemicals found on the tomato plant and soaked cotton swabs with the tomato perfume. The dodder plants grew toward the cotton swabs that smelled like tomatoes.

- Another example of a unique plant, or group of plants, is the living stone plants from South Africa, which look like small stones. These plants have windows, or transparent membranes, which are an adaptation for plants growing in lower-light environments. These plants are found in areas with lots of cloudy days; the windows let in more light on a cloudy day.

BENEFITS TO STUDYING PLANTS

- With scientific advances, there are practical benefits to studying plants. The following are examples of when a little knowledge about plants goes a long way.

- Trees consume carbon dioxide and produce oxygen. But some trees can actually contribute to climate change by emitting volatile organic carbons (VOCs). When VOCs meet nitrogen oxides produced by our cars in the presence of sunlight, they help the formation of tropospheric ozone.

- Tropospheric ozone occurs in the troposphere, where we are. The thinning of the ozone layer is occurring in the stratosphere, particularly over Antarctica, but potentially over the Arctic, as well. This ozone protects us from harmful ultraviolet radiation, which can cause skin cancer and cataracts. But the ozone in the troposphere, which also goes by the name of smog, triggers asthma in children and is the primary component of air pollution in cities.

- Not all trees produce the VOCs that contribute to smog, but some do, and in large amounts. According to a study done by the U.S. Forest Service, sycamores, locust trees, and oaks were among the greatest VOC contributors. A little botanical knowledge can help citizens make good decisions about which trees will help improve air quality.

- Another example where botanical knowledge comes in handy pertains to rubber. Henry Ford, who bought a rubber plantation in 1927, used rubber for his car's tires as well as many of the hoses and connections within the vehicle.

- The scientific name for the rubber tree is *Hevea brasiliensis*. As the name suggests, it's native to Brazil, where it grows sporadically in the rain forest. Ford bought his plantation in the Amazon, where this tree is found natively. But within a few years, the plantation failed.

- In nature, rubber trees never grow in clusters, but in the plantation, they were all planted close together—so close that a disease, known as South American leaf blight, or *Microcyclus ulei*, was able to spread through the entire plantation.

RUBBER TREE
Hevea brasiliensis

- Most rubber today is grown in Southeast Asia. The success of rubber there has to do with the fact that the climate of Southeast Asia is agreeable to rubber trees, but also that the South American leaf blight has not yet made its way to Southeast Asia. However, most botanists think it's just a matter of time before this blight makes it to Southeast Asia.

- Like Henry Ford's plantation, this could be disastrous for the plantations in Southeast Asia. There may be trees in Brazil that are naturally resistant to the blight, or scientists may be able to develop a variety of rubber tree that has resistance to the blight, but this isn't happening. When the blight hits Southeast Asia, the research on these fronts will most likely begin in full force.

- Another example where a knowledge of botany is useful is bananas. All bananas that are in the supermarket are not only the same variety, but they are exact clones of each other—genetically identical. Just as there are different varieties of apples, there are different varieties of bananas, but the one everyone eats is called Cavendish.

- The seeds inside are too immature to grow, so Cavendish bananas are effectively seedless and unable to mutate. When Tropical Race 4, a banana fungal disease called *Fusarium oxysporum* f. sp. *cubense*, reaches Latin America, the banana plantations will be susceptible to total failure.

Fusarium oxysporum f. sp. *cubense*

- Tropical Race 1 overwhelmed a previous clone banana everyone was eating in the 1950s—that was what led to the more fungus-resistant Cavendish. A more colloquial name for these Tropical Race diseases is Panama disease, which attacked bananas in 1874 in Australia. By 1890, the earlier form had made its way to Panama, where it earned its nickname. So far, research has identified possible varieties of bananas that appear resistant to Tropical Race 4.

NAMING, CLASSIFICATION, AND IMPORTANCE OF PLANTS

- Like people, plants have a first name and a last name. Their first name is the genus name, so it represents a larger group, sort of like our last name. Their second name is called the specific epithet, and those 2 names—the genus name and the specific epithet—make up the scientific name of each species of plant.

- There are more than 300,000 different species of plants—with a few thousand more discovered every year—and even the best botanists in the world don't know the scientific names for all of the species. There are unknown plants, and there are different traits to consider when deciding how to define a species.

- Although this course is not strictly about plant identification, you will be introduced to many plants and the ways that they are classified. A field guide is simply a book to identify organisms in their natural habitat. There are field guides for birds, mammals, and plants. These guides are meant to be used in the field, and they are often organized by flower color, so you don't need to know a lot of botany to use them.

- Field guides to trees are often organized by leaf shape and bark appearance. It's like a matching game. Anyone with careful observation skills can identify a common plant or tree using a field guide, but it's helpful to find the most local field guide you can. Most states have their own field guides for common trees or common wildflowers. Larger states, such as California, will have guides for particular parts of the state.

- Field guides are great for common species, but they won't include everything. That's where a manual or flora comes in: These are books that will identify all plants of a region down to the species, using a dichotomous key. A manual might be for a particular group of plants—for example, the grasses; a flora is meant to be an exhaustive account of all the plant species in a specific geographic region.

- Plants form the majority of our economic system—either as food, medicine, perfume, raw material for building, fuels, paper, clothing, or furniture—and we don't even have to pay for plants' top contribution to our survival: oxygen. According to a 2015 estimate by Royal Botanic Gardens, Kew, 10% of the world's plant species already have documented uses.

- The number-2 traded commodity in the world is a plant: coffee. Because coffee is grown and consumed on such a wide basis, the way coffee is grown can be a boom or bust to an ecosystem.

- Although plants are important to our survival, there is a real, though avoidable, phenomenon known as plant blindness, a term first coined by Dr. James H. Wandersee and Dr. Elizabeth Schussler. We go about our business, but we don't really see or know about the plants around us.

- A hundred years ago, identifying what a plant is played a more widespread role in education and daily life, whereas only recently have plant scientists been developing a much deeper understanding of how plants do what they do.

- Not only is plant knowledge useful, necessary, and life-saving, but it's also a good way of life. Celebrating the secrets of nature provides joy and satisfaction.

READINGS

Mabey, *Cabaret of Plants*.

If you want to go deeper with plant identification, pick up a local field guide. The Sierra Club, the National Audubon Society, and Peterson Field Guides all have good series. If you're already familiar with local field guides, perhaps you're ready to move on to the local flora, or a book that will identify all plants of a region to species using a dichotomous key, which you can obtain from the Native Plant Society website for a particular state. The U.S. Forest Service explains this here:

https://www.fs.fed.us/wildflowers/features/books/

This site also has links to several parts of the country where more specific books are listed.

QUESTIONS

1. How much do you already know about botany? Do you know the names of 5 plants in your yard? Would you know the name of 5 native plants near your home?

2. What was most surprising about plants for you in this lecture?

LECTURE 1 TRANSCRIPT
THE JOY OF BOTANY

Botany is the study of plants, everything from biochemical reactions to whole forests. What's so interesting about plants? As my dad commented when I first said I was thinking of studying plants, he said, "But you can't take a plant on a date." He's a dentist, so he works with people, and I knew what he meant. At first glance, plants seem, well, a little boring perhaps: they don't talk, they don't move, they may not ignite our compassion like animals. And yet, I would argue that they do talk in a language that's different than ours, but it's communication all the same. They do move—some even move faster than we do. Plants have a quieter way of igniting our passion because we all know that we're dependent on plants. It's easy to take them for granted, but wherever we look, I think we can be surprised at what's out there in the plant kingdom. Let me show you what I mean with a few examples.

To give you a picture of some of the incredible diversity of some 300,000 species of plants, let me introduce you to a few interesting and very different plants. Consider the dodder. The dodder plant is a parasitic plant, which means it gets all of its nutrition from other plants. Dodder is the common name, and it really refers to an entire genus of plants, and that genus is called *Cuscuta*. Interestingly, *Cuscuta* is an old medieval term for one who dodders, which means one who shakes or trembles, especially with old age.

Because this plant is parasitic, it will oftentimes have no chlorophyll or leaves of its own, and its leaves are so tiny they're easily overlooked and have a bright orange color. It's found all over the United States, so it's clearly found a good way to make a living. I'm not sure why it has the common name of something that shakes and trembles—maybe it's the effect it has on other plants. Research with the dodder plant shows that it's anything but trembling; in fact, it doesn't attack a host randomly—it actually searches out for a host it prefers.

Research into chemical signaling in plants and animals by Dr. Consuelo De Moraes was able to show that dodder vines grow toward plants; they don't grow toward empty pots. What's more, if the dodder plant is given a choice between a tomato plant or a wheat plant, it will always choose the tomato plant. Apparently, the wheat plants produce a substance that the dodder plants can't stand. So, Dr. De Moraes wanted to see just how strong the dodder's responses were to tomato plants. In fact, according to a 2012 article in *Scientific American*, she wanted to test the hypothesis that the dodder plants could smell the tomato. We normally don't think of plants as being capable of smelling, since they lack noses, but what if the definitive marker for a sense of smell was just a response to an olfactory cue?

If that's the case, could dodder respond to the smell or the chemical signals from tomato? To test this hypothesis, Dr. De Moraes actually made a tomato perfume concocted from the chemicals one would actually find on the tomato plant. Anyone who's grown tomatoes knows that these plants have a lot of sticky fluid all over the leaves and stems, and it's pretty pungent. She then soaked cotton swabs with the tomato perfume to see if dodder plants would take the bait, and they did. They grew toward the cotton swabs that smelled like tomatoes. So, plants can be surprising in their abilities. Perhaps you never thought a plant could smell you, but maybe they can.

Another example of a unique plant, or a group of plants, are the living stone plants from South Africa. These plants literally look like small stones. If you didn't know they were plants, you might walk right over them. A fascinating trait of these plants is that they have windows. I'll bet you didn't know that plants had windows. And, for those of you who know that leaves absorb light for photosynthesis, you're probably thinking, why would a plant have transparent tissue? It's not exactly like cellophane, more like a translucent membrane. The stone plant looks a bit like a cactus, though it isn't—it's a whole different family. Cactuses are found natively in the New World, and the living stone plants, known as *Lithops*, are native to the Old World. So, as I said, these plants are mainly below ground with just the very top of the plant protruding above the soil. In some of these plants, the part protruding is transparent.

At first thought, the reason for the transparency may seem obvious: the window allows light to penetrate deep into the plant for photosynthesis. But, as I tell my students, botany isn't rocket science, it's more complicated. Why would I say that? I mean no disrespect to rocket science, but the math and physics of rocket science always stay the same, no matter where you are on Earth—even in space, where the force of Earth's gravity does change, but everything is still governed by the same principles.

Botany, on the other hand, is not like math or physics—there are very few rules. The rules that do exist always seem to have exceptions. Everything about plants comes in a lot of unique forms and functions. For example, I can't say all plants have flowers. Most plants—over 85% of all plant species—do have flowers, and there are some 40,000 other species of plants that have no flowers at all. Nor can I say all plants have seeds, since non-seed plants are almost as numerous as non-flowering plants. So, I hope you see what I mean: very few rules in botany.

So, back to our leaf windows in the living stone plant. It may seem obvious that the window lets light penetrate. But, when scientists started working with the plant, they found that high light levels actually decreased photosynthetic rates. Just like us, plants can get too much sun. When they do, the photosynthetic machinery can't keep up, and the rate of photosynthesis declines.

The experimenters found a sweet spot with the amount of light, and that sweet spot was the amount of light typical for a cloudy day. Sure enough, when the investigators looked at the distribution of living stone plants with windows, they found them occurring in areas with lots of cloudy days. The window lets in more light on a cloudy day, but can't let in too much light on a sunny day. So, the window adaptation is an adaptation for plants growing in lower light environments—not what you might think when you first hear about windows in leaves.

And with so many scientific advances, there are also more and more practical benefits to studying plants, too. I want to give you three specific examples where a little knowledge about plants goes a long way. It's my hope that this course will give you more than a little knowledge about plants and that, by the end of the course, you'll come to see plants in a whole new light. The first example comes from trees. Everyone knows that trees consume carbon dioxide and produce oxygen. So, if you lived in a city looking to improve its air quality, you might want to plant a bunch of trees. But what if some trees could actually contribute to climate change? This turns out to be true. Scientists have long known that some trees emit volatile organic compounds, or V When these VOCs meet nitrogen oxides produced by our cars in the presence of sunlight, they help the formation of tropospheric ozone.

Now, you might say, "Oh good, that ozone will help to repair the ozone hole," and it is the same kind of molecule, but the ozone I'm talking about occurs in the troposphere, where we are. The thinning of the ozone layer—it's not really a hole, per se—is occurring in the stratosphere, particularly over Antarctica, but potentially over the Arctic as well. This ozone is in the stratosphere and protects us from harmful ultraviolet radiation, which can cause skin cancer and cataracts. But, the ozone in the troposphere is not so good. In fact, tropospheric ozone goes by the unflattering name of smog, and this smog triggers asthma in kids, and it is also the primary component of air pollution in cities. Smog is what gives my hometown of Denver the brown cloud on high ozone days.

Now, back to the trees that can contribute to this smog. Not all trees produce the VOCs that contribute to smog, but some do, and in large amounts. According to a study done by the U.S. Forest Service in Brooklyn, sycamores, locust trees, and oaks were among the greatest VOC contributors. Maybe all that air pollution in L.A. is just a result of all those California sycamores? Unlikely—there are a lot of cars in the L.A. basin. But the Blue Ridge Mountains in Appalachia get their name from the blue haze caused by isoprene, a VOC emitted from trees there. However, not all trees are emitters. Some trees emit very little isoprene, like redbuds, witch hazel trees, and tulip trees. The point here is not that planting trees is bad; the point is that a little botanical knowledge can help citizens make good decisions about which trees will help improve air quality.

Another example where botanical knowledge would come in handy pertains to rubber. Henry Ford was well known for trying to control every part of production for his automobiles. One aspect of the control was over rubber. The rubber just wasn't used for tires, it was—and still is—used for many of those hoses and connections within the vehicle. Henry Ford bought a rubber plantation in 1927. The scientific name for the rubber tree is *Hevea brasiliensis*. As the name suggests, it's native to Brazil, where it grows sporadically in the rain forest. Ford bought his plantation in the Amazon where this tree is found natively, but within a few years, the plantation failed. Why? In nature, rubber trees never grow in clusters, but in the plantation, they were all planted close together—so close, that a disease was able to spread through the entire plantation, and that disease was South American leaf blight, or *Microcyclus ulei*, which is also native to Brazil. When trees grow far apart, the chance of the disease getting from one tree to another is small, but if the trees are close together, it's like putting a toddler in a candy store: bonanza.

Interestingly, most rubber today is grown in Southeast Asia. There's also a rubber resurrection, and more things are being made with natural rubber. For example, Patagonia started making wetsuits from natural rubber instead of neoprene. The success of rubber in Southeast Asia has to do with the fact that the climate of Southeast Asia is agreeable to rubber trees, but also that the South American leaf blight has not yet made its way to Southeast Asia. However, most botanists think it's just a matter of time before this blight makes it over to Southeast Asia. Like Henry Ford's plantation, this could be disastrous for the plantations in Southeast Asia. There may be trees in Brazil that are naturally resistant to the blight, or scientists may be able to develop a variety of rubber tree that has resistance to the blight, but this isn't happening. When the blight hits Southeast Asia, the research on these fronts will most likely begin in full force.

Let's consider what's happening with bananas. All bananas that are in the supermarket are not only the same variety; they are exact clones of each other. Just as there are different varieties of apples—for example, Granny Smith, Red Delicious, and my favorite, Braeburn—there are different varieties of bananas, but the one everyone eats is called Cavendish. We don't really think about varieties of bananas because our stores only carry the one type, and all of those individuals are genetically identical. The seeds inside are too immature to grow, so Cavendish bananas are effectively seedless and unable to mutate.

So, when Tropical Race 4, a banana fungal disease called *Fusarium oxysporum* f. sp. *cubense* reaches Latin America, the banana plantations will be susceptible to total failure. What's interesting is that f. sp., which is a term botanists use to say this is a different form of this disease, that abbreviation f. sp. could be used for plants, too. And it means that this new form of the disease, or plant, doesn't need to be its own subspecies, but we need a way to classify it separately from other forms.

Tropical Race number one overwhelmed a previous clone banana that everyone was eating in the 1950s. That was what led to the more fungus-resistant Cavendish. A more colloquial name for these Tropical Race diseases is Panama disease, which attacked bananas in 1874, oddly enough in Australia. By 1890, the earlier form has made its way to Panama, where it earned its nickname. At least the rubber trees weren't genetically identical. So far, research has identified possible varieties of banana that appear to be resistant to Tropical Race 4. Botanists are working hard on this.

So, these are three stories about plants that give you some insight into the variety that encompasses botany, and the variety we will investigate in this course. I next want to tell you a bit about how I got into botany. I want to tell you this to inspire you that it's possible to be a botanist without being born with a deep love for plants. I did spend a lot of time outside as a kid, but I never considered the plants. I enjoyed finding frogs and turtles in the creek behind my house. My first real memory of having anything to do with plants, other than watching dad mow the lawn, was building a lean-to at day camp. We ravaged that forest for branches, and picked grass and leaves to throw on top. Fast forward to college, and I was premed, like pretty much everyone else studying biology.

My college required that all students take a plant class in order to earn a degree in biology. I didn't really want to take a plant class; I wanted more human anatomy or physiology. I took the plant class, it was OK, but it was a lot of taxonomy and memorization, and the magic of plants, the wonder of them, just didn't grab me at that time. So, in this course, I've tried to give you the wonder of plants, their stories, and the interesting things about them.

After that first plant class, I graduated and applied to medical school. The summer after college graduation, I got a job as a wilderness guide that involved some backpacking along with other adventures in the West. While hiking high in the mountains, I was noticing all the flowers in the alpine and thinking about how beautiful they were. I especially noticed a plant called the alpine forget-me-not. Its scientific name is *Eritrichium nanum*. Of course, I didn't know that scientific name at the time, I was just captivated by what I think we can all agree is one of the most beautiful colors to be found in flowers: deep blue. There aren't very many true blue flowers.

According to David Lee, a retired professor of biology at Florida International University, fewer than 10% of all flowering plants are blue. Dr. Lee also wrote the book *Nature's Palette: The Science of Plant Color*. In this book, he explains that plants don't have a blue pigment. Green is from chlorophyll; yellow and orange are from beta-carotene; red and pink are the anthocyanins. In order to get blue, Dr. Lee explains, plants must modify their anthocyanins by using pH shifts and mixing pigments, molecules, and ions. It's a bit like mixing up watercolors: a combination of yellow paint and green paint makes blue paint.

Of course, I didn't know anything about anthocyanins staring into the yellow center of the brilliant blue alpine forget-me-not. That plant made me think that maybe I wanted to experience and understand more stuff like this. But who am I kidding? Get paid to hike around the mountains and just look at plants? Jobs like that don't exist. I decided not to go to medical school, and, instead, I would take some more classes and think about my next move. I took a plant identification course, and in that identification course, I discovered that I actually liked plants—a lot. I thought, maybe I'll go to graduate school.

Now look, graduate school isn't all roses and chocolates, but I had a blast. I was falling in love, and when you fall in love, it takes time. Love at first sight is just really liking the way a person looks and thinking that maybe you could love them. When I saw those alpine plants, I had a feeling like that—maybe I could love these plants. And when you get to know something really intimately, then you can begin to love them.

George Washington Carver, the botanist who studied alternative crops to cotton, such as peanuts and sweet potatoes, has a lovely quote that sums up my feelings here. He wrote:

> Anything will give up its secrets if you love it enough. Not only have I found that when I talk to the little flower or to the little peanut they will give up their secrets, but I have found that when I silently commune with people they give up their secrets also—if you love them enough.

So, plants are like people. When we really give our attention to something, we can't help but begin to love it. When I started to study plants, really look at them, really learn their stories, it was the beginning of a lifelong relationship. It's a relationship I hope you will all develop as well.

Because I want you to have a relationship with plants, I will sometimes use their scientific names. When you meet someone for the first time, you typically ask their name. And, once you develop a relationship with that someone, you know their full name, and maybe one or more of their nicknames. You don't say, "Hi, tall guy with brown hair, how are you?" And so it is with plants. Like us, plants have a first name and a last name. Their first name is the genus name, so it represents a larger group, sort of like our last name. Their second name is called the specific epithet, and those two names—the genus name and the specific epithet—make up the scientific name of each species of plant. Now, there are over 300,000 species of plants, and even the best botanists in the world don't know the scientific names for all 300,000.

But, just as there are over 7 billion people on the planet, and you don't know everyone's name, you still make an effort to learn names, and we will make that effort here. At first, the names may seem cumbersome and difficult to pronounce. Some awkwardness is always present when one learns a new language. I think you will begin to see, though, that scientific names will sometimes be very useful in describing something about the plant. After all, while every single person on the planet is a member of the species called *Homo sapiens*, there are an estimated 300,000 different species of plants with a couple thousand more discovered every year. There are even estimates about how many plants remain to be discovered, and this is why many of the number of species of plants will be estimates. There are unknown plants, and there are different traits to consider when deciding how to define a species.

Although this course is not strictly about plant identification, I will introduce you to many plants and the ways that they are classified. A field guide is simply a book to identify organisms in their natural habitat. There are field guides for birds, mammals, and plants. These guides are meant to be used in the field, and they are often organized by flower color, so one needn't know a lot of botany in order to use them. Field guides to trees are often organized by leaf shape and bark appearance. It's a lot like a matching game. Anyone with careful observation skills can identify a common plant or a tree using a field guide, but it's helpful to find the most local field guide you can. Most states have their own field guides for common trees or common wildflowers. Larger states like California will have guides for particular parts of the state.

The field guides are great for common species, but they won't include everything, and that's where a manual or a flora comes in. These are books that will identify plants of a region down to species, using a dichotomous key. A manual might be for a particular group of plants—say, the grasses—whereas a flora is meant to be an exhaustive account of all the plant species in a specific geographic region.

Many of us tend to forget that plants form the majority of our economic system, either as food, medicine, perfume, raw material for building, fuels, paper, clothing, furniture, and we don't even have to pay for plants' number one contribution to our survival: oxygen. According to a 2015 estimate by Royal Botanic Gardens, Kew, 10% of the world's plant species already have documented uses. For example, the number two traded commodity in the world is a plant: coffee. The number one traded commodity, if you're wondering, was plants—now fossilized plants, or oil. Because coffee is grown and consumed on such a wide basis, the way coffee is grown can be a boon or a bust to an ecosystem.

In addition to wanting you to develop a closer relationship with plants—all kinds of plants—I have a few other goals for you in this course, as well. I'd like you to understand the fascinating differences and similarities between plants and animals. I'd like you to get better acquainted with the various parts of the plant and what they do. I want you to understand the basic process of photosynthesis and evolution.

And after we've established this foundational knowledge, we'll take an evolutionary walk through plant development, and I hope you'll gain an appreciation for the adaptations that plants have developed over time, from mosses to ferns, then conifers, and then flowering plants. Once we reach flowering plants, we'll take a biome approach, looking more closely at plants in specific environments, from the ocean to the mountains. Along the way, we'll also learn about physiology in plants, and we'll meet some other organisms that aren't plants, but are either similar to plants in their makeup, or necessary for plants to survive.

And, when you complete this course, you'll be a botanist, and botanists make great protagonists. *E.T.* was the earliest film I remember as an example. You may not remember, but E.T. and his friends were on planet Earth collecting plants. Do you remember the female lead in the first *Jurassic Park* in 1993? Dr. Ellie Sattler was a paleobotanist, and in the movie, she correctly diagnosed a dinosaur that was sick from eating a poisonous plant. I think we can all agree that the sequel to *Jurassic Park*, *Jurassic World*, isn't nearly as good because the lead protagonist is not a botanist. Of course, many people have seen the movie *Avatar*, and you may recall that Sigourney Weaver was, in fact, a botanist.

Even Matt Damon famously saves himself while stranded on Mars because he is a botanist. In a great example of truth following fiction, scientists in Australia named a new species after the character's name, Mark Watney. The new species is called *Solanum watneyi*. This was an especially good choice for the plant to be named after the character because this plant is the same family as potato, *Solanum tuberosum*, which is what Mark Watney grows on Mars to save himself. Now, granted, I did my doctorate at UCLA, so I have a special penchant for Hollywood botanists, but I also think their presence in so many feature films is evidence of what we all know: botany will save our lives.

And yet there is a real, though avoidable, phenomenon known as plant blindness. This term was first coined by Dr. James H. Wandersee who was at Louisiana State University, and Dr. Elisabeth Schussler, now a professor at University of Tennessee, Knoxville. We go about our business but we don't really see or know about the plants around us. Even as children, we focus on animals. Everyone knows how a giraffe differs from an elephant, but how does the century plant, *Agave americana*, differ from *Araucaria araucana*, the national tree of Chile? How does the rare flower of *Amorphophallus titanum*, the titan arum, differ from *Musa sapientum,* the tree that gives us bananas? What separates living stone plants from the giant Amazon water lily, *Victoria amazonica*? We know the difference between cardinals and robins, but how does an elm differ from a birch? The answer is in the leaf base. The part of the leaf that attaches to the stem, or the petiole of the leaf, is uneven in elms, and straight across in a birch. What schoolchild learns this?

Ironically, a hundred years ago, identifying what a plant is played a more widespread role in education and daily life, whereas only recently have plant scientists been developing a much deeper understanding of how plants do what they do. Not only is plant knowledge useful, necessary, and lifesaving, it's also a good way of life. Celebrating the secrets of nature, something I call *Natura Revelata*, provides joy and satisfaction even if other pursuits aren't so fulfilling. The revelry we can experience at knowing the names and stories of the plants around us really is a field party. My hope in this course is to open your eyes to the full science of botany, old and new, that is all around us—the what and the how. Once we pay attention, we see a world that is too wonderful to ignore.

LECTURE 2

PLANTS ARE LIKE PEOPLE

In 1975, Jerry Baker published a gardening book called *Plants Are like People* that compares people and plants. For example, plants like the same temperature of water that we like. They like about the same amount of light that we do; with individual variations, we enjoy a sunny day, and so do plants. Plants have a circadian rhythm of rest and activity, and so do we. We drink water; they need water, too. In this lecture, you will discover some of the ways that plants are like people.

PLANT INTELLIGENCE

- A very hot topic of research is plant intelligence. This topic is a bit controversial because there was a book called *The Secret Life of Plants*, first published in 1989, that claimed, among other things, that certain types of music encouraged plant growth. Even though this has never been confirmed in a repeatable lab study, the book has been a wild success.

- In some ways, the book created a rift in botany that persists to this day. Scientists interested in the ideas of plant communication or plant behavior tend to be labeled on the fringe by more mainstream botanists. But there are also attempts to rethink plant behavior from a more mainstream perspective.

- Monica Gagliano, an Italian botanist who studies plant communication, published a 2012 paper with colleagues entitled "Towards Understanding Plant Bioacoustics," which investigates the perception of sound in plants. The authors argue that many animals lack eardrums yet can still hear, so perhaps plants can hear through some organ or mechanism we don't yet understand.

- They make a good argument that there are constant sounds in substrates such as soil. Animals who live underground still hear, though their hearing is generally reduced. The authors outline an experiment where the roots of young corn are shown growing toward a sound emitted in water.

- Roots need water, so it makes sense that roots will grow toward sources of water. It's even been reported that roots will grow toward pipes with water in them, even if the pipes are sealed and not leaking any water. This also suggests that the roots are somehow sensing the sound.

- Gagliano has also been experimenting with *Mimosa pudica*, also known as the sensitive plant, to test its memory. It's a small shrub-like, or frutescent, herb that is native to the tropics of North and South America. There are about 430 species of *Mimosa*, but only a few species display the special trait that *Mimosa pudica* displays: When touched, its leaves collapse immediately.

SENSITIVE PLANT
Mimosa pudica

- The *Mimosa* plant itself doesn't move, but the movement of the leaves is rather instant. How do the leaves move so fast without muscles?

- The *Mimosa* also has a special thickening at the base of the leaf called a pulvinus. These cells are what control the drooping of the leaves.

- Consider that the leaf is fully turgid, a botanical term that means the opposite of wilting. The leaves are flat, but when you touch them, they droop or wilt immediately. When touched, the plant produces an electrical signal that causes potassium channels in the cell wall of the pulvinus to open.

- When all the potassium rushes out of the cell, the water will follow it by osmosis—the biologic principle by which water moves from an area of high water concentration to an area of low water concentration—thus causing the pulvini to wilt, or become flaccid, which makes the leaves droop. (The more general process of diffusion is the movement of anything from a high concentration to a low concentration.)

- A flaccid cell also causes plasmolysis, which is when the cell shrinks due to lack of water and the cell membrane pulls away from the cell wall. A flaccid cell doesn't always have plasmolysis, but if it remains flaccid long enough, plasmolysis will occur.

- In testing the memory of the plants, Dr. Gagliano would drop the *Mimosa* plants, and the leaves would collapse. She would repeat this several times until the *Mimosa* stopped collapsing their leaves. You could say that the potassium channels can't reboot or make some sort of physiological argument, but then she would drop the plants a week later, and the leaves would not collapse. To test them, she would give them some other stimulus, and the leaves would droop immediately because they had not been conditioned for a response.

PLANTS VERSUS ANIMALS

- There are many ways that the cells of plants and animals are similar, but there are a few ways that they're different.

- Plant cells have 3 important structures that animal cells don't have: a cell wall, a vacuole, and chloroplasts.

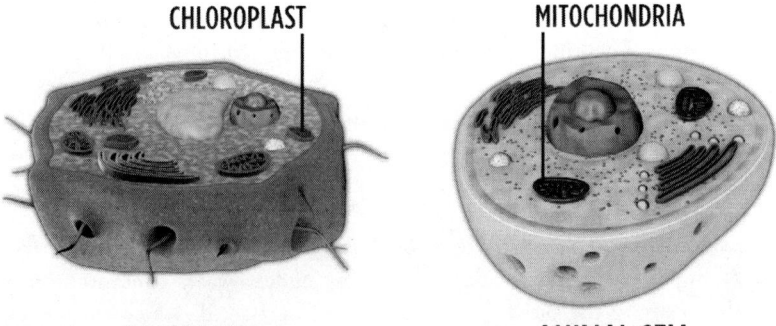

- All cells, including animal cells, have some kind of boundary between the cell and everything else, the extracellular fluid or matrix. Both plants and animals have a cell membrane, but plants also have a cell wall. The cell wall is tougher, usually thicker, and has more structure than a cell membrane.

- In plants, the cell membrane is called the plasma membrane, and it is inside the cell wall. Put another way, on top of the plasma membrane is the primary cell wall. This cell wall is made up of polysaccharides, which is a fancy word for a lot of sugars stuck together.

- One of these polysaccharides is cellulose. We can't digest cellulose, even though it's just a chain of glucose molecules; instead, cellulose is a component of dietary fiber. Cellulose is also the most abundant macromolecule on Earth.

- These cellulose molecules are long strings of glucose that crisscross each other to add structure and strength. The cellulose molecules are called microfibrils, and embedded around these are pectin molecules. Pectin provides flexibility to the cell wall.

- Primarily, the cell wall keeps the cell from bursting under turgor pressure. Most plant cells are turgid, or full of water to maintain their shape. An animal cell membrane would burst under this pressure. Plants are able to get such high pressures in their cells because the cell wall holds in the water.

- Plant cell walls also have small pores in them, which allow for water, and a few small nutrients, to pass into the cell. Osmosis is the movement of water through a semipermeable membrane, and the cell wall is an example of a semipermeable membrane.

- Cell walls can be either primary cell walls, which are still growing and living, or secondary cell walls, which are mostly no longer alive but contain a substance called lignin to provide structure. In animal cells, the cell membrane is important because it receives many signals from the rest of the body.

◇ The cell membrane contains receptors on the surface of the cell membrane that receive signals from the rest of the body. If the cell needs some glucose, it has a glucose receptor on the cell membrane that is open and therefore receives glucose. The evidence that plant cell walls regulate responses in the same way is not that strong yet, and this is a big interest in botanical research.

◇ The second unique component, for nonwoody plant cells, is the vacuole. The vacuole is an organelle, or cellular organ, whose main purpose is to store water. The vacuole stores water, but this water is also the primary component that gives cells their structure.

◇ The third unique component that plants possess is the chloroplast. The chloroplast is an organelle that contains chlorophyll and is the main site of photosynthesis. In some ways, chloroplasts resemble the mitochondria of an animal cell, which is the main site of energy production.

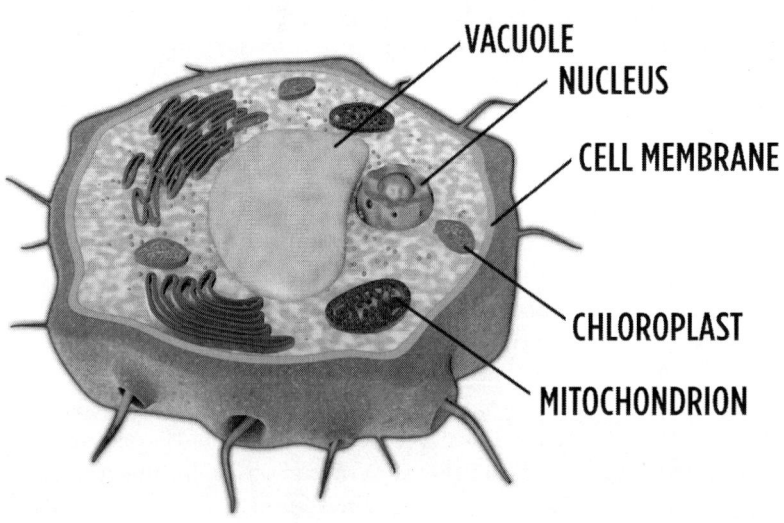

- In fact, the chemical reactions are exactly opposite in the chloroplast and the mitochondrion. The chloroplast takes in sunlight, carbon dioxide, and water and creates glucose, oxygen, and energy. The mitochondrion takes in glucose and oxygen and creates carbon dioxide and water, a process known as cellular respiration.

- All plant cells also have mitochondria, just like animals, to break down the glucose produced during photosynthesis. But mitochondria are especially important for cells that don't photosynthesize, such as the roots.

- Both plant cells and animal cells contain a nucleus. The nucleus contains the deoxyribonucleic acid (DNA). The genetic information, or genes, are contained within the DNA. DNA works pretty much the same in plants as it does in animals. Plants and animals not only share the same DNA building blocks, but they also share many of the same DNA sequences, or genes.

- A group of cells together makes up a tissue. Because plants have such different cell types from animals, they also have different tissue types than animals.

- There are 3 main tissue types in plants, and these are categorized by the thickness of the cell wall. Tissue with the thinnest cell wall is called parenchyma, the tissue that makes up much of the internal structure of the leaves of plants.

- Collenchyma is tissue with irregularly thickened cell walls. These are structural cells, though they are not found in the roots. Typically, they are most common in stems of herbaceous (nonwoody) plants, such as celery.

- Sclerenchyma is tissue that contains cells with thickened secondary walls that may contain lignin, which is a substance that provides thickness. Sclerenchyma serves a protective or supportive function in the plant.

- Plants and people have some genes in common and use the same basic building blocks of DNA to make proteins. Plants also have tissues, but plant tissue types depend on the thickness of cell walls, which our animal tissue types don't have, so the form and function of plant tissues are different.

◇ Most important—and most surprising, given these differences—plants are able to sense their environments and respond. The open question is whether plant sensation is akin to sensation in animals and whether plant responses to their environment are akin to animal behavior and communication.

READINGS

Bock, et al, *Identifying Plant Food Cells in Gastric Contents for Use in Forensic Investigations.*

Chamovitz, *What a Plant Knows.*

Karban, *Plant Sensing and Communication.*

Trewavas, *Plant Behaviour and Intelligence.*

QUESTIONS

1   Do you think that plants can communicate like humans? Why do you think that botanists have been so slow to study plant behavior? Why would it be difficult to accept that plants are communicating?

2   What are cellular parts that plants have that animals don't have, and what are their functions?

3   Suppose that you had a red-flowered plant with dominant alleles for red, RR (one allele came from the mother and one came from the father). You are going to cross this red-flowered plant with a plant that has white flowers and is recessive with the alleles, rr. If color is completely dominant, what is the predicted outcome of the cross? How many flowers will be red, and how many will be white?

LECTURE 2 TRANSCRIPT
PLANTS ARE LIKE PEOPLE

In what ways are plants like people? In 1975, Jerry Baker published a book called *Plants Are like People*. It's a gardening book that is very cute in terms of comparing people and plants. For example, what temperature of water do plants like? They like the same temperature you would like. How much light would they like? They like about the same amount we do, with individual variations—we enjoy a sunny day, and so do plants. They have a circadian rhythm of rest and activity, and so do we. We drink water; they need water, too.

And there's more. A very hot topic of research is plant intelligence. Yes, plant intelligence. This topic is a bit controversial because there was a book called *The Secret Life of Plants*, first published in 1989. Definitely written for the lay enthusiast, the book has chapters entitled "Plants Can Read Your Mind" and "The Mystery of Plant and Human Auras." This was the book that claimed certain types of music encouraged plant growth. Surely this must be one of the most often repeated elementary school science fair experiments of all time, but it's never been confirmed in a repeatable lab study. Even if plants are responsive to sound, it's the people taking care of plants who are more likely to be affected by human music.

I think most botanists thought the book was kooky in a New Age sort of way. However, that book was a wild success, still ranking number 20 on Amazon's top selling botany books, almost 3 decades after it came out. And, in some ways, the book created a rift in botany that persists to this day. Scientists interested in the ideas of plant communication or plant behavior tend to be labeled on the fringe by more mainstream botanists.

But there are also attempts to rethink plant behavior from a more mainstream perspective. An Italian botanist named Monica Gagliano is one scientist studying plant communication. With colleagues, she published a 2012 paper entitled "Towards Understanding Plant Bioacoustics." This is not a paper about the sounds that plants make. By the way, if a tree falls in the forest, does anyone hear it? OK, it's not that kind of paper.

The authors state, "Here, we present a rationale as to why the perception of sound and vibrations is likely to have also evolved in plants." The perception of sound—these authors are investigating what plants hear. It's pretty convincing stuff. They argue that many animals lack eardrums, yet they can still hear. Snakes hear through their jaws, and insects hear through their antennae. Why, they argue, could plants not hear through some organ or mechanism we don't yet understand?

They make a good argument that there are constant sounds in substrates such as soil. Animals who live underground still hear, though their hearing is generally reduced. The authors outline an experiment where the roots of young corn are shown growing toward a sound emitted in water. Pretty much everyone knows that roots need water, so it makes sense that roots will grow toward sources of water. It's even been reported that roots will grow toward pipes with water in them, even if the pipes are sealed and not leaking any water. This also suggests that the roots are somehow sensing the sound.

This same botanist, Dr. Monica Gagliano, has been experimenting with *Mimosa pudica* to test its memory. After all, it's the sensitive plant. It's a small shrub-like herb, something botanists call *frutescent*, which means shrub-like. This plant is native to the tropics of North and South America. There are about 430 species of mimosa, but only a few species display the special trait that *Mimosa pudica* displays. Its remarkable trait is, when touched, its leaves collapse immediately. In fact, the scientific name *pudica* is Latin for "shy."

Before I tell you about the work with mimosa's intelligence, let's explore the mechanism of this leaf movement. Most plants don't move like this, and in fact many people think that plants don't move at all. It's true that the mimosa plant itself doesn't move, but the movement of the leaves is rather instant. How do the leaves move so fast without muscles?

In order to understand this, we need go over a basic biologic principle called osmosis. Now, you've probably heard people say something like, "I'll learn it by osmosis." This really irritates me because osmosis is only possible when there's a movement of water. They are actually thinking about a more general process of diffusion, which is the movement of anything from a high concentration to a low concentration. So, if I bring a cup of coffee into your room, you're likely to smell it before too long, because the scent molecules of coffee are moving from a high concentration, which is just above the coffee, to a low concentration, which is everywhere else in the room. So, when someone says they'll pick it up by osmosis, they mean they'll pick it up by diffusion, from a high concentration of knowledge to a low concentration of knowledge, which would be their brains.

So, osmosis is a special case, involving the diffusion of water. This process will also be important when we think about movement of water through the plant. So, in osmosis, water will move from an area of high water concentration to an area of low water concentration. Now, you're probably wondering, how can you have low water concentration? You have water or you don't. Well, that's not exactly true. Water has more water in it than, say, Kool-Aid. In Kool-Aid, there are other things in the water, and we call them solutes—sugar, flavorings, color. So, if water moves from a high concentration to a low concentration, think about which way the water will move if we have two chambers separated by a membrane that is only permeable to water. In one of the chambers is Kool-Aid, and in the other chamber is pure water. Remember, the membrane separating the chambers is only permeable to water. And water, like everything else, will move from a high concentration to a low concentration. Which way will the water move?

If you said the water will move from the water side into the Kool-Aid side, you're right. The concentration of water was greater in the pure water than the concentration of water in the Kool-Aid. Now, what does this have to do with plants? The membrane that is permeable only to water in our example is much like a plant's membrane or cell wall. So, back to our mimosa example. The mimosa also has a special thickening at the base of the leaf called a *pulvinus*. These cells are what control the drooping of the leaves.

Now, remember back to our osmosis example. Consider that the leaf is fully turgid—that's a botanical term that is the opposite of wilting. So, the leaves are flat, but when you touch them, they droop or wilt immediately. When touched, the plant produces an electrical signal that causes potassium channels in the cell wall of the pulvinus to all open. When all the potassium rushes out of the cell, the water will follow it by osmosis, thus causing the pulvini to wilt, or become flaccid, which makes the leaves droop. A flaccid cell also causes *plasmolysis*, which is when the cell shrinks due to lack of water, and the cell membrane pulls away from the cell wall. A flaccid cell doesn't always have plasmolysis, but if it remains flaccid long enough, the plasmolysis will occur.

Now, back to our Italian scientist looking for plant intelligence in mimosa. As we just learned, this plant will respond to touch by drooping its leaves. It would probably do this as a response to herbivory—a trick, as if to say, "I'm not a good plant to eat." So, in this study, Dr. Gagliano is testing the memory of the plants. Yes, their memory. How well can the mimosa plant remember something that happens to it?

Her experiment went something like this. She would drop the mimosa plants, and the leaves would collapse, just as we discussed. She would repeat this several times until the mimosa stopped collapsing their leaves. OK, the potassium channels can't reboot, or some sort of physiological argument. But then she would drop the plants all week a whole week later, and the leaves would not collapse. To test them, she would give them some other stimulus, and the leaves would droop immediately because they had not been conditioned for a response. So, if plants can respond to sound, respond to touch, and perhaps even have memory, how else are plants like people? And in what ways might they be different? We can build this discussion from the bottom up by thinking about plant and animal cells.

There are a lot of ways that the cells of plants and animals are similar, but there are a few important ways that they're different. Let's look at the differences first and then outline some of the similarities. Plant cells have three important structures that animal cells don't have. Plants have: number one, a cell wall; number two, a vacuole; and number three, plants have chloroplasts.

Let's start with the cell wall. All cells, including animal cells, have some sort of boundary between the cell and everything else, the extracellular fluid or matrix. Both plants and animals have a cell membrane, but plants also have a cell wall. This is just the way it sounds; the cell wall is tougher, usually thicker, and has more structure than a cell membrane. In plants, the cell membrane is called the plasma membrane, and it's inside the cell wall. Put another way, on top of the plasma membrane is the primary cell wall, and this cell wall is made up of polysaccharides, which is a fancy word for a lot of sugars all stuck together.

One of these polysaccharides is cellulose. We can't digest cellulose, even though it's just a chain of glucose molecules. Instead, cellulose is a component of dietary fiber. Cellulose is also the most abundant macromolecule on Earth. These cellulose molecules are long strings of glucose that crisscross each other to add structure and strength. We call the cellulose molecules microfibrils, and embedded around these are pectin molecules. Pectin is the primary component in jams that gives them their viscosity, and the same here with the cell wall. The pectin provides flexibility.

Primarily, the cell wall keeps the cell from bursting under turgor pressure. Most plant cells are turgid, or full of water, to maintain their shape. An animal cell would burst under that kind of pressure, but plants are able to get such high pressures in their cells because the cell wall holds in the water. Plant cell walls also have small pores in them, which allow for water to pass into the cell, and a few small nutrients. Remember, earlier in this lecture, we talked about osmosis as being the movement of water through a semipermeable membrane. The cell wall, even though it's a wall, is an example of a semipermeable membrane. Cell walls can be either primary cell walls, which are still growing and living, or they can be secondary cell walls, which are mostly no longer alive but contain a substance called lignin to provide structure. In animal cells, the cell membrane is important because it receives many signals from the rest of the body. Let's think of the cell membrane as a cell phone. You get it? Cell phone—it's sending and receiving signals.

OK, seriously though, the cell membrane contains receptors on the surface of the cell membrane that receive signals from the rest of the body. Let's say the cell needs some glucose. It has a glucose receptor on the cell membrane that is open, and so it receives glucose. The evidence that plant cell walls regulate responses in the same way is just not that strong yet, and this is a big interest in botanical research right now.

The second unique component for non-woody plant cells is the vacuole. The vacuole is an organelle, which is sort of the way is sounds, a cellular organ, or an organ for the cell—that's an organelle. The vacuole's main purpose is to store water. When you look at a cartoon of a plant cell, about 90% of that cell is taken up by the vacuole and everything else is sort of squeezed around that organelle. The vacuole stores water, but this water, as we've mentioned, is also the primary component giving cells their structure.

Both vacuoles and cell walls can be used to clean up toxic waste in a process called phytoremediation. *Phyto* is Latin for "plant," and remediation means to restore. For example, the mining industry sometimes leaves residual heavy metals in soils and waterways. Certain plants, called hyperaccumulators, can draw heavy metals up through their roots and store these metals in their tissues. For example, rapeseed, or *Brassica napus*, is a plant in the mustard family that gives us rapeseed oil, which is also canola oil. This *Brassica* plant is also a good accumulator of metals like mercury, lead, and chromium. Generally, the heavy metals can be stored in the cell walls or in the vacuole of a plant.

The third unique component that plants possess is the chloroplast. The chloroplast is an organelle, which, as we said, is a mini-organ in the cell. Chloroplasts contain chlorophyll, and they are the main site of photosynthesis. In some ways, chloroplasts resemble the mitochondria of an animal cell. The mitochondria is the main site of energy production in the animal cell. In fact, the chemical reactions are exactly opposite in the chloroplast and the mitochondria. The chloroplast takes in sunlight, carbon dioxide, and water and creates glucose and oxygen and energy. The mitochondria takes in glucose and oxygen and creates carbon dioxide and water, a process known as cellular respiration.

All plant cells also have mitochondria. Why? Like animals, just to break down the glucose produced during photosynthesis. But mitochondria are especially important for cells that don't photosynthesize, like the roots. It also implies that these cells need oxygen. This is one reason we aerate our lawns, so that the roots of the grass can get oxygen. Both the mitochondria and chloroplasts are thought to have originated from free-living bacteria that became parasitic, or somehow lived symbiotically with other cells. Over time, these bacteria changed to become mitochondria of modern plant, fungal, and animal cells, while others became the chloroplasts of modern plant cells. This process of bacteria becoming symbiotic organisms living inside other cells is called the theory of endosymbiosis.

Both plant cells and animal cells contain a nucleus. The nucleus contains the DNA, or deoxyribonucleic acid. The genetic information, or genes, are contained within the DNA. Perhaps unsurprisingly, DNA works pretty much the same in plants as in animals. This is unsurprising because this code was developed very early on in evolutionary history, so it has been retained over many millions of years.

In order to understand how DNA and genes work, it helps to understand the structure. Deoxyribonucleic acid is a chain of nucleotides. A nucleotide consists of a 5-carbon sugar—that is, the deoxyribose—a phosphate group, and a nitrogen-containing base. This last part, the nitrogen-containing base, can either be adenine, thymine, guanine, or cytosine. These get abbreviated A, T, G, and C. Movie buffs may remember the science fiction film *Gattaca*. Note that the title uses only A, T, G, and C, which is the genetic code of DNA. When I first learned about the genetic code, it seemed sort of unimpressive that the entire blueprint for life came down to these four nucleotide bases, but it does. It's helpful to think of it this way: the entirety of the English language comes down to 26 letters.

The similarity is they each form words. The four nitrogen-containing bases form codons, and these are sort of like words. Each three nitrogen bases forms a codon. That codon will code for a particular amino acid. Now, here's some math review. If there are four bases and three bases per codon, how many codons are possible? That would be 4^3, or 64. Interestingly, as I said, the codons will code for an amino acid, but there are only 21 amino acids essential for life, so all our essential amino acids might have been possible with only 3 bases, except for one thing: the complementarity of DNA would be lost. Complementarity means that the nitrogen bases bind to each other, and that's how they create the double helix we all know and love. A binds to T and C binds to G.

The way proteins are made in plants and animals is the same. It's due to the complementarity of the DNA. To make a protein, first the DNA is read by messenger RNA in the nucleus. Second, the messenger RNA takes the message out of the nucleus to the ribosomes, where it's translated by transfer RNA. Third, the transfer RNA then assembles the amino acids in the sequence that is dictated by the messenger RNA. There is a special codon called a stop codon to tell the transfer RNA that the protein is complete—no more amino acids need to be added. The amino acids strung together make up a protein. This process of creating proteins from the recipe in the genes is happening in the cells all the time. Whether or not a gene is being activated is known as gene regulation.

So the DNA contains the genes, the genetic information from the parents. Humans are diploid organisms, which means that they have two copies of all the DNA: one from mom, and one from dad. Sometimes, these two copies have the same information for a certain gene and sometimes they have different information for a certain gene. That bit of information for a certain gene is called an allele. I like to think of alleles as flavors. If the gene is ice cream, then the flavors are chocolate, vanilla, strawberry. There might be many alleles in a population, but any diploid organism will just have two flavors of alleles: one from mom, and one from dad. Most animals are diploid, and many plants are, too, but plants can also have more copies, which is one way we get different varieties of plants. Your apple, for example, might be diploid, triploid, or tetraploid, depending on the variety of apple.

Plants and animals not only share the same DNA building blocks, they share many of the same DNA sequences or genes. For example, the human genome, or collection of all the human genes, and the rice genome share about 25% of the same genes. However, when hearing such a statement, it's also important to realize that shared genes are expressed via alleles that are very different for humans compared with plants. On the other hand, some proteins such as cytochrome C, which is important for respiration, are found in all plants and animals.

We owe a lot of basic information about genetics to a monk living in the 19th century. Gregor Mendel did his classic pea experiments in what is now the Czech Republic during the 1850s. He actually developed these ideas about inheritance without knowing anything about genes, alleles, or DNA. I love this example because I think it demonstrates how science is a useful construct even when there isn't that much that's understood.

In Mendel's time, people thought inheritance was all blending. If your mom had blonde hair and your dad had dark brown hair, you would have lightish hair. Sometimes that's true, but a lot of times it isn't. By using pea plants as a model organism, Mendel was able to meticulously breed pea plants and make observations about the outcomes. They have male and female parts, so they can be self-fertilized within the plant, or they can be cross-fertilized—that is, between different plants. Mendel did his crosses, or pollination, with a paintbrush. He would get pollen from one flower and then put it on the pistil of another flower. He would use the pollen and pistil from the same flower if he wanted a self-pollination cross.

He then had to wait for the fruit to form before he could tell what he got. Pea plants take about 3 weeks to form fruit after the plant flowers. This is one reason why botanists don't use pea plants as model organisms much anymore. But for Mendel, they worked well, because there are a number of visible differences called phenotypes that can be observed from various crosses. The seven traits that Mendel observed were seed color, seed shape, seed coat color, pod color, pod shape, flower position, and stem length.

So, even though Mendel didn't know it at the time, the seven traits that he looked at were seven different genes. The physical trait of the gene was the allele, so, looking at seed shape as the gene, plants could have a smooth allele or a wrinkly allele. So, to determine if blending was occurring, Mendel crossed two homozygous parents. Homozygous means that a given parent had the same two alleles, so the wrinkly parent had two wrinkly alleles, and the smooth parent had two smooth alleles. When he did this, he found that all of the offspring were smooth. He had disproven the idea of universal blending because none of the seeds were a blend of smooth and wrinkly. He also observed that one allele, smooth, was dominant over the other.

He then wanted to know if these traits could be inherited independently, so he did another cross of two parents that were both heterozygous for the wrinkly trait, which means that each parent had one wrinkly allele and one non-wrinkly allele. In genetics, the dominant allele is usually written as a capital letter and the recessive allele is written in lower case. He was able to show that the allele for the wrinkly could be inherited separately from the allele for color, so that there were all sorts of combinations that were possible: green-smooth, green-wrinkly, yellow-smooth, and yellow-wrinkly. Even though there were only 4 phenotypes that resulted, there were 10 different genotypes.

Thus, Mendel was able to show some pretty important features of genetics, and these basic features are true for plants and people. First, he was able to show that blending doesn't always occur, and this is because the alleles segregate from their parents. This happens when parents form gametes, or egg and sperm. The egg and sperm only get one copy of an allele, and these segregate during gamete formation. Mendel's second finding was the idea of independent assortment, which means that genes don't have to be inherited together. Mendel's third observation was that of dominance, meaning that some alleles are dominant over others.

Some textbooks call these Mendel's laws, but this is inaccurate, I believe. Certainly, segregation occurs during gamete production, but some genes are in fact linked, and will be inherited together. Also, some alleles are codominant, meaning that both alleles can be expressed. An example of codominance is seen in the human blood type AB. This blood type means that an A allele from one parent and a B allele from another parent would both show up in the blood phenotype. An example in plants would be a pink carnation, which would result from the breeding of a red carnation with a white carnation. Both colors get expressed and pink is the result. So, Mendel's ideas are good models for how much of genetics works, but it's not how genetics works all the time.

So, we know that the basic genetic principles are the same in plants and animals, but we learned that there are three unique features of plant cells: the cell wall, the vacuole, and the chloroplast. A group of cells together makes up a tissue. Because plants have such different cell types from animals, they also have different tissue types than animals.

There are three main tissue types in plants, and these are categorized by the thickness of the cell wall. Tissue with the thinnest wall is called parenchyma. Thinking about form and function, where might plants want to have the thinnest cell walls? Certainly somewhere there might be a lot of exchange—the leaves. And parenchyma is the tissue that makes up much of the internal structure of the leaves. Collenchyma is a tissue with irregularly thickened cell walls. These are structural cells, though they are not found in the roots. Typically, they are most common in the stems of herbaceous or non-woody plants, like celery.

Sclerenchyma is tissue that contains cells with thickened secondary walls that may contain lignin, with is a substance that provides thickness. Sclerenchyma serves a protective or supportive function in the plant. This tissue type is easy to remember when one thinks of arteriosclerosis, which means a hardening of the arteries. "Sclero" comes from the Greek *scleros*, meaning hard. Anyone who has eaten a pear has intimate knowledge of sclerenchyma because pears contain a special cell called stone cells that have lots of sclereids. The grainy or gritty taste in pear flesh comes from the sclereids. In fact, under a microscope, these cells are so recognizable they've been used to solve crimes. Chromoplasts from red bell pepper are another example.

Forensic botany is the use of botanical material to solve crimes. Dr. Jane Bock and colleagues first published a lab manual about this for the U.S. Department of Justice in 1969, with information about plant tissue types and plant cell walls that have been used to solve dozens of cases. Sclereids of pears are so recognizable that one can still find them in the gastric contents of victims. It turns out that there are a number of recognizable cells in plants that have been used to detect where and when last meals were eaten.

So, plant intelligence may be controversial, but there's no doubt that plants can solve crimes. Plants and people definitely have some genes in common and use some of the same basic building blocks of DNA to make proteins. Like people, plants have tissues, but plant tissue types depend on the thickness of the cell walls, which our animal tissue types don't have at all, so the form and function of plant tissues is different. Most important and most surprising, given these differences, plants are able to sense their environments and respond. The open question is whether plant sensation is akin to sensation in animals, and whether plant responses to their environment are akin to animal behavior and communication.

LECTURE 3

MOSS SEX AND PEAT'S ENGINEERED HABITAT

Mosses and the other bryophytes live in places that later plants on Earth have been less eager to inhabit. They can thrive in low light and nearly anaerobic conditions. And while they do need nearly constant water to thrive, many can bounce back from even a long dry spell within hours. They aren't large, but they offer sights—and insights—available nowhere else in the plant kingdom.

MOSS

- The plant kingdom is divided into a dozen phyla, also known as divisions. The division that includes moss is known as Bryophyta. The bryophytes are thought to be one of the oldest groups of plants that first made it to land. This means that they developed new ways to breathe, preserve moisture, and reproduce. These 3 traits are what distinguish all plants from algae, even from algae that can do photosynthesis.

- A few smaller plant divisions were traditionally lumped in with the bryophytes: the liverworts (phylum Marchantiophyta), with upward of 6000 species; and the hornworts (phylum Anthocerotophyta), which have only a few hundred species.

- All 3 phyla of bryophytes, in this broader sense, were the first plants to colonize land. They also happen to be quite different, in shared ways, from all other plants. This is why they have all traditionally been grouped together and called bryophytes in a looser sense.

LIVERWORT
Marchantiophyta

HORNWORT
Anthocerotophyta

UMBRELLA LIVERWORT
Marchantia polymorpha

◆ They all lack true leaves and true roots; in fact, they lack vascular tissue altogether. This essentially means that they have no way to conduct sugars or water through the plant. It all basically happens by osmosis or diffusion. Without dedicated tissues to move water, minerals, and sugars through the plant, and without true roots to absorb water and minerals, bryophytes are restricted to being small.

◆ Just because mosses arrived early and aren't big doesn't mean that we have to consider them as evolutionary runts or remnants. They are still very well adapted to the habitats in which they live, such as the side of tall trees or on rocks in a stream. Mosses are necessary for many animals and form the basis of whole ecosystems, such as peat bogs. Mosses are also really good at being able to handle low-light conditions. Mosses do require habitats that are wet, at least some of the time, and this has to do with the way they reproduce.

◆ Across all 3 bryophyte phyla, there are only around 15,000 bryophytes, which is about 5% of all plant species, but they live in remarkable ways, in many habitats, and have a unique way of reproducing.

◇ How is a moss different from its ancestors? Green algae are a lot like bryophytes—they photosynthesize and need water to reproduce—but there are 2 important differences. The most important is that the mosses, like all other true plants, will have a protected place for the embryo to develop. By contrast, algae release their sperm and eggs directly into the water, where they fuse to form zygotes, which can then travel through the water to settle somewhere else as a developing embryo.

◇ A second partial distinction is that all algae can reproduce sexually or asexually, while only some plants can reproduce asexually, such as mosses. What separates mosses and other asexual plants more clearly from algae is that the asexual propagule in algae can be dispersed. But in plants, any asexual reproduction—when it's possible—is generally anchored to the parent plant.

◇ The other major difference between algae and bryophytes is that the bryophytes have a special waxy cuticle all over that keeps water in their cells. This cuticle is present in all plants. By contrast, only a few red algae have a cuticle. Because algae live in water, they don't need a cuticle. These 2 developments in the bryophytes—protection for the developing embryo and a waxy cuticle to protect the entire plant—were the most important adaptations to land.

GOUTY-MOSS
Oedipodium griffithianum

MOSS SEX

- Both algae and mosses can reproduce sexually or asexually. They do something pretty unusual: They alternate between a haploid and diploid generation. Diploid means 2 copies of the chromosomes; haploid means 1 copy of the chromosomes. Algae also alternate generations. Animals don't have a haploid life stage. The only haploid thing in animals is the sperm and the egg, and these things are not free-living as whole generations for any animal.

- The production of the sperm and the egg is pretty straightforward—only in mosses, it's a whole gametophyte generation that produces sperm and egg, which means that the whole free-living plant is either male or female with only 1 copy of chromosomes (it's haploid).

- Then, the sperm swim to the egg, which is located in a special chamber, the archegonium. This swimming is why most mosses are found in areas that will have occasional water, so that the sperm can swim to the egg.

- When the egg and sperm fuse together, the fertilized egg, or zygote, develops in the archegonium. But that fertilized egg is now diploid, with 2 copies of the chromosomes—1 copy from the egg and 1 copy from the sperm. It is called the sporophyte, and it produces spores.

- Spores are another important characteristic of the bryophytes. Not only do bryophytes lack true leaves, true roots, and vascular tissue, they don't even have seeds. They have spores instead.

- A spore is haploid. In the sporophyte, a process called meiosis occurs, which changes the diploid cells to haploid. This is the same process that occurs in our bodies to make gametes, meaning our eggs or sperm.

- Seeds are diploid. They have all the information to make a diploid organism. But in plants that don't have seeds, spores will germinate to form new haploid plants, and this haploid plant is called the gametophyte.

- The gametophyte plant then produces egg and sperm, each of which are also haploid, and returning to the beginning of the life cycle. This process is called alternation of generations because the haploid and diploid generations are alternating. Even though they're alternating, they are still attached, because the sporophyte, the diploid generation, will grow out of the gametophyte.

- Although there are similarities in algal and bryophyte reproduction—in the alternation of haploid and diploid life stages—when bryophytes became successful on land, they had not only the waxy cuticle and a dedicated place for the fertilized egg to develop, but they also had a more sophisticated approach to alternation of generations.

- For algae, alternation of haploid and diploid generations produces no important change in the body of the organism. But for bryophytes, the haploid body is very different from the diploid body, with the haploid body being better adapted to independent life on land.

- In most mosses, the haploid body looks like a short, leafy structure that stays close to the ground. This is so that water can accumulate within the moss and the sperm can swim to the egg. The diploid part of the moss is a delicate stalk on which the spore-releasing capsule perches to disperse the spores.

- All the bryophytes have 2 unique adaptations that enabled them to live on land: a protected place for the embryo to develop and the waxy cuticle. The mosses have one more adaptation, the development of stomata. These are small pores in the leaf that allow for the exchange of gases, meaning that carbon dioxide goes in and oxygen goes out and thus the gases are exchanged. This greatly enhanced the ability of the mosses to live on land.

PEAT MOSS

- *Sphagnum* is a genus of a few hundred mosses that are commonly known as the peat mosses. This group of mosses is economically important because of their water-holding capacity, which comes from the fact that sphagnum includes many dead cells surrounded by live cells.

PEAT MOSS

Sphagnum sp.

- Sphagnum grows in watery, treeless bogs. But it doesn't just live in bogs; it can actually create these bogs. Because sphagnum moss can hold so much water, it can basically flood out other plants. In fact, sphagnum moss can hold roughly 20 times its weight in water.

- This amazing water-holding capacity will make it very difficult for tree roots to access oxygen, because wetland soil is lower in oxygen. Tree roots need oxygen because the cells in the roots undergo respiration, and those roots will drown in too much water because they can't get oxygen. Over time, trees, shrubs, and many other plants will die out, but the sphagnum moss will remain, robust in its engineered habitat.

- The lack of oxygen also slows down the growth of bacteria, and without bacteria, decomposition is also slow. Because of slow decomposition, the dead sphagnum accumulates over time; this is the peat moss with which gardeners are familiar.

- This slow decomposition also means that there aren't as many minerals being released back to the soil, so bogs are limited in nutrients. The bog of sphagnum is also highly acidic; over time, the sphagnum moss changes the acidity by lowering the pH.

- The process of having some sphagnum in a pond or lake to a whole bog with roughly 2 meters of sphagnum and other bog plants takes time—hundreds or even thousands of years. Because a peat bog takes so long to accumulate, harvesting peat for burning isn't a sustainable resource, at least not on the timescale of trees. Burning peat, which holds a lot of carbon, isn't great for the atmosphere.

- The International Peatland Society is working on a sustainable practices certification and guidelines so that some peat harvesting can take place and gardeners can continue to use peat moss.

- Peatlands occupy only about 3% of Earth's surface, yet they store perhaps twice as much carbon as all the forests on Earth put together. So, preserving peatland may be an especially efficient way to keep carbon locked away, while degrading peatland may have especially serious effects.

READINGS

Gilbert, *The Signature of All Things*.

Kimmerer, *Gathering Moss*.

QUESTIONS

1 What is a distinguishing factor of a plant? For example, how are plants distinguished from algae?

2 How is sphagnum moss an ecosystem engineer?

LECTURE 3 TRANSCRIPT
MOSS SEX AND PEAT'S ENGINEERED HABITAT

In the movie *Avatar*, there are bioluminescent plants that glow in the dark. But did you know there really is a glow-in-the-dark plant? It's a moss that lives mostly in caves and dark places. In normal environments, this moss gets outcompeted by other mosses, but around cave entrances and in other very shady places, it can thrive. In North America, its range loops from Alberta, Canada to Montana, Oregon, and Washington, northward through British Columbia to Alaska. The species is circumboreal, which means it's found throughout the higher latitudes of the Northern Hemisphere, but below the polar latitudes.

This species is called *Schistostega pennata*, but it has several common names: luminous moss, cave moss, and dragon's gold. In Japan, there is a national monument for this moss, and also a 1954 novella called *Luminous Moss* by Taijun Takeda, which was made into an opera with the same title in 1972. I won't spoil the story, but the protagonists end up living in a cave, where luminous moss makes a big impression.

To understand how and why luminous moss glows the way it does, we need to understand some things about mosses generally. The plant kingdom is divided into a dozen phyla, also known as divisions, and we'll start with the division that includes moss, known as Bryophyta. The prefix "bryo" means moss or moss-like. The Greek root *phyta* means—you should know this one—*phyta* means plant. The Bryophytes are thought to be one of the oldest groups of plants that first made it to land. This means they developed new ways to breathe, preserve moisture, and reproduce. These three traits—new ways to breathe, reproduce, and preserve moisture—are what distinguish all plants from algae, even from algae that can do photosynthesis.

There are a couple of smaller plant divisions that were traditionally lumped in with the bryophytes. These are the liverworts—Phylum Marchantiophyta, with upward of 6000 species—and the hornworts—Phylum Anthocerotophyta—which have only a couple hundred species. All three of these bryophyte phyla, in this broader sense, were the first plants to colonize land. They also happen to be quite different, in shared ways, from all other plants. This is why they have all traditionally been grouped together and called bryophytes in a looser sense. First, they all lack true leaves and true roots—in fact, they lack vascular tissue all together. This essentially means that they have no way to conduct sugars or water through the plant. It all basically happens by osmosis or diffusion. Without dedicated tissues to move water, minerals, and sugars through the plant, and without true roots to absorb water and minerals, bryophytes are restricted to being small, some only a millimeter. A bryophyte will never be as tall as a tulip because it lacks the upward plumbing to get water up that high, or the downward pipes to get sugars down so low.

Just because mosses arrived early and aren't big, doesn't mean that we have to consider them as evolutionary runts or remnants. They are still very well adapted to the habitats in which they live, such as the side of tall trees or on rocks in a stream. A traditional botany course will start with the mosses and then proceed through evolution until the flowering plants in a way that can leave the impression that flowering plants are somehow better than mosses. But this is not true. Mosses are necessary for many animals and form the basis of whole ecosystems, like peat bogs. And in terms of being able to handle low light conditions, mosses are really good, while most flowering plants are pretty bad, with ferns falling roughly in between.

Mosses do require habitats that are wet, at least some of the time, and this has to do with the way they reproduce. Are we really going to talk about moss sex? Yes we are. Moss sex is interesting because it's so different than the way most other plants do it. Across all three bryophyte phyla, there are only around 15,000 bryophytes, which is about 5% of all plant species, but they live in remarkable ways, in many habitats, and have a unique way of reproducing.

So, how is a moss different from its ancestors? Green algae are a lot like bryophytes—they photosynthesize, they need water to reproduce—but there are two important differences. The most important is that the mosses, like all other true plants, will have a protected place for the embryo to develop. In fact, everything that we now call a plant is known as Embryophyta, to mark the central importance of a protected embryo. By contrast, algae release their sperm and eggs directly into the water, where they fuse to form zygotes, which can then travel through the water to settle somewhere else as a developing embryo.

A second partial distinction is that all algae can reproduce sexually or asexually, while only some plants can reproduce asexually, such as mosses. What separates mosses and other asexual plants more clearly from algae is that the asexual propagule in algae can be dispersed. But in plants, any asexual reproduction, when it's possible at all, is generally anchored to the parent plant.

The other major difference between algae and bryophytes is that the bryophytes have a special waxy cuticle all over that keeps water in their cells. This cuticle is present in all plants. By contrast, only a few red algae have a cuticle. And because algae don't live in water, they don't need a cuticle. These two developments in bryophytes—protection for the developing embryo, and a waxy cuticle to protect the entire plant—were the most important adaptations to land.

Let's get back to our moss sex. As I said, both algae and mosses can reproduce sexually or asexually. To think about this, we need to look at chromosomes, and we need to unpack a few terms of botanese, the language of botanists. The most familiar chromosomes are diploid. The prefix "di" means two and "ploid" refers to chromosomes. Thus, diploid means two copies of chromosomes. The contrast word we need to understand here is haploid—this is one copy of the chromosomes. These words are important in showing how mosses and algae do something pretty unusual: they alternate between a haploid and diploid generation. To put this in perspective, it would be like my eggs just sort of living life on their own, doing their own thing outside my body for a whole generation. Crazy! In order to understand this unique life cycle, we have to start somewhere in the circle. This is literally a chicken and egg type thing here, except the egg is much more independent.

Which came first? It's hard to say, since algae also alternate generations as well. What's also so interesting is that animals just don't do this; animals don't have a haploid life stage. The only haploid thing in animals is the sperm and the egg, and these things are not free-living as whole generations for any animal.

While we're on the subject of the haploid stage, in bryophytes it's the actually the haploid stage of the gametophyte that is truly independent, while it's the diploid sporophyte that remains attached and dependent. This would be like me being dependent on my own egg.

So, let's start where we're probably all comfortable: the production of sperm and egg—pretty straightforward, only in mosses it's a whole gametophyte generation that produces sperm and egg, which means that the whole free-living plant is either male or female with only one copy of the chromosomes: it's haploid. Then the sperm swim to the egg, which is located in a special chamber, not the womb, but the *archegonia*. The singular is *archegonium*. This swimming is why most mosses are found in areas that will have occasional water, so that the sperm can swim to the egg. When the egg and sperm fuse together, the fertilized egg, or zygote, develops in that archegonium. But, that fertilized egg is now diploid—two copies of the chromosome: one from the egg and one copy from the sperm. So it gets a special botanese name: the sporophyte. The sporophyte produces spores.

Aha, spores. This is another important characteristic of the bryophytes. Not only do they lack true leaves, true roots, and vascular tissue, they don't even have seeds. They have spores instead. What, you may ask, is a spore? And why is the spore not a seed? A spore is haploid. What, back to the single copy of genes lifestyle all over again? Yes. In the sporophyte, a process called meiosis occurs, which changes the diploid cells to haploid. This is the same process that occurs in our bodies to make what we call gametes, meaning our eggs or sperm. So, the spores are haploid. Seeds are diploid; they have all the information to make a diploid organism. But in plants that don't have seeds, spores will germinate to form new haploid plants, and in botanese, this haploid plant is called the gametophyte.

Our gametophyte plant then produces egg and sperm, each of which are also haploid, and now we are back to the beginning of the life cycle again. This whole process is called alternation of generations because the haploid and diploid generations are alternating. Even though they're alternating, they are still attached, because the sporophyte—the diploid generation—will grow out of the gametophyte. Remember, the sporophyte developed from the fertilized egg, which was located in the archegonium of the gametophyte.

Looking at the algal alternation of generations, it's clear that there are parallels between algae and bryophytes in the alternation of haploid and diploid life stages. But why would something like this have evolved? Students really like to ask this question. A typical answer often says because this is type of reproduction was probably present in the ancestor to the plants, a sort of green algae. Over plant evolution, there was a gradual move toward a dominant diploid stage, most likely to mask deleterious genes that would show up in a haploid organism.

But this is also a great time to mention that natural selection isn't always the most efficient process. Remember, whatever organisms are successful are the ones that reproduce, and that's the way generations of offspring will continue to reproduce unless there's a mutation that manifests an even better reproductive success. Thus, although there are similarities in algal and bryophyte reproduction, when bryophytes became successful on land, they had not only the waxy cuticle and a dedicated place for the fertilized egg to develop; they also had a more sophisticated approach to alternation of generations.

For algae, alternation of haploid and diploid generations produces no important change in the body of the organism. But for bryophytes, the haploid body is very different from the diploid body, with the haploid body better adapted to independent life on land. In most mosses, the haploid body looks like a short, leafy structure that stays close to the ground. We know that this is so water can accumulate within the moss and the sperm can swim to the egg. The diploid part of the moss is a delicate stalk on which the spore-releasing capsule perches in order to disperse the spores.

How do we know that bryophytes were the first plants on land? We have a few pieces of evidence. The first comes from fossils. By dating the rock layer surrounding a fossil, scientists can determine about when that fossil was formed. This sort of dating is called relative dating—for some of us, maybe sort of like dating in middle school. Not that kind of dating. Fossils themselves can also be dated using absolute dating. In both cases, the dating occurs using radioisotopes. Some isotopes are stable, which is when an atom has a different number of neutrons, but it doesn't degrade over time, but in radioisotopes the neutrons do leave the atom over time. This is called radioactive decay, and this is what makes radioactive dating possible. The time it takes for half of the isotopes in a sample to decay to the normal type of atom is called a half-life. By comparing the number of radioactive isotopes left in a sample, and knowing the half-life, the age of the sample can be determined.

Another technique to determine when an organism evolved is to use a molecular clock, which takes advantage of the fact that mutations seem to come at roughly constant rates over time. Let's suppose a common ancestor begins with a certain sequence in the DNA. After 25 million years, there is a base pair mutation in each of two daughter lineages. After another 25 million years, there is another mutation in the base pairs, to form two more daughter lineages. So, using this fossil evidence, we calibrate our molecular clock to say there is a base pair mutation every 25 million years. Known facts about mutation are translated into a rate of mutation, which acts as a clock by which time from divergence can be estimated.

Since neither the radioactive dating nor the molecular clock technique is perfect, both of these techniques are used together to provide an estimate. Using the estimates, the first plants likely appeared on land about 425 million years ago. Some estimates put that date back to about 475 million years ago. These early land plants probably evolved in an area of shallow water from something like a green algae. The most likely first plant was probably something more like a liverwort, our *Marchantia*. The unfortunate name liverwort comes from the Old English word "wyrt," which means herb. The liver part comes from the fact that the general shape of the plant might resemble the shape of a liver.

This name also brings up an interesting tidbit about the naming of plants and phyla. When I went to school, the liverworts were Phylum Hepatophyta. That was nice, because the prefix "hepato" refers to the liver and the *phyta* part means plant. But this name was problematic because there wasn't actually a genus or family in the phylum that had the term hepato in it. In other words, that name "hepato" wasn't actually represented anywhere in the phylum. So, the new phylum name is Marchantiophyta. The namesake is liverwort genus *Marchantia*, which is the most common and best known of the liverworts.

The liverwort *Marchantia* has a flattened gametophyte with no obvious leaflike structures, so it's easy to see how it could have been derived from green algae. The haploid gametophyte actually forms specialized structures called *gameophores* or *gametangiophores*, which are the regular flattened gametophyte called the *thalli*, just sort of rolled up and growing perpendicular to the thalli. These gametangiophores produce male *antheridiophores* and female *archegoniophores* that produce the sperm and egg separately. When the sperm fertilizes the egg, the sporophyte grows out from the archegonia and releases the spores. On top of all these unique structures, the liverworts can also reproduce asexually by the formation of small balls of tissue called *gemmae*, which are produced in gemmae cups, which really are small cups on the base of the thallus.

I have here a bryophyte. This is a *Marchantia*, or a liverwort, and it sort of looks like a liverwort, maybe, the lobed shaped of the thallus, which is the green sort of leafy-like material that you can see here: that's the gametophyte. You can't see the sporophyte on this particular organism, but what you can see are the gemmae cups, and the gemmae cups are where the asexual propagule will be produced. But they're really small, so I need a hand lens—this is my 10× hand lens—and I need this to actually see them. But I know kind of what I'm looking for, so let's see if we can find it. Oh, I see one; there's one right there. So, it can really be a lot of fun, but you've got to take your time, and it really helps to have a magnifying lens or a hand lens so that you can see up close.

The point here is that liverworts are uniquely different from the rest of the bryophytes, particularly the mosses, because of their flattened shape and the way that they reproduce. The hornworts are also very different, primarily in the way their spores are displayed. Hornworts have no capsule—the spores are attached along a horn-like structure, which is the diploid sporophyte.

We already mentioned that all the bryophytes have two unique adaptations that enabled them to live on land: a protected place for the embryo to develop, and the waxy cuticle. The mosses have one more adaptation: the development of stomata. These are small pores in the leaf that allow for the exchange of gases, meaning carbon dioxide goes in and oxygen comes out—thus, the gases are exchanged. This greatly enhanced the ability of mosses to live on land.

But back to our cave moss for a minute. Now that we know something about the growth form of mosses, we can begin to understand why this particular moss glows in the dark. The beginning, immature growth of a moss gametophyte is called the *protonema*, singular; or the *protonemata*, plural. The protonemata in cave moss grow in a lens formation so that the chloroplasts, which appear as bright green lumps, can sit at the focal point of the lens-shaped cell and receive more light. Thus, the chloroplasts gather light more efficiently, while the lens-shaped cell also reflects some of this light, which makes them glow in the dark. The immature haploid gametophyte making this show possible most closely resembles a partly upright green algae.

So, we see that the mosses have a unique type of reproduction, unique forms that go along with this type of reproduction, and are likely the oldest land plants. Despite all this uniqueness and goodness, mosses are sometimes referred to as the lower plants. Most botanists don't really adhere to this sort of naming anymore, as it implies there is something inferior about the mosses. Although these bryophytes may be limited in size and stature, and confined to areas where there's moisture, they're still extraordinarily well adapted to their environments.

On the other hand, some things that look mossy may not in fact be a moss. Such is the case with so-called Spanish moss, which is neither Spanish nor a moss. This plant, often seen draping off of live oaks in the Southeastern United States, is actually a bromeliad, a family of flowering plants that grow in the canopies of other trees. It has tiny flowers—green, brown, or yellow—that are easy to overlook. Spanish moss is native to the U.S., not Spain. In fact, it wouldn't be found in Spain at all unless it was cultivated there, because Spain lacks the humidity for the plant to grow naturally. Legend has it that the name Spanish moss was derived as an insult. The Native Americans used to call it tree hair. The French explorers then turned it into *barbe espagnole*, or Spanish beard, perhaps to insult their rivals in the exploration of the New World. The name Spanish moss just stuck.

Likewise, reindeer moss is not a moss, either. This important source of food for reindeer and caribou is actually a lichen, which is neither a moss nor even a plant. A lichen is a symbiotic mutualism of algae and fungus. Textbooks have traditionally stated that it takes a total of two species to make a lichen. On this basis, scientists tried for years to grow lichens in the laboratory, but in 2016, Toby Spribille from the University of Graz in Austria, now working at the University of Montana, discovered that there are actually two fungal partners, plus one algal species. Evidently, for lichens, it takes three to tango. All this time, the second species of fungus involved in any lichen symbiosis had gone unnoticed. Researchers now think the missing ingredient is a small yeast. There are always new things to discover in botany.

So, I like to think of lichen as an algae sandwich on fungus bread. The fungus grows as long threadlike strands called *hyphae*, while the algal cells inside photosynthesize, and so they provide the fungus with sugars. This is a symbiosis where all three organisms are benefitting, but the result is not a plant, and it's not a moss. A somewhat well known example that is a moss is sphagnum moss. Sphagnum refers to a whole genus of a few hundred mosses, which are commonly known as the peat mosses. This group of mosses is economically important because of their water-holding capacity, which comes from the fact that sphagnum includes many dead cells surrounded by live cells.

I think this is best described by bryologist Robin Kimmerer, a botanist who studies mosses and is a professor at the State University of New York College of Environmental Science and Forestry. Dr. Kimmerer wrote a lovely book called *Gathering Moss: A Natural and Cultural History of Mosses*. In it, she writes:

> What amazes me most about Sphagnum is that most of the plant is dead. Under the microscope you see that every leaf has narrow bands of living cells which border the patches of dead cells, like green hedgerows around empty pastures. Only one cell in twenty is actually alive. The others are merely dead cell walls, skeletons surrounding the open space where the cell contents used to be.

This is a beautifully written and informative book, and it was also part of the inspiration for Elizabeth Gilbert's novel *The Signature of All Things*, which is a sweeping book centered on a female bryologist, who studies mosses and then life more generally. Although *The Signature of All Things* has not proved as popular as Gilbert's *Eat, Pray, Love*, I enjoyed it a great deal more, maybe because of the plant parts.

So, at first glance, sphagnum may not seem terribly exciting. I mean, it's a mostly dead, small moss that grows in watery, treeless bogs. The exciting thing is that sphagnum doesn't just live in bogs; it can actually create these bogs. The idea that one species can manipulate or create its own habitat shouldn't be all that surprising to us—that's exactly what humans do. But other creatures do it, too. Think of a beaver. A beaver moves into an area that has a stream and, if the stream is small enough, a dam can be created overnight. What was a stream will now be a pond. The beavers have engineered their own habitat, and so they are what ecologists call ecosystem engineers.

It turns out that sphagnum moss, too, is an ecosystem engineer. Because sphagnum moss can hold so much water, it can basically flood out other plants. In fact, sphagnum moss can hold roughly 20 times its weight in water. This amazing water-holding capacity will make it very difficult for tree roots to access oxygen, because wetland soil is lower in oxygen. Tree roots need oxygen because the cells in the roots undergo respiration, like we do, and those roots will drown in too much water because they can't get oxygen. Over time, trees, shrubs, and many other plants will die out, but the sphagnum moss will remain, robust in its engineered habitat.

The lack of oxygen also slows down the growth of bacteria, and without bacteria, decomposition is also slow. This is why several bog bodies have been discovered, because there are fewer bacteria and the decomposition process is slow. Because of slow decomposition, the dead sphagnum accumulates over time, and this is the peat moss with which gardeners are familiar. This slow decomposition also means that there aren't as many minerals, like nitrogen, being released back to the soil, so bogs are limited in the nutrients, which is probably the main reason why carnivorous plants like Venus flytraps live in the surface of bogs. There may not be much nitrogen in the bog below, but the there are bountiful bugs in the air, each one a delicious, protein-powered snack.

Now, as if this weren't enough engineering, the bog sphagnum is also highly acidic. The sphagnum moss changes the acidity through cation exchange. Chemistry tells us that a cation is an atom with a positive charge. These cations are found in the soil and water surrounding the sphagnum in low concentrations. The main cations present are calcium, Ca^{2+}; sodium, Na^+; potassium, K^+; and magnesium, Mg^{2+}. The 2+ or + represents the charge. These cations have a positive charge because they are missing an electron, one +; or missing two electrons, 2+.

Because nutrients are limited in the bog, sphagnum must employ some special tactics to absorb them. The cation exchange occurs when sphagnum pumps out one kind of cation—in this case a simple hydrogen ion, which is also just a proton without an electron—in exchange for the cations it wants, the calcium, sodium, potassium, and magnesium. What's left outside is an accumulation of H+, which causes acidity. This process can leave a pond that once didn't have sphagnum to lower the pH down to about 3.5 once sphagnum has taken over.

Of course, the process from some sphagnum in a pond or a lake to a whole bog with roughly 2 meters of sphagnum and other bog plants takes time—hundreds or thousands of years. Because a peat bog takes so long to accumulate, harvesting peat for burning isn't a sustainable resource, at least not on the timescale of trees. Burning peat, which holds a lot of carbon, isn't so great for the atmosphere. The International Peatland Society is working on a sustainable practices certification and guidelines so that some peat harvesting can take place and gardeners can continue to use peat moss.

Because of the slow decomposition and the low pH, peat moss was actually used as a wound dressing during World War I, because cotton supplies were limited because of the war with Egypt. Perhaps most impressive, peatlands occupy only about 3% of Earth's surface, yet they store perhaps twice as much carbon as all the forests on Earth put together. There is some overlap—wetlands with mangrove forests have trees but also produce a lot of peat. So, preserving a peatland may be an especially efficient way to keep carbon locked away while degrading peatland may have especially serious side effects.

We know that bogs take a long time to accumulate, but how fast is other moss growth? Apparently, it's quite slow, too, usually more than a decade. Meanwhile, the horticultural trades still collect moss for all sorts of arrangements. According to Dr. Patricia Muir, the craft and floral trades collected some 17 million pounds of moss for horticultural use in 2003, and much of that was not permitted. Much of this moss is collected from the canopies of trees. At first glance, maybe moss poaching doesn't seem like a big deal, but it is. Mosses are super important bioindicators. This means that if moss is abundant, everything else is probably doing OK. A study in 2016 in Portland, Oregon, reliably used moss growth to make detailed pollution maps of the city. Moss are good bioindicators because they lack true roots—

they pick up most of what they need from the air, especially epiphytic mosses, which are those that grow in trees. Mosses also provide habitat and food for other animals.

So maybe we can just grow mosses, like a crop. That's exactly the idea that Dr. Nalini Nadkarni had. Dr. Nadkarni is a professor of biology at the University of Utah, which is not a hotbed of moss research, but before that she was a faculty member at The Evergreen State College in Olympia, Washington, which is a moss hotbed. In thinking about the question of how to cultivate moss, she writes:

> No protocols exist for growing mosses in large quantities. To learn how best to grow them, I needed help from people who (1) have long periods of time available to observe and measure the growing mosses, and (2) have access to extensive space to lay out flats (shallow plastic trays) of plants, and (3) fresh minds to put forward innovative solutions. All three can characterize incarcerated individuals.

Dr. Nadkarni's idea became the Moss-in-Prisons project. That is, although mosses seemed too slow growing for the commercial trade, Dr. Nadkarni invited prisoners to participate in the research, and incidentally earned Dr. Nadkarni several award and accolades for marrying justice and conservation. Plants save the day, again.

Mosses and the other bryophytes live in places that later plants on Earth have been less eager to inhabit. They can thrive in low light and nearly anaerobic conditions. And while they do need nearly constant water to survive, many can bounce back from even a long dry spell within hours. They aren't large—indeed, their rhizoids and other structures are sometimes only a single cell thick—but they offer sights, and insights, available nowhere else in the plant kingdom. So, get yourself a field glass. Mosses and their non-vascular relatives are the Old Ones, and they're always worth another look.

LECTURE 4

FERN SPORES AND THE VASCULAR CONQUEST OF LAND

Even though ferns have been on the planet for a very long time, their ancestry is still a bit of a mystery, and they do a lot of unusual things. Even though we may not use them much economically, at least not yet, ferns still have a lot to tell us about themselves and the rest of the plant kingdom.

FERNS

- Our understanding of ferns isn't as deep as it is for other plants. We don't use ferns for much economically. They don't have seeds, flowers, or fruits, so there isn't much to eat, though there are some edible species and parts. We don't use ferns for much beyond horticultural beauty.

- Even botanists find it difficult to figure out their phylogeny, or ancestral tree. The names of the phyla, which is the second largest category after kingdom, have changed over time, and fern species often look alike, even to botanists, but they have different genes, which prevents cross-fertilization. Plus, ferns are among the oldest land plants, so they've had a long time to mutate back and forth.

- There are many gaps in our evolutionary understanding of ferns. As the phyla now stand, there are 4 different phyla of ferns that are extinct, which means that these phylogenies get figured out for whole groups of plants using only what can be seen in the fossil record.

- Ferns are very good at invading an area after a disturbance, which, in ecology, is anything out of the usual that will disrupt the growth of plants. Disturbances can be small, such as a herd of deer coming through to graze, or big, such as a volcanic eruption. Hawaii has had several volcanic eruptions in the last century. When lava flows over the existing plant life, everything underneath burns up. Eventually, the lava hardens. Even though there is nothing but hardened lava to begin with, eventually plants find their way onto this lava to colonize it.

- About 6 different types of plants can colonize these lava fields. How long it takes these ferns to move in depends on the amount of water. On the dry side of the islands, it takes much longer for colonization to occur. If it's wet enough, the National Park Service reports that sword ferns can appear on flows after about a year. Apparently, this fern is intolerant of shade but very tolerant of sun and areas of low soil quality.

- In addition to needing water to grow, ferns, like mosses, need water to reproduce, because, like mosses, they still have swimming sperm. Like mosses, ferns need alternation of generations to reproduce, but ferns have a slightly different reproductive pattern. Still, there are some similarities with mosses, such as a free-living gametophyte stage.

- This haploid stage—meaning one copy of the chromosomes—is heart-shaped in ferns and is called the prothallus. This word is also used to describe the gametophyte of the mosses.

- This haploid prothallus contains both the archegonia, which will produce the egg, and the antheridia, which will produce the sperm. The prothallus also contains rudimentary rootlike structures called rhizoids, but it has no leaves. The prothallus grows flat on the ground, and it's small—about 2 to 5 millimeters wide.

- Then, as in mosses, the motile sperm must swim through at least a thin layer of water to the egg. The fertilized egg, or zygote, will then develop into the sporophyte, which grows right out of the archegonia to become the plant we recognize as a fern. In ferns, the gametophyte generation is still present, but it's much reduced. This trend will continue as we make our way to the seed plants.

- Like mosses, ferns produce spores instead of seeds. The spore is haploid and will produce the gametophyte. When we get to seed plants, there is no free-living gametophyte stage. Still, this way of life, with a free-living gametophyte stage, is very successful for ferns and mosses.

- Each fern plant can produce billions of spores, and each spore is about 20 times lighter than the lightest seed, and much smaller. This means that spores can travel massive distances on the wind. Fern colonization of the lava fields is probably more a result of getting there before seeds might be able to get there, rather than a specialized adaptation to germinating specifically on lava fields.

FERN SPORES

◇ This ease of dispersal is one reason why many ferns are invasive. Especially in the tropics, a field of ferns growing in full sunlight is an indication of a disturbance—usually, ground was cleared for gardening or grazing. These ferns form such dense thickets that trees cannot get established, and the forest cannot regrow without some assistance from people or another kind of disturbance, such as fire.

◇ Although ferns can travel anywhere, they can't survive everywhere. Unlike mosses, which thrive in the high latitudes, most ferns don't withstand cold very well, though both the Arctic and Antarctic possess several species.

DISTINGUISHING FEATURES

- Worldwide, there are anywhere from 10,000 to 15,000 species of ferns and what botanists call the fern allies, which are phyla other than the main phyla of ferns, which is Monilophyta. That name has changed over time to try to form a monophyletic group, which is a group that includes one ancestor and all of its descendants.

- Four other phyla generally included in the allies of ferns are all extinct. Those extinct phyla included the so-called seed ferns, which didn't so much die off as evolve into later seed plants. There are 2 phyla now that are extant, or still living: Lycopodiophyta, also known as the lycophytes or the club mosses, which are not actually mosses; and Monilophyta.

- Another phylogeny of ferns groups them into 4 classes, which is the next taxonomic level down from phyla. These classes seem to make the most sense. Plant phyla end in the suffix -phyta, classes end in -opsida, orders end in -ales, and families end in -aceae.

- Despite the difficulty in determining the fern phylogeny, all living ferns reproduce by spores. These spores are often the way you can determine if a plant is indeed a fern, or a fern ally.

- The spores are formed in organs called sporangia, and all the sporangia are typically grouped together in lines or small dots on the leaves of the fern. (Leaves for a fern are usually called fronds.) One of these small dots will contain numerous sporangia, and together these sporangia will form a sorus, which is one dot, or a group of sori, which is many dots.

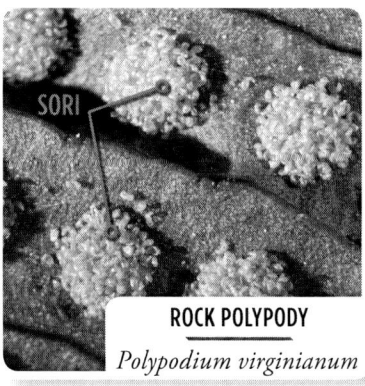

ROCK POLYPODY
Polypodium virginianum

◇ Another way to distinguish a fern is to determine where the leaves, or fronds, start. In ferns, all the fronds come from the base, and this is even true in tree ferns.

◇ Vascular tissue is the main adaptation that separates ferns from mosses. The development of this tissue cannot be overstated in its importance to life on land. Vascular tissue is equivalent to a circulatory system in animals. The difference between an animal that has to rely on diffusion for transport within the body versus an animal with a circulatory system is the same as the difference between mosses and ferns. All the other plants in this course have vascular tissue, and ferns got there first.

◇ All vascular plants have 2 main types of vascular tissue: the xylem and the phloem. The xylem transports water and minerals up from the ground, and the phloem transports sugars from photosynthesis in the leaves to the rest of the plant. The phloem typically travels along the inside of the bark in woody plants. Xylem and phloem are found in ferns, and no ferns are woody, not even tree ferns.

SOFT SHIELD FERN
Polystichum setiferum

◇ Like other ferns, tree ferns don't have branches. All the fronds come from the top of the trunk-like stem. Also, like other living ferns and fern allies, tree ferns have lignin in the cell walls of the xylem. Lignin is a component of the cell wall in many vascular plants. Like cellulose, lignin is a polymer, but it is a polymer of phenyl propene basic units. Propene is a hydrocarbon, and phenyl signifies a 6-carbon ring with 5 hydrogens attached. This complex branching structure provides a lot of durability and strength to the cell wall.

- Together, vascular tissue and lignin provided the structure for all ferns to stand up and live on land, but tree ferns got even bigger. Tree ferns were very dominant growth forms, from about 200 million years ago, up until the extinction event that wiped out the dinosaurs. They were the quintessential dinosaur food. Most tree ferns living today are in the order Cyatheales.

LIVING IN THE UNDERSTORY

- While some tree ferns and the ferns that colonize lava fields can withstand full sunlight, many ferns are well adapted to living in the understory.

- Dr. Fay-Wei Li provided evidence about how ferns might have become adapted to living in the understory. It seems that many ferns that are adapted to full sun don't have a special photoreceptor called neochrome. A photoreceptor is a chemical within a plant that absorbs light.

- Most photoreceptors will only absorb light in a specific wavelength, but neochrome is special because it absorbs light in both the red and the blue. This means that ferns have a way to see in the dark understory. What's even more fascinating is that this trait wasn't just passed down from fern to fern in the 400 million years that ferns have been on the planet.

- Dr. Li showed that neochrome has only been present in ferns for about 179 million years. Dr. Li hypothesizes that genes were transferred between a hornwort and a fern about 179 million years ago and this is how some ferns are thriving in low-light levels. They borrowed the gene from a nonvascular plant, which perhaps got it from cyanobacteria.

- Fern dominance in low light is one reason why ferns make good houseplants. In fact, most of our common houseplants are understory forest plants that don't require a lot of light.

READINGS

Large and Braggins, *Tree Ferns*.

Mehltreter, Walker, and Sharpe, eds., *Fern Ecology*.

Moran, *A Natural History of Ferns*.

QUESTIONS

1   How are ferns different from mosses?

2   In what kinds of habitats would you expect to find ferns?

LECTURE 4 TRANSCRIPT
FERN SPORES AND THE VASCULAR CONQUEST OF LAND

Nineteenth-century Britain had a mania for ferns. Botanists were making new discoveries; Victorian collectors were fanatically collecting and, in some cases, decimating rare species from all over the British Isles; and Victorian graphic artists were putting fern motifs on all sorts of decorations and commemorative displays. By the 20th century, the fern craze had ferned out, with only fainter echoes reaching the U.S. Ferns subsided into something like a go-to cliché. In fact, if you were to search the internet now for ferns in the news, you might get an episode of *Between Two Ferns*, an ironic comedy show whose only connection with plants is that it takes place, literally, between two ferns.

So, when it comes to ferns, we think we know them, but we don't. We might think of ferns as delicate tropical plants for example, but they are much more than this. I think a couple of things may account for the fact that our understanding of ferns isn't as deep as it might be for other plants. First, we don't use ferns much economically—they have no seeds, no flowers, no fruits, so there doesn't seem to be much to eat, though in reality there are some edible species and parts, such as fiddleheads from ostrich ferns, and ancient humans may have eaten fern tubers. These days, we don't use ferns for much beyond horticultural beauty.

I think another reason ferns have become so overlooked is that even botanists find it difficult to figure out their phylogeny, or their ancestral tree. The names of the phyla, which is the second largest category after kingdom—remember, King Philip Came Over From Good Spain—have all changed since I learned them in school. The phyla have changed—phyla. It happens all the time that families and genera change, but the whole phyla?

Part of what makes ferns trickier is that fern species often look alike, even to botanists, but they have different genes, which turn out to prevent cross-fertilization. Another difficulty comes from the fact that ferns are among the very oldest of land plants, so they've had a long time to mutate back and forth. Together, what this means is that there are a lot of gaps in our evolutionary understanding of ferns. As the phyla now stand, there are four different phyla of ferns that are extinct, which means that these phylogenies get figured out for whole groups of plants using only what one can see in the fossil record.

Your general thought of a fern might be something growing on the forest floor, a plant that doesn't get too tall and was adapted to growing in the understory of a forest where light levels are low. A general biology textbook might put it something like this: "Ferns live in shady places that provide enough moisture, such as forests, fields, swamps, and areas near streams." What you may not realize about ferns, though, is that they are very good at invading an area after a disturbance.

In ecology, a disturbance is anything out of the usual that will disrupt the growth of a plant. So, a thunderstorm isn't considered a disturbance, but a lightning that causes a forest fire is considered a disturbance, because the fire is a change in business as usual. Although many plants are adapted to fire, it's still a disturbance to the plant community as a whole. Disturbances can be big or small. A small disturbance might be a herd of deer coming through to graze, sort of like mowing the lawn, while a big disturbance might be something big, like a volcanic eruption.

The Hawaiian Islands are no strangers to volcanic eruptions. As part of the Pacific Ring of Fire, where geologic activity promotes volcanoes and tsunamis, Hawaii has several volcanic eruptions in the last century. When lava flows over the existing plant life, everything underneath burns up. Eventually, the lava hardens. In Hawaii, there are two types of hardened lava: the *pahoehoe* and the *a'a*. These are differentiated by the sound of walking on lava in bare feet: *a'a*—"ah ah ah"—because this kind of lava is spiky and rough, it's difficult to walk on; and the *pahoehoe* lava is smoother.

Even though there is nothing but hardened lava to begin with, eventually plants will find their way onto this lava and colonize it. The process of that occurring is called primary succession. The vegetation that recolonizes an area that has been disturbed, but not disturbed to where there is no plant life, is called secondary succession. This all becomes important to ferns because about six different types of plants can colonize these lava fields, regardless of the type of lava. You might have thought it would be difficult to colonize the smooth *pahoehoe*, but apparently the type of lava doesn't change the colonization rate. How long it takes these ferns to move in depends on the amount of water. On the dry side of the island, it takes much longer for colonization to occur. If it's wet enough, the National Park Service reports that sword ferns can appear on lava flows after about one year. Apparently, this fern is intolerant of shade, but very tolerant of sun and areas of low soil quality, to say the least. These plants all need water, too.

It's not just that these plants need water to grow, they do, but like the mosses, the ferns also need water to reproduce because, like the mosses, they still have swimming sperm. Like mosses, ferns need alternation of generations to reproduce, but ferns have a slightly different reproductive pattern. Still, there are some similarities with mosses, such as the free-living gametophyte stage. This haploid stage—remember, haploid means one copy of the chromosomes—is heart-shaped in ferns, and it gets a special name: the *prothallus*. The word thallus, or thalli for plural, comes from the Greek word *thallos* for "green shoot," so this word was also used to describe the gametophyte of the mosses.

This haploid prothallus contains both the archegonia and the antheridia. Do you remember which produced what? Remember, the G generally means girl: it's the archegonia that will produce the egg, and the antheridia will produce the sperm. The prothallus also contains rudimentary root-like structures called rhizoids, but it has no leaves, and the prothallus grows flat on the ground, and it's small—about 2–5 millimeters wide. Then, as in mosses, motile sperm must swim through at least a thin layer of water to get to the egg. The fertilized egg, or zygote, will then develop into the sporophyte, which grows right out of the archegonia to become the plant we recognize as fern. So, in ferns, the gametophyte generation is still present, but it's much reduced. This is a trend that will continue as we make our way to the seed plants. But the seeds aren't there yet; ferns produce spores, not seeds.

So, what's so great about a spore anyway, and why isn't it a seed? Remember that mosses, too, produced spores and the spore's haploid, and will produce the gametophyte—that's the big difference. When we get to seed plants, there is no free-living gametophyte stage—it disappears. Still, this way of life, with a free-living gametophyte stage, is very successful for ferns and mosses.

Each fern can produce billions of spores, and each spore is about 20 times lighter than the lightest seed, and much smaller. This means spores can travel on the wind massive distances. This ability to disperse is one reason why there are over 50 endemic ferns in Hawaii. Endemic means that they are found no place else on Earth. So, spores got to Hawaii millions of years ago and then evolved into something different from species on the mainland.

What's so interesting about the ferns colonizing the lava fields is that it's probably more a result of getting there at all rather than a specialized adaptation to germinating specifically on lava fields. In over 30 years of research on Mount St. Helens, Dr. Roger del Moral determined that which plants were on a site was determined by who got there, not necessarily by who grew there best. So, it's not necessarily that these ferns that colonize lava beds have some unique adaptation to grow there; it's that their spores are really good at getting places and getting there first before seeds might be able to get there.

This ease of dispersal is one reason why many ferns are actually invasive. Especially in the tropics, a field of ferns growing in full sunlight is an indication of a disturbance, usually ground that was cleared for gardening or grazing. These ferns form such dense thickets that trees cannot get established, and so the forest cannot regrow without some assistance from people or another sort of disturbance like a fire. But, again, even if there is a fire, the fern spores are likely to make it to the burned up field first, and then the process begins again. In fact, one of the ferns that forms these thickets is a relative of one of the ferns that can colonize the lava beds. Again, the secret is in the dispersal—the lightweight, super small spores that can travel anywhere. And ferns are resilient outside lava beds as well. The so-called resurrection fern, *Pleopeltis polypodioides*, can lose 75% or more of its water. Some estimates even say 97% of its water. It looks fairly dead and brown during dry times, but when it's rehydrated by rains, it appears green again.

Although ferns can travel anywhere, they can't survive everywhere. Unlike mosses, which thrive in high latitudes, most ferns don't withstand cold too well, though both the Arctic and Antarctic possess several species. Still, the center of diversity for ferns is in the tropics. For example, Costa Rica in Central America has about 900 species of ferns, more than found in all of the United States.

Worldwide, there are anywhere from 10,000 to 15,000 species of ferns and what botanists can call the fern allies. I really like that term: ferns and fern friends—it sounds so pleasant. The fern allies, of course, are the other phyla that are the main phyla of ferns, which is Monilophyta. Now, that name has changed a lot. When I was an undergraduate student, the main ferns were the Pteridophyta, and the horsetails got their own phyla, the Equisetophyta. However, the names changed to try to form a monophyletic group, which is a group that includes one ancestor and all of its descendants. It's just a good way for botanists to try to organize life, but it means changes when new data is available. Ferns have a long history, almost equal to that of mosses, so paleobotanists always have new jigsaw pieces on hand, and many others have yet to be discovered.

Four other phyla generally included in the allies of ferns are all extinct. Those extinct phyla included the so-called seed ferns, which didn't so much as die off as evolve into later seed plants. So, thank you, seed ferns. Anyway, there are two phyla now that are extant, or still living: the Lycopodiophyta, also known as the lycophytes or the misnomer club mosses, since they aren't mosses; and the other fern phyla is the Monilophyta.

Another phylogeny of ferns groups them into four classes, which is the next taxonomic level down from phyla— King Philip Came Over From Good Spain. These classes seem to make the most sense, and this is a good time to point out the endings in plant taxonomy. All the plant phyla we have discussed will all end in the suffix "phyta"; all the plant classes we will discuss end in the suffix "opsida"; and all the plant orders will end in "ales"; and all the plant families will end in "aceae." For now, we'll stick to phyla and classes.

Despite the difficulty in determining the fern phylogeny, one thing is clear about all living ferns: all ferns today reproduce by spores. These spores are often the way you can determine if a plant is indeed a fern or a fern ally. The spores are arranged in organs called *sporangia*, and all the sporangia are typically grouped together in lines or small dots on the leaves of the fern, although leaves of the fern are usually called fronds. One of these small dots will contain numerous sporangia, and together these sporangia will form a sorus, which is one dot, or a group of sori, which is many of these dots. The word sorus comes from the Greek word *soros*, meaning "a heap." So, these sori are the main way to determine if the plant is a fern.

Another way to distinguish a fern is to determine where the leaves, or fronds, actually start. In ferns, all fronds will come from the base, and this is even true in a tree fern. A group of ferns called the whisk ferns will branch, but even these ferns lack true leaves—they look more like green stems. This group also lacks true roots, having simpler rhizoids that anchor them. But this group reproduces by spores and has vascular tissue, so they're included with the ferns. And vascular tissue is the main adaptation that separates ferns from mosses.

The development of this tissue cannot be overstated in its importance to life on land. Vascular tissue is equivalent to a circulatory system in animals, so the difference between an animal that has to rely on diffusion for transport within the body versus an animal with a circulatory system is the same as the difference between mosses and ferns. All the other plants we'll meet in this course will have vascular tissue, and ferns got there first. All vascular plants have two main types of vascular tissue: the xylem and the phloem. The xylem transports water and minerals up from the ground and the phloem transports sugars and photosynthates in the leaves to the rest of the plant.

The word xylem comes from the Greek *xylos* meaning "wood." Most wood is actually dead xylem cells. The word phloem comes from the Greek word *phloios*, which means "bark of trees." Indeed, as we'll see when we get to trees, the phloem typically travels alongside the inside of the bark in woody plants. Now, a non-woody plant might have xylem and phloem, too. Just because these words are derived from wood terms, doesn't mean that xylem and phloem are only found in wood. Xylem and phloem are found in ferns, and no ferns are woody, not even the tree ferns.

Like other ferns, tree ferns don't have branches—all the fronds will come from the top to the trunk-like stem. Also, like other living ferns and fern allies, tree ferns have lignin in the walls of the xylem. Lignin is a component of the cell wall in many vascular plants, and, like cellulose, lignin is a polymer, but it is a polymer of phenylpropene basic units. Propene is a hydrocarbon, and phenyl signifies a six-carbon ring with five hydrogens attached. This complex branching structure provides a lot of durability and strength to the cell wall.

Together, vascular tissue and lignin provided the structure for all ferns to stand up and live on land, but the tree ferns got even bigger. Tree ferns were very dominant growth forms from about 200 million years ago right up until the extinction event that wiped out the dinosaurs, so they were the quintessential dinosaur food. Most tree ferns living today are in the order Cyatheales. Do you remember which taxonomic group that was? Was it a phylum? No, no—plant phyla end in phyta. What's the next taxonomic group: King Phylum Came Over Class? Was it a class? That's a hard one. We haven't talked much about classes, but plant classes end in "opsida." Next grouping: King Philip Came Over—what is the O for? Order, yes, it's an order. Plant orders end in "ales."

So, as we discovered, the tree ferns are a bit of a misnomer. They aren't true trees in the sense that they don't actually produce wood like we know from a tree. However, they do produce a *pith*, which is a large area in the center of the trunk, and surrounding the pith is a cortex, which contains vascular tissue. Since these ferns don't contain true wood, they don't grow outward like a tree trunk, they only grow upward. In botany, we call this upward growth primary growth.

Because the tree fern only shows primary growth, it's difficult to tell how old it is. Some tree ferns are estimated to be about 200 years old, but this isn't all that old for a tree. Even though we think of tree ferns as tropical, there are some that live on subantarctic islands off the coast of New Zealand. Probably, these ferns were some of the first plants to arrive, thanks to the advantage of the spore, and they've been able to keep out other forms. There are forests of New Zealand that are dominated by tree ferns. Interestingly, there are not forests of tree ferns in the subtropics of the northern hemisphere, which may have to do with the inability of the tree ferns to handle the freezing winters of the northern hemisphere mid-latitudes.

Because tree ferns are so common in New Zealand, the Maori people there use them as construction materials. Even today, there are fences made out of tree ferns in New Zealand. And the fern is such a strong symbol of New Zealand that there was a vote to change the New Zealand flag to one with a fern on it. The 19th-century fern maniacs would have been so proud. Well, in 2015, New Zealand voters were asked to select their favorite alternative flag. The winning alternative flag had a frond of a silver fern on it, but it was still defeated by the traditional flag in early 2016. The silver fern, *Cyathea dealbata*, is a tree fern, and to my knowledge, it would have been the only flag of a country in the world with a fern leaf. Even though the silver fern doesn't appear on the New Zealand country flag, it does appear on the flag of the New Zealand rugby team, the All Blacks.

While some tree ferns and ferns we met earlier colonizing lava fields can withstand full sunlight, many ferns are well adapted to living in the understory. Given that ferns were the dominant plant form on land during the Carboniferous period 360–290 million years ago, what changed? I mean, when they were dominant, they were clearly well adapted to living in light. We know some ferns can live in full light, but why are many ferns adapted to shady places?

Dr. Fay-Wei Li of Duke University provided evidence about how ferns might have become adapted to living in the understory. It seems that many ferns that are adapted to full sun don't have a special photoreceptor called neochrome. A photoreceptor is a chemical within the plant that absorbs light. Most photoreceptors will only absorb light in a specific wavelength, but neochrome is super special because it absorbs light in both red and blue. What this means is that ferns have a way to see in the dark understory. What's even more fascinating is that this trait wasn't just passed down from fern to fern in the 400 million years that ferns have been on the planet. Dr. Li was able to show that this neochrome has only been present in ferns for about 179 million years. So where did it come from?

This is the super cool part. Remember how we talked about moss sex and fern sex? Remember how mosses and liverworts, hornworts, and ferns all have a free-living gametophyte in their life cycle, and that this gametophyte is haploid, meaning it has only one copy of the chromosome? Well, Dr. Li hypothesizes that genes were actually transferred between a hornwort and a fern about 179 million years ago, and this is how some ferns are thriving in low light levels. They borrowed the gene from a non-vascular plant, which perhaps got it from a cyanobacteria.

This is all astounding because most gene transfer occurs from parent to offspring. Sure, bacteria can transfer genes between species, but plants and animals don't do this very often. This process is called horizontal gene transfer because genes aren't transferred between parent and offspring, which would be vertically; the genes are transferred between two individuals of a different species—it's so fascinating. Then, all of the ferns that descended from the fern that received the gene from the hornwort became well adapted to exploit the area beneath the trees. This is how some ferns have become dominant in low light environments.

Fern dominance in low light is one reason why ferns make good houseplants. In fact, most of our common houseplants are understory forest plants that don't require a lot of light. A common house fern that most people are familiar with, and a fern that can be grown in some amount of sunlight, is the Boston fern. The Boston fern is definitely a fern, and the drooping version now popular as a houseplant began as a tropical plant mutant that was first noticed and commercialized in Boston. The fern was grown by a Southerner living in Boston, and the mutation allowed it to grow in the drier and cooler climates of northern households. The Boston fern is most likely descended from the sword fern of Florida.

It turns out that it might actually be difficult for mutations to show up in ferns, maybe a bit more difficult than in other plants, and the reason for this has to do with polyploidy. Remember, earlier we said that the gametes—the sperm and the egg—are haploid: they have one copy each of the chromosomes. When the sperm and egg fuse, they form an embryo that is the diploid, and this is also called the sporophyte.

Now, to get down with ferns, we need to get comfortable with the very cool fact that plants can actually have more than two copies of the chromosomes, and that's OK. In humans, you may have heard of trisomy 21, or Down syndrome. This occurs when there is a third copy of the 21st chromosome. In what botanists call polyploidy, there are third, fourth, fifth, sixth, seventh, and even eighth copies of all the chromosomes, and the plant does just fine. Research in the past decade has revealed that anywhere from 47% to 100% of flowering plants have polyploidy somewhere in their ancestry. The ferns, however, tend to be polyploid even more than flowering plants, and this polyploidy can even lead to speciation, or the formation of new species.

Let's say for example, that you're an organism with just three chromosomes. Keep in mind that humans each have 46 chromosomes—23 from dad and 23 from mom. Also keep in mind that the number of chromosomes doesn't relate to the complexity of the organism. As we'll see, many ferns have hundreds of chromosomes—a lot more than we do. Plants are fine carrying around extra copies, and some of the ferns have been around for a very long time.

OK, so say you are a three-chromosome organism. You go to form your sex cells, but instead of just giving one copy of each of the three chromosomes, you give two copies, or six chromosomes all together. If your gamete is still viable, as happens in plants quite often, then your offspring will now be triploid, meaning there will be three copies of each chromosome. This is all assuming that your partner gamete was produced normally, or with just one copy. If your partner gamete accidentally received two copies of the chromosome, like your gamete did, then the resulting offspring would have four copies of the chromosome. How many would that be? If you said 12, you're right. An organism that has 4 copies of 3 chromosomes would have 12 chromosomes all together. We would call this organism a tetraploid. The tetra would imply four copies of the chromosomes. Now that you're an organism with four copies of the chromosomes, you might not be able to reproduce with organisms with just one copy or even with two copies of the chromosomes, and so a new species would be formed through polyploidy.

There is still some debate in the botany literature about the advantages of polyploidy. Having an extra copy of a gene could come in handy if your copy gets damaged, but it becomes harder for mutation to show up in the organism. If mutations don't show up, meaning they don't get physically expressed, then there isn't anything for nature to select for or against, and so natural selection and adaptation may slow down. Yet, if so many plants have polyploidy somewhere in their history, it's unlikely that it's a big disadvantage. After all, frequent use of polyploidy is itself another form of mutation.

The fern life cycle of sporophyte and gametophyte generations has another unusual aspect. The haploid spores can germinate in three different ways: into the male gametophyte, the female gametophyte, or a hermaphrodite gametophyte. Which one the spore becomes can actually be determined by chemicals in the environment. In a 2014 paper in *Science*, Dr. Junmu Tanaka at Nagoya University in Japan explored this question. She found that spores that land first are more likely to germinate to female. These female gametophytes then release a chemical into the environment that causes the other spores that haven't yet germinated to develop into male gametophytes.

What Dr. Tanaka actually discovered was the chemical that causes this change. The female gametophytes release an inactive plant hormone called *gibberellin*. This same hormone, gibberellin, is found in lots of different kinds of plants across the plant kingdom, where it usually stimulates stem or other growth. But in ferns, the gibberellin hormone controls sexual difference. The hormone is inactive when the females release it, but when the inactive form is picked up by other developing spores, enzymes in those spores activate the gibberellins, and the active gibberellin will make them male. The trick is that the inactive form is absorbed by the developing spore much faster than the active form, which is why the females secrete the inactive form.

So, we've seen that even though ferns have been on the planet for a very long time, their ancestry is still a bit of a mystery, and they do a lot of unusual things. Even though we may not use them much economically, at least not yet, ferns still have a lot to tell us about ourselves and the rest of the plant kingdom.

LECTURE 5

ROOTS AND SYMBIOSIS WITH NON-PLANTS

The biggest, tallest, heaviest, oldest living things depend on roots, even though we do not see most roots, because they are underground. Many people think of roots as going deep underground, but most roots are surprisingly shallow. Some plants are connected by a single system of roots and underground stems to form a single larger plant. All plants exchange friendly and unfriendly chemicals with each other through their roots, sometimes even across the species barrier, on a constant basis. Besides interacting with other plants, underground is also where most plants interact with bacteria to obtain nitrogen and where plants have symbiotic, often mutualistic, relations with fungi to increase the surface area of the plant root systems.

THE IMPORTANCE OF ROOTS

- Roots are so important that they are the first evidence of germination in the seed. When a seed has germinated, it has sent out its new root, which is called a radicle. Roots are where plants absorb water, but what else do they do? And how does that water get all the way to the top of very tall trees?

- One redwood *Sequoia*, named Hyperion, was measured at approximately 379 feet tall. Roots for such a tall tree are surprisingly shallow, only a few feet deep, but they can spread horizontally for 100 feet in all directions.

- Although trees have root pressure, due to osmosis, root pressure is not enough to make water travel that high. Osmosis is the movement of water through a semipermeable membrane, and plant cells are semipermeable, especially to water. So, if a root has a lot of solutes in it, then the water will move from the soil, where the concentration of water is higher, into the root, where the concentration of water is lower. So, water flows down its concentration gradient into the root, but that's not enough pressure to send water up to the leaves.

GIANT SEQUOIA

Sequoiadendron giganteum

- For the water to get that high, it is drawn up by evaporation in the leaves. This means that the water concentration decreases as one goes up the tree. Because water is always moving from a greater concentration to a lesser concentration, it's moving up the tree to get to lesser concentration.

- The air outside the leaves will always have an even lower concentration of water than the leaves themselves, so evaporation is always occurring through the leaves. Moving water through the plant is called transpiration, so this particular process of water moving up and out of the plant is called evapotranspiration.

- But roots have much more to offer than just the uptake of water. Plants make their own sugars during photosynthesis, and for photosynthesis to occur, plants need other raw materials from the ground, including nitrogen, phosphorous, and potassium.

- The most important of these 3 elements is nitrogen, which is needed to make chlorophyll and DNA. Although the air we breathe is about 70% nitrogen, this is atmospheric nitrogen, or N_2, which is a form that neither we nor plants can use. Plants need their nitrogen in a form where the nitrogen atoms can be removed and used, and N_2 is too energetically expensive for plants to break.

- The chemical process to make nitrogen fertilizer, the Haber-Bosch process, requires extreme heat and pressure to make ammonia, or NH_3, which is a form of nitrogen that plants can use. Plants can also use NH_4^+, which is the ammonium ion.

- Plants can also use nitrate, NO_3^-, which can be produced when lightning flashes through the sky, breaking N_2 bonds, and then rain brings the nitrate and other nitrogen oxides down to Earth. All of these molecules contain nitrogen, and because the bonds aren't as strong, plants can extract the nitrogen and use it.

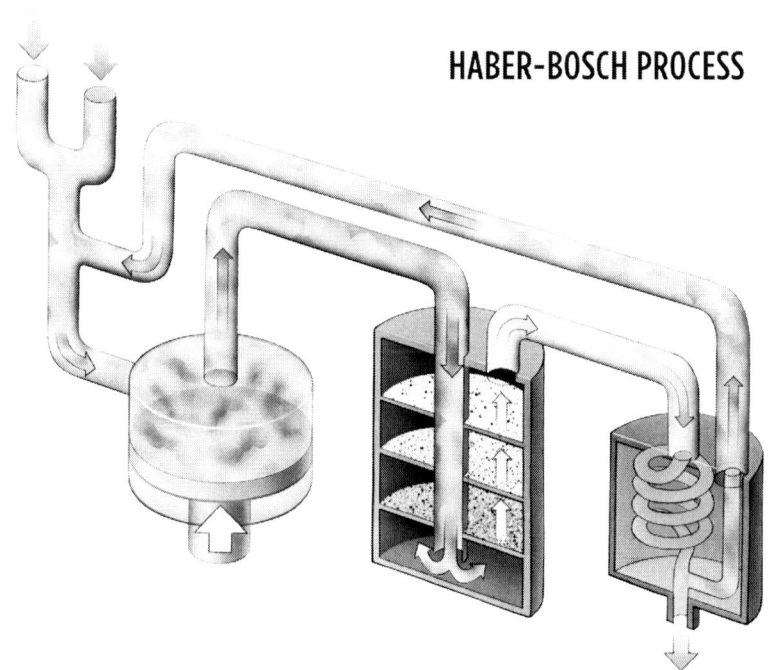

HABER-BOSCH PROCESS

NITROGEN

- All plants need a source of nitrogen, and most plants typically get it from bacteria.

- Bacteria don't have nuclei. They are in a group of organisms called prokaryotes, which means "before the nucleus." Plants and animals are eukaryotes, which means "true nucleus." Given that prokaryotes don't have a nucleus, they must have first evolved a long time ago, perhaps as early as 3.4 billion years ago, and some of these early bacteria were photosynthetic, though not exactly in the way that plants are today.

- Bacteria also have cell walls, though these are not like cell walls in plants. They have a completely different structure, and they are typically much smaller than a plant or animal cell.

◇ Bacteria do have chromosomes and genes, like plants and animals, but they are not contained in a nucleus. Bacterial genes are still on chromosomes, but they are free-floating in the cell.

◇ Bacteria also have extra pieces of DNA called plasmids, which are one reason why bacteria can change so quickly. Bacteria don't reproduce sexually; they simply divide. Mutation can occur and bacteria can exchange plasmids, but the genetic information of the bacteria can change quickly.

◇ Many people tend to think of bacteria only in pejorative terms. We call them germs and fight them with antibiotics. That's changing, and expressions like "probiotics" and "friendly bacteria" are more commonly understood. Plants have a lot to teach us about how bacteria can be very good, and the leading way bacteria are good is by providing usable nitrogen to plants.

◇ Bacteria can change atmospheric nitrogen into usable nitrogen in a process called nitrogen fixation. An organism that can fix nitrogen is a called a diazotrope. Nitrogen in the atmosphere is pretty much inert because the 2 nitrogen atoms are joined together in a triple bond, which takes a lot of energy to break. Bacteria have special enzymes called nitrogenase that will break this bond and convert the nitrogen into ammonia. Other bacteria help decompose organic material to make nitrogen available to plants.

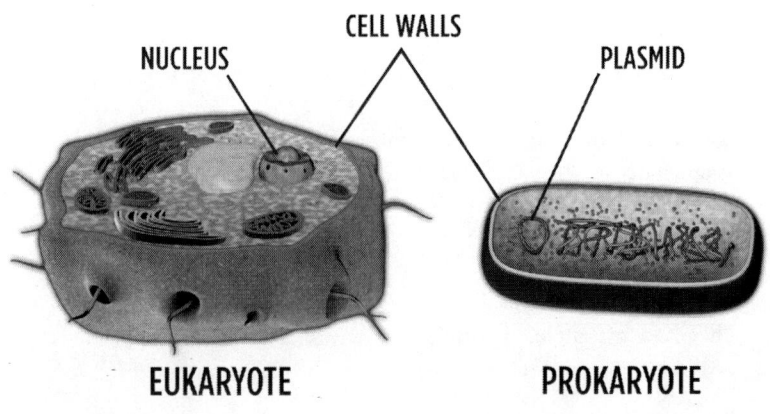

- There are about a dozen different bacterial phyla that can fix nitrogen. Some of these bacteria are free-living in the soil, using ammonia to make their own amino acids. Other bacteria are symbiotic with plant roots. The word "symbiotic" simply means living together. A symbiont doesn't have to be beneficial. In the case of nitrogen fixing bacteria, these symbionts are beneficial. When both organisms of a symbiosis benefit, ecologists call this a mutualism.

- This mutualistic symbiosis occurs in the roots, and the process is best known in the legumes—the pea or bean family, called Fabaceae. This family has many species, and it has many species that we use for food, such as soybeans and peanuts. These beans and peas have the symbiotic bacteria rhizobia fixing nitrogen in their roots.

- Rhizobia infect the roots of the bean plant early on, so the seeds aren't already inoculated with the bacteria when they germinate. In fact, the legume plant releases flavonoids, which are chemical want ads for rhizobia, into the soil. When the rhizobia sense the flavonoids, they release a chemical nodulation factor, which encourages the root to grow small nodules, which look like small peas attached to the roots.

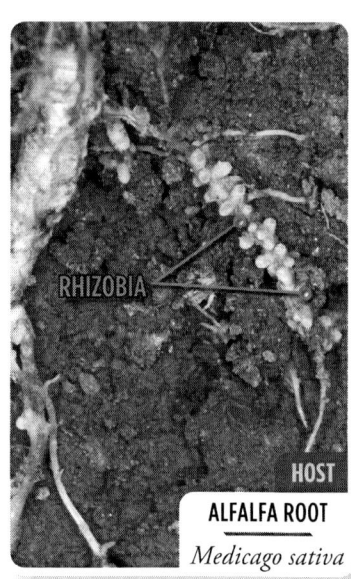

ALFALFA ROOT
Medicago sativa

- The bacteria form an infection thread to move in through the root hairs and fix nitrogen in their new home. The bacteria get a good habitat, and the plant gets a built-in source of nitrogen. The plant also feeds the bacteria sugars. Interestingly, although rhizobia can live in soil just fine, it can't fix nitrogen when it's free-living in the soil. It has to have a host.

ROOTS AND SYMBIOSIS WITH NON-PLANTS

◈ Not all legumes have associations with rhizobia. This leads to questions about how this mutualistic symbiosis came to be. It's possible that it evolved early on but then was lost by later descendants or that it evolved several times within the family. The mutualism is more common for legumes found in low-nitrogen environments. Rhizobia aren't the only bacteria that form symbiotic mutualisms with plants.

◈ Another genus of bacteria called *Frankia* form nitrogen-fixing root nodules with several types of woody plants. Even a few grasses get in on the action: Sugarcane, maize, and rice associate with different nitrogen-fixing bacteria that do not produce nodules. Bacteria have a variety of associations with plants. Some are very tight, forming nodules, and others are looser, with bacteria living in the area around the roots, called the rhizosphere.

◈ One exciting area of botany research involves seeing if other species can be coaxed into this bacteria mutualism. Because fertilizer is expensive to make and because plants only use about half of what gets applied, using bacteria to fix nitrogen across many kinds of crop species is an exciting idea.

OTHER MINERALS

◈ In addition to nitrogen, the other minerals that plants need are phosphorous and potassium. There is a host of macro- and micronutrients—including carbon, hydrogen, oxygen, phosphorous, potassium, nitrogen, sulfur, calcium, iron, and magnesium—that get into the plant the same way that water would, moving from a high concentration to a low concentration. Because it doesn't involve the movement of water, this process is called diffusion instead of osmosis.

◈ To pick up these minerals, plants need to have a lot of root area to exploit as much of the soil as possible. Overall, roots have 3 main functions: to support or anchor the plant, to absorb water and minerals, and to store carbohydrates. The first function of support can be met with the primary root system, which can be one big taproot or several fibrous roots. Most woody plants have a taproot first and then send out spreading roots.

◇ In addition to taproots and fibrous roots, there is another type of root called an adventitious root, which is derived from tissue other than root tissue. Adventitious roots often grow right out of the stem. A special example of adventitious roots are prop roots, which help prop up a plant, or provide extra support.

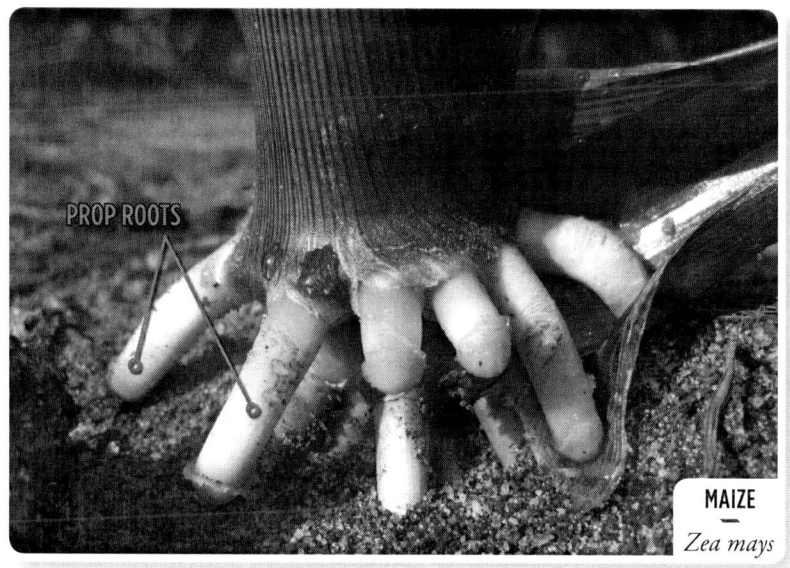

MAIZE
Zea mays

FUNGUS

◇ Another symbiotic mutualism that plants have to help increase the surface area of their roots occurs with fungus. Special types of fungus, called mycorrhizae, actually grow into the plant roots. The mycorrhizae are small threadlike structures of the fungus, called hyphae, that spread throughout the soil. The mycorrhizae increase surface area and therefore pick up extra nutrients, and the plant provides sugars to the mycorrhizae.

◇ Although they're special, mycorrhizae are very common; it's estimated that about 80% of all flowering plants and all of the gymnosperms have mycorrhizal associations.

◇ Scientists have found that the mycorrhizal associations beneath the forest soil connect trees of different species. Studies have also shown that trees can send chemical signals to each other via mycorrhizal connections.

READINGS

Capon, *Botany for Gardeners.*

Hodge, *Practical Botany for Gardeners.*

Kourik, *Understanding Roots.*

QUESTIONS

1   The beginning of the lecture talked about Pando, the giant clonal organism in southeastern Utah that might be the oldest and largest organism on Earth. Many plants are connected underground with underground stems or roots to form clones. What is a clone? What are the advantages to a clonal life-form? When would a clonal way of life not be so good?

2   What are mycorrhizae? Why are they so important for plants? What does the mycorrhizae get from growing with the plant?

LECTURE 5 TRANSCRIPT
ROOTS AND SYMBIOSIS WITH NON-PLANTS

Let's talk about roots by talking about extreme plants. In particular, let's talk about the biggest, tallest, heaviest, oldest living things, which it turns out depend on roots, even though we do not see most roots, because they're underground. Now, it's not that plants are like icebergs with, say, 90% or more of their mass below the surface. In fact, the biggest organism, as far as we know, is so large because of connections made by the roots, and this discovery was made in a paper published in *Nature* in 1992. The lead author, Mike Grant, was my first biology professor at CU Boulder in 1989, so this story is rather near and dear to my heart. In the letter to the editor of *Nature*, Dr. Grant acknowledges that some people believe the largest organism was a fungus, *Armillaria bulbosa*, which was estimated to weigh 10,000 kilograms and have an area of 15 hectares. That's about 15 football fields, or 37 acres.

But another candidate for the largest organism is a giant sequoia, or *Sequoiadendron giganteum*, which weigh in at just under 200,000 kilograms, which is 20 times the mass of the fungus, though not as much area or volume. Dr. Grant then reports finding an organism both more massive in weight and larger in area. It is—drumroll, please—an aspen tree. Now, many of you are probably familiar with the aspen tree, *Populus tremuloides*. After all, aspen, along with its sister Eurasian species, *Populus tremula*, has the largest range of any tree in the world. So, you've probably seen aspen trees, though never one as massive as a giant sequoia, or as large as the fungus. But, what you may not know about aspen is that aspen are clonal, and this means that when you look at a whole grove of aspen trees, they are all genetically identical. The clones are connected underground by their roots. When a new tree is born, it's just a sucker, or a shoot, coming off of one of the roots of an existing tree.

This doesn't mean that aspen are incapable of producing sexually—they can; aspen actually have male and female trees. In this sense, they're more like humans, and most trees have both sexes represented in one tree. In botanical language, we call trees with both sexes in one individual *monoecious*. Aspen are *dioecious*. That word is Greek for "two houses": a house or individual that is male, and another house or individual for female. Both the male flowers and female flowers are referred to as catkins. The name catkin refers to any sort of arrangement of small unisexual flowers in a pendulous stalk, and they're invariably wind-pollinated and hang down such to receive the wind. It used to be thought that aspen sexual production was rare, but it seems to be fairly common in that most stands of aspen, called *genets*, are related for about 50 meters, or half a hectare. The individual trees within the genet are called *ramets*.

Let's get back to Dr. Grant and his claim that an aspen genet is both the heaviest, largest, and possibly oldest organism on Earth. Of course, not all aspen genets are equally big, but Dr. Grant and his team, Dr. Jeffry Mitton and Dr. Yan Linhart, who was my genetics teacher by the way, found a genet in southwestern Utah that surpassed all records. This aspen clone was estimated to contain 47,000 individual ramets, or individual trees. If each one of those ramets was estimated conservatively at 100 kilograms, and the estimated root structure for each tree at an additional 30 kilograms, the clone would come in somewhere near 6 million kilograms, which is triple the sequoia and 60 times the mass of the fungus.

Later work showed the area of this clone to be about 43 hectares, or 107 acres, much larger than the 15 acres estimated for the fungus. What's more, as mentioned, these ramets are not just genetically identical; they're attached underground by their root system. If size weren't impressive enough, scientists speculate that this may be the oldest organism on the planet, and certainly the individual ramets aren't that old, but aspen are well adapted to disturbances. This means the root system can resprout after an avalanche, or a mudslide, or a forest fire. In this way, for such a clone to be so large, it must have resprouted many times in its life. Although we don't have the right scientific tools to date it exactly, these scientists speculate that it may be 80,000 years old.

Because the clone was the largest, heaviest, oldest organism on the planet, Dr. Grant named it Pando, which is Latin for "I spread." I think this is a wonderful attribute. As I said to you before, having a relationship with plants, you should call them by their name. We are *Homo sapiens*, but we all have names. The clone is *Populous tremuloides*, but its name is Pando. And, if you've remembered that aspen are dioecious, which means separate male and female plants, some of you may be wondering if Pando is a he or a she. Pando is a he, which is probably good because if he were a she, that might set off proposals to change Pando to Panda, which would be bad Latin, and might cause some confusion.

Clearly, having a large, spreading root system is an incredible asset here in creating the heaviest, largest, and possibly oldest organism on the planet. Roots are so important that they are the first evidence of germination in the seed. When a seed has germinated, it has sent out its new root, which is called a radical, and roots are where the plants absorb water. But what else do they do? And how does that water get all the way to the top of a very tall tree? One redwood sequoia named the Hyperion was measured at 379.3 feet tall. The roots for such a tall tree are surprisingly shallow, only a few feet deep, but they can spread horizontally for 100 feet in all directions.

If you've ever enjoyed a beverage through a straw, that's the right mechanism for thinking about how water gets to the top of a tree. Although trees have root pressure due to osmosis, root pressure is not enough to make water travel that high. Osmosis is the movement of water through a semipermeable membrane, and plant cells are semipermeable, especially to water. So, if a root has a lot of solutes in it, then the water will move from the soil, where the concentration of water is higher, into the root, where the concentration of water is lower. So, water flows down its concentration gradient into the root, but that's not enough pressure to send the water up to the leaves.

In order for water to get that high, it's drawn up by evaporation in the leaves. What this means is that the water concentration decreases as one goes up the tree. Since water's always moving from a greater concentration to a lesser concentration, it's moving up the tree to get to lesser concentrations, and the air outside the leaves will always have an even lower concentration of water than the leaves themselves, so evaporation is always occurring through the leaves. Moving water through a plant is called transpiration, so we call this particular process of water moving up and out of the plant evapotranspiration.

But roots have much more to offer than just the uptake of water. We know that plants make their own sugars during photosynthesis, but in order for photosynthesis to occur, plants need other raw materials from the ground. Sometimes these raw materials are referred to as nutrients, but this can be vague and confusing, because for human nutrition we might think of carbohydrates or sugars as nutrients. For this reason, I prefer to talk about specific elements, starting with nitrogen, phosphorous, and potassium, and instead of the vague word nutrients. I also like these words better because they have scientific abbreviations from the periodic table: N for nitrogen, P for phosphorous, and K for potassium. Talking about NPK is also the gardeners' language for plant fertilizers, but it's a useful construct for botany generally because all plants need these elements.

The most important of these elements is nitrogen, which is needed to make chlorophyll and DNA. Although the air we breathe is about 70% nitrogen, this is atmospheric nitrogen or N_2, which is a form that neither we nor plants can use. Plants need their nitrogen in a form where the nitrogen atoms can be removed and used, and N_2 is too energetically expensive for plants to break. In fact, the chemical process to make nitrogen fertilizer, the Haber-Bosch process, requires extreme heat and pressure to make ammonia, or NH_3, which is a form of nitrogen that plants can use. So, plants also use NH_4^+, which is the ammonium ion. The ion means that it has a charge on it, in this case a positive charge due to an extra hydrogen atom. Plants can also use nitrate, which is NO_3^-, which can be produced when lightning flashes through the sky, breaking N_2 bonds, and then rain brings nitrate and other nitrogen oxides down to Earth. All of these molecules contain nitrogen, and because the bonds aren't as strong, plants can extract the nitrogen and use it.

Chemical fertilizers obviously aren't available to most plants growing in nature. Chemical fertilizers account for about 25% of all nitrogen fixation on Earth, while lightning contributes roughly another 10% of all nitrogen fixation. But all plants need a source of nitrogen, and most plants typically get it from bacteria. Now bacteria are amazing, but we can't see them without a microscope, and they are really different. Bacteria don't even have a nucleus. So, I want to take a bit of time here to explain what they are, so that we can understand how bacteria help plants.

Bacteria are in a group of organisms called prokaryotes, which means before the karyote, or before the kernel, which means before the nucleus. Plants and animals are eukaryotes, which means true kernel, or true nucleus. Now, given that prokaryotes don't even have a nucleus, they must have first evolved a long time ago, perhaps as early as 3.4 billion years ago, and some of these early bacteria were photosynthetic, though not exactly in the way that plants are today.

Bacteria also have cell walls, though again these are not like cell walls of plants. They have a completely different structure, and they're typically much smaller than a plant or animal cell. Bacteria do have chromosomes and genes, like plants and animals, but they're not contained in the nucleus. Bacterial genes are still on chromosomes, but they're just free-floating in the cell. Bacteria also have extra pieces of DNA called plasmids, and these are one reason why bacteria can change so quickly. Bacteria don't reproduce sexually; they simply divide. But, mutation can occur and bacteria can exchange plasmids, so the genetic information of the bacteria can change quickly.

Many people tend to think of bacteria only in pejorative terms—we call them germs; we fight them with antibiotics. Of course, now that's changing, and expressions like probiotics and friendly bacteria are more commonly understood. Plants have a lot to teach us about how bacteria can be very good, and the leading way bacteria are good is by providing usable nitrogen to plants.

Bacteria can change atmospheric nitrogen to usable nitrogen, and this process is called nitrogen fixation. An organism that can fix nitrogen is a called a *diazotrope*. Nitrogen in the atmosphere is pretty much inert because the two nitrogen atoms are joined together in a triple bond, and this bond takes a lot of energy to break. Bacteria have special enzymes called *nitrogenase* that will break this bond and convert the nitrogen to ammonia. Other bacteria help decompose organic material to make it available for plants. There are about a dozen different bacterial phyla that can fix nitrogen. Some of these bacteria are free-living in the soil, using ammonia to make their own amino acids. Other bacteria are symbiotic with plant roots. The word symbiotic simply means living together. A symbiont doesn't have to be beneficial. In the case of nitrogen-fixing bacteria, these symbionts are beneficial. When both organisms of a symbiosis benefit, ecologists call this a mutualism.

This mutualistic symbiosis occurs in the roots, and the process is best known in the legumes, which is a French word meaning vegetable, but we might know legumes best as the pea or bean family. Botanists call this family the Fabaceae. The family level is below kingdom, phylum, and class and order, but botanists focus a lot on families because they usually have similarities in the flower or fruit structure. I like to tell my students that beans and peas are fabulous. And they are pretty fabulous, too. For one thing, this family has lots of species, and it has a lot of species that we use for food, such as soybeans and peanuts. On top of all this, these beans and peas have the symbiotic bacteria rhizobia fixing nitrogen in their roots. Clover are another plant using this same bacteria.

It turns out that rhizobia infect the roots of the bean plant early on, so the seeds aren't already inoculated with the bacteria when they germinate. In fact, the legume plant will actually release flavonoids into the soil, which are chemical want ads for rhizobia. When the rhizobia sense the flavonoids, they release a chemical nodulation factor, which encourages the root to grow small nodules which look like, well, peas attached to the roots. The bacteria form an infection thread to move in through the root hairs and live happily ever after fixing nitrogen in their new nodule home. Get it? Nodule home, like module home. So the bacteria get some good real estate, or habitat, and the plant gets a built-in source of nitrogen. The plant also feeds the bacteria sugars. Interestingly, rhizobia can't fix nitrogen when it's free-living in the soil; it has to have a host. Rhizobia can live in the soil just fine, but it won't fix nitrogen till it's in a comfy nodule with the promise of sugars.

Not all legumes have associations with rhizobia. This leads to the question about mutualistic symbiosis and how it came to be. It's possible that it evolved early on, but then was lost by later descendants, or that it evolved several times within the family. The mutualism is more common for legumes found in low-nitrogen environments. Rhizobia aren't the only bacteria that form symbiotic mutualisms with plants. Another genus called *Frankia* form nitrogen-fixing root nodules with several types of woody plants, such as alder, the *Alnus* genus; or bayberry, the *Myrica* genus; and *Casuarina*, a Mediterranean tree genus. *Frankia* grow really slowly in culture, and they require special nutrition, which means that they may be specialized symbionts—that is, they perform better when they're in a mutualism.

Even a few grasses get in on the action: sugarcane, *Saccharum officinarum*; maize, *Zea mays*; and rice, *Oryza sativa*, associate with different other nitrogen-fixing bacteria that do not produce nodules. These bacteria include *Glucoacetobacter, Azospirillum, Herbaspirillum,* and *Azoarcus*. What these examples demonstrate is the variety of associations that bacteria have with plants. Some are very tight-forming nodules, and others are looser, with bacteria living in the area around the roots, called the rhizosphere.

One exciting area of botany research involves seeing if other species can be coaxed into this bacteria mutualism. Because fertilizer is so expensive to make, and because plants only use about half of what gets applied, using bacteria to fix nitrogen across many kinds of crop species is an exciting idea.

So, we've seen how plants get nitrogen, but, remember, the other minerals plants need are phosphorous and potassium. There are actually a whole host of macro and micronutrients, which are easy to remember by the following mnemonic: C. Hopkins Café, Mighty Good. The CHOPKNS CaFe Mg stands for carbon, hydrogen, oxygen, phosphorous, potassium, nitrogen, sulfur, calcium, iron, and magnesium. All of these elements are plant macronutrients, except iron, which is a micronutrient. A small amount of iron is needed to make chlorophyll and mitochondria, but too much quickly becomes toxic.

All these minerals get into the plant the same way that water would, moving from a high concentration to a low concentration, only this isn't osmosis because it's not the movement of water. Anything moving from a high concentration to a low concentration is the process of diffusion. In order to pick up these minerals, though, plants need to have a lot of root area to exploit as much of the soil as possible.

Overall, roots have three main functions: to support or anchor the plant, to absorb water and minerals, and to store carbohydrates. The first function of support can be met with the primary root system, which can be one big taproot or several fibrous roots. Most woody plants will have a taproot first, and then send out spreading roots. Many of us have seen the spreading roots of trees coming up through sidewalks. Why would tree roots want to come up through a sidewalk? One reason is that sidewalks are impermeable to water, so water runs off the sidewalk, so the area around the sidewalk can have greater moisture.

Some trees are more prone to spreading their roots along the surface of the soil. Some trees won't cause your sidewalk much harm because their roots aren't as prolific near the surface of the soil. These trees are bur oak; hedge maple, *Acer campestre*; amur maple, *Acer ginnala*; sweetbay magnolia, *Magnolia virginiana*; crabapple, *Malus baccata*; and some varieties of Japanese crape myrtle, *Lagerstroemia fauriei*. Trees to avoid planting near sidewalks because they develop large surface roots include Norway maple, *Acer platanoides*; and red maple, *Acer rubrum*; beech, or *Fagus* species; sweetgum—and I love this name—*Liquidambar styraciflua*; and weeping willow, *Salix babylonica*. Also, keeping the crown pruned will slow down the root growth, but the best solution is to give trees plenty of water, or plant smaller trees.

So, we mentioned taproots and fibrous roots, but there is another type of root called an adventitious root, which is derived from tissue other than root tissue. Adventitious roots will also grow right out of the stem. A special example of adventitious roots are prop roots, which are just like they sound, they help prop up the plant—they provide extra support. Some of you may be wondering when I'm going to talk about potatoes or turnips or other so-called root crops. So let's be clear: the carrot is a root vegetable, but many of our so-called root vegetables are botanically stems called tubers, or bulbs. The difference is that the roots absorb water and minerals, and potatoes and turnips simply store starch, but they don't absorb water or minerals.

Aside from carrots, there are a few roots that we eat or use for various purposes, and the most important of these is probably cassava. Remember that roots absorb water and minerals, and provide support for the plant, but roots also store starch for the plant as well. The cassava plant is native to South America, probably Brazil, but it's cultivated throughout the tropical and subtropical world. The cassava forms large adventitious roots that store a lot of starch, and this starch is an important food source for many countries. Cassava is reported to have amongst the greatest yield of starch per acre of crop than any other in the world, often exceeding 20 tons of roots per acre.

However, cassava comes in two varieties: a sweet type and a bitter type. Both types contain a glycoside of hydrocyanic acid, or HCN, and when ingested, this hydrocyanic acid can shut down the respiratory system. There was actually a case of food poisoning in the Philippines where 27 children died after eating cassava cakes that were improperly made. The hospital was 20 miles away and that delayed the treatment. All these root types also have smaller growth off of them called root hairs. They do in fact look like small hairs, and their function is to increase the surface area of the roots so that more absorption of minerals and water can take place.

There is yet another symbiotic mutualism that plants have to help increase the surface area of their roots. This mutualism occurs with fungus. There are special types of fungus that actually grow into the plant roots, and these are called mycorrhizae. The "myco" part means fungal and the "rhizae" part means root, so literally fungus root. The mycorrhizae are small, thread-like structures of fungus called hyphae that spread throughout the soil. The mycorrhizae increase surface area and so pick up extra nutrients—remember, NPK, not sugars—and the plant provides sugars to the mycorrhizae. Although their special, mycorrhizae are very common. It's estimated that about 80% of all flowering plants and all of the gymnosperms have mycorrhizal associations.

There are two basic types of mycorrhizae: ectomycorrhizae and endomycorrhizae. Both of these types grow into the root, but the ectomycorrhizae stay outside the outer cells of the root, while the endomycorrhizae typically grow into the cells of the root. Ectomycorrhizae are more commonly associated with woody species, and they can actually match the total mass of the root. Endomycorrhizae are more commonly associated with herbaceous plants, and they don't have nearly the amount of mass.

Now, let's back up for a minute and talk about fungus. What do you know about fungi? Do you know why a mushroom always gets invited to the party? Because he's a fun guy. Sorry, I had to. Fungi are not plants. They're often included with plants because they're not animals either, and they are definitely multicellular, but when you look at the genetic evidence, fungi are actually more closely related to animals than they are to plants. I'm sort of annoyed that fungi get grouped with the plants. It's not that I don't love fungi, I do, but I think they deserve a little more respect. I mean, the world would be covered in rotting ooze if it weren't for decomposers like fungi and bacteria. Also, fungi are much older than land plants, probably 1.3 billion years old. The common ancestor of plants didn't make the move to land until about 700 million years ago.

Still, plants and fungi have some superficial similarities. They both don't really move all that much. They both have cell walls, though the structure of the fungal cell wall is different from that of plants. The fungi cell wall is made of chitin, which is the same hard substance in the exoskeletons of insects. Chitin is a long chain of carbohydrates, but it's much more rigid than the cellulose found in plant cell walls. Plants and chloroplasts photosynthesize sugars. Fungi are heterotrophs, meaning they can't make their own food. Most fungi are decomposers, which means that they use enzymes to break down external sources of carbon.

Most fungi live below the ground in those long thread structures called hyphae, and a large group of hyphae all together is the mycelium. And the mycelium can sometimes form the reproductive structure that we know as a mushroom. So, when you pick a mushroom from the ground, you aren't killing the fungus, because most of it is below ground; you're destroying its reproductive structure. But it may have already released its spores, which look a little like pollen but are whole new fungi in tiny pollen-sized packets, ready to be dispersed and settle somewhere new.

Even people can use mycorrhizae. The delicious chanterelle, *Cantharellus cibarius*, mushroom is the fruiting body—the mushroom, if you will—of the mycorrhizae. Also, the hearty *Boletus edulis* forms mycorrhizae associations among oaks, but it's also delicious. Scientists have even found that the mycorrhizal associations beneath the forest soil are actually connecting trees of different species. In a 1997 paper in *Nature*, Suzanne Simard and colleagues used stable isotopes to trace the movement of carbon between trees. Stable isotopes are useful in botany because, as their name suggests, they are stable, a reminder from chemistry that isotopes are molecules that have more or less neutrons in the nucleus of the atom than normal. So, a normal carbon atom has 12 protons and 12 neutrons, but a stable isotope of carbon has 12 protons and 13 neutrons.

This extra neutron makes it easy to detect, so the isotope can label where the carbon is moving. If you enclose plants and give them a labeled source of carbon dioxide with this isotope, you can trace where the carbon-13 moves. Using this technique, Dr. Simard and her colleagues determined that carbon was moving in two directions between two tree species: paper birch, or *Betula papyrifera*; and Douglas fir, which is *Pseudotsuga menziesii*. The paper even showed that the Douglas fir was getting the better end of the deal, receiving more net carbon than the paper birch in the trade-off.

Further studies have shown that trees can send chemical signals to each other via mycorrhizal connections. Plants that were infected with harmful fungi in the lab sent chemical signals to their neighbors. Plants that had received the chemical from the mycorrhizae were less likely to get an infection even after exposed to the harmful fungus. Researchers hypothesize that this carbon exchange would help seedlings get started as well.

Many people think of roots as going deep underground in search of water, but most roots are surprisingly shallow. Instead, most of the action is just below the surface, where an individual plant becomes far more than the sum of its individual parts. Some plants, such as aspens and poplar trees, are literally connected by a single system of roots and underground stems to form a single larger plant, and some plants exchange friendly and unfriendly chemicals with each other through their roots, sometimes even across the species barrier, on a constant basis.

Besides interacting with other plants, underground is also where most plants interact with bacteria to obtain nitrogen, and underground is where plants have symbiotic, often mutualistic, relations with fungi to increase the surface area of the plant root systems. In fact, the underground network is so vast, connected, and important that some scientists call it the Wood Wide Web. Just as the internet ties people together and makes much of our life possible, something similar seems to be true for the roots for plants.

LECTURE 6

STEMS ARE MORE THAN JUST THE IN-BETWEEN

Stems—from the long stem of a rose to the amazing cork of *Quercus suber*—all seem to share 2 functions: gas exchange; and the transport of water up from the ground and the movement of sugars down to the roots. But there are exceptions. In early spring, the direction of movement is reversed, until leaves can be formed. Modified stem types exist that can store sugars. Stems can be rhizomes, below ground, where they look like roots. The stem is the vascular core of the plant, and arguably the overlooked heart of what it means to be a plant.

CORK

◈ Cork is the outermost layer of a stem. Many people would call this the bark, but the bark actually includes the cork. It also includes the phloem, which is the tissue that carries sugars from the leaves to the root. If all layers of the bark are removed, all the way around the tree—a process called girdling—then that will kill the tree.

◈ When cork is harvested, the inner bark is left intact. It's also important that the cork cambium is left intact. This is a layer just inside the cork and outside the phloem that will produce the new layer of cork. Any tissue called a cambium will produce other tissues.

◈ The cork is made up of cells that have cell walls that are filled with suberin. Sort of like lignin, the complex substance that provides toughness and allows ferns to stand tall, suberin is a complex substance that provides waterproofing. In fact, suberin is so complex that the molecule's exact structure is still being determined.

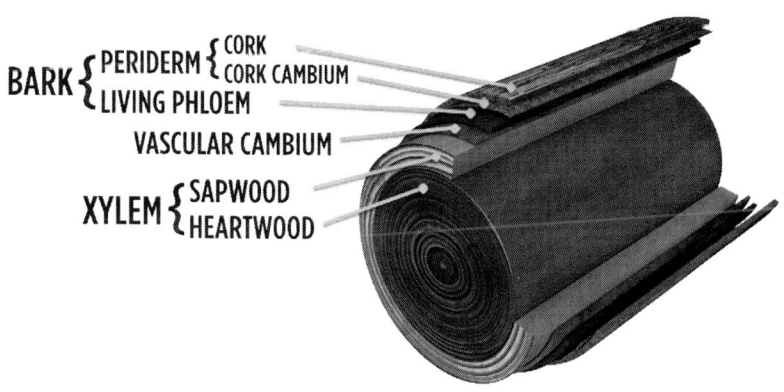

- All woody plants have cork. The *Quercus suber*, found throughout the Mediterranean coastlands, produces a lot of cork, which has a lot of suberin. Another oak, such as *Quercus rubra*, or red oak, doesn't produce nearly as much cork.

- The Mediterranean region, where cork grows, has hot, dry summers, perfect for fire. So, having a big cork means that the inside of the tree can be protected from the fire. Cork is fire resistant, so the fire, especially if it's a ground fire that doesn't get too hot or into the canopy of the tree, will just burn around the cork tree, leaving it unharmed.

CORK OAK
Quercus suber

- On the outside, any woody stem has this structure: The cork is on the outside, and surrounding the cork is cambium (where new cork is made), which together are called the periderm. Inside the periderm is the phloem, which first appeared in ferns but is fundamentally the same for all vascular plants. The phloem transports the sugars from fronds or leaves down to the roots. There may be other parts of the plant that need sugars, too, such as developing fruits, and the phloem would transport sugars there.

- Interior to the phloem and xylem of the stem is the vascular cambium, which makes new vascular tissue. Cells within the vascular cambium differentiate to phloem on the side that is closest to the outside of the trunk, and cells nearest the inside become the xylem cells. The cells in the vascular cambium start off as undifferentiated stem cells, and as they move into the xylem or phloem, they develop traits of those tissue types. The accumulation of these xylem cells creates wood. Lignin is the substance responsible for toughness in the plant cell, and xylem cells are heavily ligninized.

- Both the phloem and the xylem don't contain just one type of cell. Like the blood of humans, numerous kinds of cells are involved. The cells are specialized in terms of their function. Many of the cells of the phloem are specialized to transport sugars. Likewise, many of the cells in the xylem are specialized for transporting water.

- The words "xylem" and "phloem" refer to all of the following: the tissue type, the cells that make up the tissue, and the fluids within the tissue. The 2 most important cells in the xylem are tracheids and vessel elements, both of which are lined with pits that open and close in very sophisticated ways to make sure that water gets to the parts of the plant where it's needed most.

- Like most other parts of the plant, stems also have 3 other basic tissue types: parenchyma, collenchyma, and sclerenchyma. Parenchyma are the cells with the thinnest walls, collenchyma has uneven walls, and sclerenchyma has the thickest walls.

- All of these cells are found throughout the xylem and phloem. Their function relates to their structure. The parenchyma cells usually act as storage cells because it's easy to get things in and out of a thin cell wall. Collenchyma cells typically provide fiber and a dynamic structure. Sclerenchyma cells provide even more structure.

TREE RINGS

- Trees can tell time. Tree rings are concentric rings in the tree as seen when the tree is in cross-section. The rings form because of the seasons. In the wintertime, the tree is growing very little if it's an evergreen, and not at all if it's a deciduous tree that has lost its leaves for the winter. Because the tree isn't taking up a lot of water and is making very few sugars to store in the wood, the cells are small, and therefore the cell walls are close together. This makes the rings darker in the winter.

◇ In the spring and summer, the trees are growing more quickly and the cells are larger, so the lighter wood between the dark rings is the wood that is laid down in the spring and summer. The dark rings put down in winter are called the latewood, and those laid down in the spring and early summer are called the earlywood.

◇ Tree rings can answer questions about climate change, fire history, and archaeology, so they are very useful. But not all trees have tree rings. Because the rings form due to the seasons, in climates that are not seasonal, trees will grow without obvious tree rings. Therefore, finding tree rings in tropical trees is very difficult. Many tropical climates have a wet season and a dry season, but for most species, the dry season is not long enough, or dry enough, to produce a visible ring.

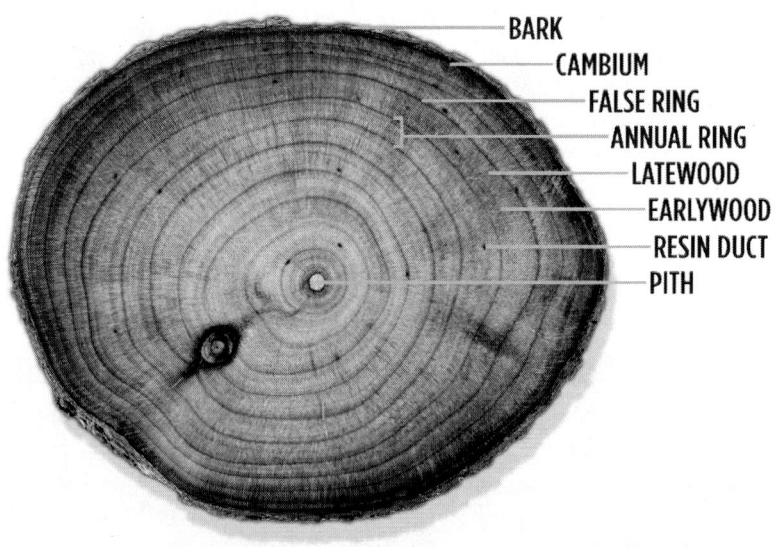

STEMS

- All stems start out as herbaceous stems, while only some become woody over time. When the stem first comes out of the seed, it's herbaceous material. The word "herbaceous" means anything that isn't woody, so this would include a grass, a buttercup, or a mint—any plant without woody growth.

- After the stem develops for a while and the cells start to differentiate, the cells in the xylem will get lignin deposited in their cell walls, and those cells will become wood. In this way, the development of the cork also comes later.

MINT
Mentha sp.

- Herbaceous stems don't have annual rings or wood, so the arrangement of the vascular tissue looks different. The arrangement of these tissues within the stem is an important clue to evolutionary relationships.

- The mosses didn't have vascular tissue, but the ferns did, so the arrangement of xylem and phloem within fern stems is considered the ancestral arrangement. The earliest vascular seedless plants have phloem that surrounds the xylem or columns of phloem dispersed within the xylem. Either of these arrangements is called a protostele.

- There are some unique types of stems. There is an important type of underground stem called a rhizome, also known as rootstock. You may wonder why a rhizome is a stem if it's growing underground, and the answer has to do with the arrangement of vascular bundles.

◇ The arrangement in a rhizome resembles the vascular arrangement in stems, not in roots. Rhizomes also lack root hair, so they are not absorbing water. The function of the rhizome is to spread the plant vegetatively. Many grasses are rhizomatous, as are many weeds that spread really well, such as bamboo. Rhizomatous trees include aspens and poplars.

◇ Another type of horizontal stem that grows on top of the ground, rather than underneath it, is a stolon, or runner. These are most familiar in strawberry. Although a stolon can have some rooting from the nodes of the stem, the main function of the stolon is the same as a rhizome, to allow the plant to spread clones.

STRAWBERRY
Fragaria sp.

◇ Other stem modifications are bulbs and corms. A bulb is often underground, but tulip bulbs and onions are both botanically stem tissue, rather than root tissue, for the same reason that rhizomes are stems.

◇ A corm is a bulb that is solid, meaning that it is swollen stem tissue that does not have leaves and has a basal plate, where roots will come from. Gladiola and crocus both have corms. A tuber, such as the potato, is another modified swollen stem that is underground, but it lacks the basal plate that corms and bulbs have.

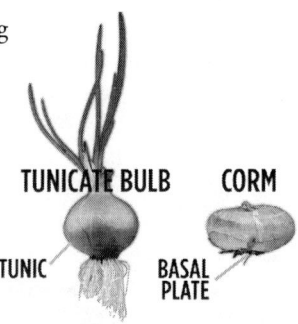

◇ All of these modified stems are for storing sugars, but they can also give rise to new plants via vegetative propagation. "Vegetative" just means clonal, with the offspring identical to the parent plant.

- Although we typically don't think of stems as the site of photosynthesis, there are a number of stems that do photosynthesize. Basically, any stem that is green is photosynthesizing. Stem photosynthesis is especially prevalent in areas where leaf size might need to be reduced, such as climates that are hot or dry or maybe very windy.

- Stems also have gas exchange, which means that they exchange their waste (oxygen) for the gas they need (carbon dioxide). This gas exchange occurs through small, somewhat spongy openings in the stem called lenticels. Lenticels are visible to the naked eye, especially on the branches of woody plants. On trees like birch, with smooth bark, they're pretty obvious, but trees with thicker, rougher bark have lenticels, too.

- They look like small splits in the wood, but they are there on purpose to exchange gasses. If the stem isn't photosynthetic, then these lenticels can take in oxygen, because cells that don't photosynthesize will respire, just like the respiration in our cells.

READINGS

Capon, *Botany for Gardeners*.

MacAdam, *Structure and Function of Plants*.

Speer, *Fundamentals of Tree-Ring Research*.

QUESTIONS

1. Describe the difference between primary and secondary growth in stems.

2. What is dendrochronology? What kinds of applications can you think of that would use dendrochronlogy?

LECTURE 6 TRANSCRIPT
STEMS ARE MORE THAN JUST THE IN-BETWEEN

When I was a kid, I used to have a book called *Ferdinand the Bull*, written by Munro Leaf. Ferdinand was a bull, but he was a sort of quiet and friendly kind of bull. In the story, Ferdinand likes to sit in the shade of the cork tree. In the book, there is this very cute picture of the cork tree showing corks dangling from the leaves like fruits. Of course, the book takes place in Spain, where Ferdinand is meant to be a bull that will fight in the ring. Through a mishap, he goes to the ring, but then he gets to come back and sit under the cork tree again. This book is great because even though the cork trees are indigenous to Spain, it never dawned on me at the time that this wasn't how cork was actually produced, since cork trees are not native to North America.

The cork tree—at least the kind of cork tree that would be used to make cork for wine bottles in Spain—is called *Quercus suber*. The *Quercus* part means that it's from the genus that has all the true oaks. This tree is found throughout the Mediterranean coastlands, but most extensively in Portugal and Spain, which is why I always think of Ferdinand, the Spanish bull, sitting under his cork tree.

Cork is a sustainable resource that we get from a stem. Sometimes we might forget that tree trunks are stems, but they are—in fact, the biggest stems of all. Portugal currently exports the most cork, and that country alone is responsible for half of the world's cork. Cork trees live about 200 years, and cork can be harvested after the first 25 years of growth. After those first 25 years, it takes about 9 years for the cork to grow back, so the cork can be harvested every 9 years for the remaining 175 years of the tree's life. A typical tree will yield around 100 pounds of cork per harvest, which is enough to cork up about 4000 bottles of wine. Cork has a lower carbon footprint compared with plastic and other bottle stopper materials, but cork isn't just used for wine bottles.

It's also used for flooring, insulation, space products, shoes, and even watches. Cork is lightweight, waterproof, fire-resistant, and recyclable. You can even sell old corks on eBay. Who knew?

What is cork and why is it so great? It actually makes botanical sense to start with the cork, because cork is the outermost layer of the stem—the cork is on the very outside. Now, many people would call this the bark, and that's fine, but given our discussion of cork being a sustainable resource that doesn't kill the tree, we should definitely understand the difference between cork and bark. The bark includes the cork, but it also includes the phloem, which is the tissue that carries the sugars from the leaves to the root. If all layers of the bark are removed all the way around the tree, a process called girdling, then that would kill the tree. When cork is harvested, the inner bark is left intact. It's also important that the cork cambium is left intact. This is a layer just inside the cork and outside the phloem that will produce the new layer of cork. Any tissue called a cambium will produce other tissues.

So, the cork is made up of cells that have cell walls that are filled with *suberin*. Sort of like lignin, the complex substance that provides a lot of toughness and allows ferns to stand tall, suberin is a complex substance that provides waterproofing. In fact, suberin is so complex that in 2016 the molecule's exact structure is still being determined. It's known that suberin is a polymer of monomers, and there are many different monomers that make up that polymer. Plants have made such a complex waterproofing compound that science has not been able to copy it, or even fully describe it.

If all woody plants have cork, why don't all trees produce as much cork as the *Quercus suber*? By the way, did you notice how that specific epithet, the last part of the scientific name *Quercus suber*, sounds like suberin? That just goes to show that sometimes scientific names can be really useful when they give us a clue about the plant. Look at this really old, unharvested *Quercus suber*—it produces a lot of cork, which has a lot of suberin. Another cork, such as *Quercus rubra*, or red oak, doesn't produce nearly as much cork.

Actually, the key to understanding why the cork tree makes so much cork lies in the region where the cork grows. The Mediterranean region where cork grows has hot dry summers, perfect for fire. So, having a big cork means that the inside of the tree can be protected from the fire. As I mentioned, cork is fire-resistant, so the fire, especially if it's a ground fire that doesn't get too hot or get into the canopy of the tree, will just burn around the cork tree, leaving it unharmed.

If fire resistance is important in cork development, we might expect other trees that grow where fire is common that they might also produce a lot of cork, and so it is with ponderosa pine. This tree forms large plates of bark with a thick cork layer, presumably to protect the tree from ground fires. By the way, ornamental cork trees, or *Phellodendron*, come from Asia, and while they do have more cork than other trees, they don't produce as much cork as the cork oak, and they're not used for commercial production of cork.

On the outside, any woody stem has this structure—the cork on the outside. Surrounding the cork is a cambium where new cork is made, and together they're called the *periderm*. Inside the periderm is the phloem, which first appeared in ferns, but is fundamentally the same for all vascular plants. The phloem transports the sugars from the fronds or leaves down to the roots. There may be other parts of the plant that need sugars, too, such as developing fruits, and the phloem would transport sugars there. It's best to think about the movement as a source and a sink. Areas that produce phloem—mainly the leaves, but some stems can be photosynthetic, too—will be the source, and areas that receive phloem will be the sinks. This is important because plants don't have a heart to pump materials, so how are these sugars moved?

The best model botanists have for how phloem moves around in the plant is called the Pressure Flow Hypothesis. Basically, this model says that osmosis and diffusion move the sugars around. Let's imagine that we are next to the phloem in the leaf where there are a lot of sugars. Even though the sugar concentration is high in the leaf, the sugar concentration is also high in the phloem, so the plant has to use energy to move sugars into the phloem. This movement is called active transport. The active part just means that the plant had to expend some energy to get the sugars into the phloem. Once there are a lot of sugars in the phloem, water from the xylem will diffuse into the nearby phloem via osmosis.

Think about it. Where would the concentration of water be higher: in the xylem—mostly water; or the phloem—mostly sugar? Since the concentration of water is higher in the xylem, the water moves along its concentration gradient into the phloem, and this creates a lot of turgor pressure. The word turgor just means positive pressure from water. And this pressure will create a flow, just like turning on the hose creates a flow from the faucet to the outlet of the hose. When the sugars get to the sink, the area that needs the sugars, they're actively moved out the phloem by active transport again. Without the sugars in the phloem, the water concentration increases, and the water will move out of the xylem, which creates the low pressure. The low pressure at the sink encourages the flow from the source.

Even though I used leaves as an example of a sugar source, they're not the only sugar source. What about deciduous plants after they lose their leaves in winter? When the springtime comes, the plant needs to grow new leaves. But without leaves, how will the plants get energy it needs to grow those leaves? It's at this time of year, in early spring, that roots become the sugar source, and the leaf buds become the sugar sink.

What this means is that even though sugars normally flow from the phloem, during the early spring, the sugars from the roots actually travel up the tree in the xylem. The rate of flow depends on the temperature. During the nighttime, the xylem is at its densest, but if the daytime temperatures get warmer, the xylem will expand. This all becomes perhaps more important when we consider that this is the process that gives us maple syrup. As the xylem in the trunk expands, then the maple sap will flow out of the tree faster. It takes about 40 liters of this sap to make one liter of maple syrup.

I just used the word sap, but that's not a word botanists use very often. Sap can refer to xylem or phloem, or even the fluid inside the vacuole inside the cell. I generally think of any fluid coming out of the tree as sap, and usually we don't know if that sap is xylem or phloem. In fact, most of what we know about phloem has come from aphids. These small insects insert their straw-like mouthpart, called a *stylet*, into the phloem of the plants to suck out the sugars. The pressure of the phloem is so strong that it moves straight through out the back end, where it's called honeydew. In order to analyze the phloem, one can anesthetize the aphid, so it won't withdraw its stylet, and then remove the aphid's body from the stylet and use a small micropipette to get the phloem out of the plant.

Interior to the phloem and xylem of the stem is the vascular cambium. Do you remember what that word cambium meant? It's the tissue that generates new tissue. Cork cambium makes new cork; vascular cambium makes new vascular tissue. Cambium is sort of like stem cells. Ha, get it? Stem cells? OK, stem cells, as a developmental stage, are undifferentiated cells that can differentiate to other kinds of cells. The cambium tissue can give rise to a few different tissue types, but there are whole regions of stems called *meristems*, and these meristems are responsible for growth. Usually, the height of the stem is controlled by the apical meristem. Apical means top, and so the apical meristem is at the top of the plant. Conversely, some plants have basal meristems, like grasses, so even when their tops are cut off, they still grow from their basal meristem at the base of the plant.

A hormone called *auxin*, produced in the apical meristem, generally suppresses branching so that the top of the plant can grow tall. This is why we prune our rose bushes. By cutting off the apical buds at the tips of the branches, more branches will be produced and, typically, more roses. So, the apical meristem is responsible for upward growth, what botanists call primary growth. Woody stems also grow outward, increasing in girth, and this growth is termed secondary growth. It's the vascular cambium that is responsible for the secondary growth. By contrast, most ferns do not have a vascular cambium, and that's one reason why ferns typically do not increase much in girth. The tree ferns begin their life with as much girth as they will have as they grow—the stem doesn't get any wider as they get taller. And, again, this is why tree ferns won't get as tall as trees with wood.

Cells within the vascular cambium differentiate to phloem on one side that is closest to the outside of the trunk, and cells nearest the inside become xylem cells. The cells in the vascular cambium start off as undifferentiated cells, and as they move into the xylem or the phloem, they develop traits of those tissue types. It's the accumulation of these xylem cells that creates wood. Remembering that lignin was the substance responsible for toughness in the plant cell, these xylem cells are heavily ligninized.

Both the phloem and the xylem don't contain just one type of cell. Like the blood of humans, numerous kinds of cells are involved, and the cells are specialized in terms of their function. Many cells of the phloem are specialized to transport sugars. Likewise, many of the cells in the xylem are specialized for transporting water. When botanists say xylem and phloem, they are referring to: one, the tissue type; two, the cells that make up the tissue; and three, the fluids within these tissues. It's sort of like saying blood vessel, where that phrase includes the blood inside. The two most important cells in the xylem are tracheids and vessel elements, and what is cool is that both of these are lined with pits that open and close in very sophisticated ways to make sure that water gets to the part of the plant where it's needed most.

Like most other parts of the plant, stems also have three other basic tissue types: parenchyma, collenchyma, and sclerenchyma. Parenchyma are the cells with the thinnest walls, collenchyma has uneven walls, and sclerenchyma has the thickest walls. So, all of these cells are found throughout the xylem and phloem, and their function relates to their structure. The parenchyma cells usually act as storage cells because it's easy to get things in and out of thin cell walls. Collenchyma cells typically provide fiber and a dynamic structure, capable of supporting primary and secondary growth.

The strings in celery are typically collenchyma cells, and although you may think of celery as a stem, it's actually a very long leaf stalk, or leaf petiole in botanese. But, even though it's part of the leaf, a celery stalk and all its petioles still have xylem and phloem, which is how materials get into and out of the leaf. If you put celery in colored water, what you see are vascular bundles of xylem and phloem in the celery stalk. Sclerenchyma cells provide even more structure. Stems for many ferns and grasses skimp on collenchyma and go straight for the more rigid sclerenchyma, and these cells actually come in two shapes: fibers and sclereids. Fibers are long cells that usually occur in bundles, whereas sclereids usually occur in clumpy masses, like the gritty cells in pears.

OK, back to our walk into the woody stem. We met the vascular cambium, which has cells differentiating to phloem on the outside, and cells differentiating to xylem on the inside. These xylem cells are generally dead at maturity, even though water is still moving through them from the roots. When we discussed leaves, we talked about the mechanism for how water moves upward, and how it's analogous to a straw. Evaporation is happening in the leaves, and this evaporation essentially pulls water up from the roots through the xylem. Similar to our straw analogy, this means the water in the xylem is under negative pressure—a vacuum, if you will.

This system works pretty well, but what if the xylem freezes or the soil is really dry? What can happen then is called a cavitation, and this is where an air bubble forms in the xylem. This is not good, but it does happen. When the water in the xylem freezes, it expands, and the air that was dissolved in the xylem is not soluble in the ice, so the air bubble forms. This is the same thing that happens in our straw when we're enjoying a beverage. If we're near the end of the beverage, and we suck up some air, the delivery of the beverage is slowed, and so it is with cavitation. As these air bubbles form, they impede the suction of water up the stem.

When I learned about plant cavitation in plant physiology, our professor brought in a small tree to demonstrate. This tree had not been watered for several days, and she put what looked like a small box on the trunk of the tree. It was a sensor like a stethoscope, and it listened to what was happening inside the tree. What was happening, of course, was a bunch of popping noises, and the air bubbles burst as the suction from the leaves got stronger. This is how botanists actually quantify cavitation rates. In this way, botanists can know when a tree is under water stress. So, if a tree cavitates in the forest and no one hears it? Especially on a hot summer day? Well, maybe they just weren't listening closely enough—there is definitely a sound.

Trees don't just tell us about themselves, though; they can also tell time. Many people are familiar with tree rings. These are concentric rings in the tree as seen when the tree is cut in a cross section. The rings form because of the seasons. In the wintertime, the tree is growing very little if it's an evergreen, and not at all if it's a deciduous tree that has lost its leaves for the winter. Because the tree isn't taking up a lot of water and is making very few sugars to store in the wood, the cells are small, and so the cell walls are close together. This makes the rings darker in the winter. In the spring and summer, the trees are growing more quickly and the cells are larger, so the lighter wood between the dark rings is the wood that is laid down in the spring and the summer. The dark rings put down in winter are called the latewood, and those laid down in spring and summer are called the earlywood.

Trees don't have to be cut to see these rings. One can take a core of a tree using a device called an increment borer. This works sort of like a soil auger or an apple corer. It just takes a small piece out of the middle of the tree. It actually takes some doing to get the borer right through the center of the tree. The core is pretty fragile once it comes out, so you have to be pretty careful not to break it. The botanists who study these rings are called dendrochronologists. Dendro means tree or branch, and chronology refers to the study of time. While a dendrochronologist could just count the rings, there are a number of computer programs that will do the analysis by using crossdating techniques. These programs have information about various years from various places and can better date the rings of the tree. Using more than one tree and other methods for analysis is called crossdating, and this method has been used to recreate the climate for about the past 1000 years.

Tree rings can answer questions about climate change, fire history, and archaeology, so they're very useful, but not all trees will have rings. Since the rings form because of the seasons, in climates that are not seasonal, trees will grow without obvious tree rings. Therefore, finding tree rings in tropical trees is very difficult. Many tropical climates have a wet season and a dry season, but for most species, the dry season is not long enough or not dry enough to provide a visible ring.

In 2006, though, botanists used the Brookhaven National Synchrotron Light Source to take X-rays of tropical trees that didn't have visible tree rings. The X-ray beams were particularly focused to investigate the amount of calcium in the trees during the main growing season versus the off-season. They found that the type of tree they were investigating did show discernible rings under X-rays. Now, botanists are looking to see if other tropical trees will show their invisible tree rings.

Although we've talked about woody stems first, all stems start off as herbaceous stems, while only some become woody over time. When the stem first comes out of the seed, it's herbaceous material. The word herbaceous means anything that isn't woody, so this would include a grass, a buttercup, or a mint—any plant without woody growth. By the way, most mints will have square stems, and you can actually feel this at the base of the stem. It's not true for all mints but it's true for a lot of them. After the stem develops for a while and the cells start to differentiate, the cells in the xylem will get lignin deposited in their cell walls, and those cells will become wood. In this way, the development of the cork also comes later.

Herbaceous stems don't have annual rings or wood, so the arrangement of the vascular tissue looks different. The arrangement of these tissues within the stem is an important clue to evolutionary relationships. Of course, the mosses didn't have vascular tissue, but the ferns did, and so the arrangement of xylem and phloem within the fern stems is considered the ancestral arrangement. So, the earliest vascular seedless plants have phloem that surrounds the xylem, or columns of phloem dispersed within the xylem. Either of these arrangements is called a *protostele*. The word stele just means central cylinder of the plant.

There are some unique types of stems that are worth discussing here, because not all stems are reaching for the sky. There is an important type of underground stem called a rhizome, also known as rootstock. You may wonder why a rhizome is a stem if it's growing underground, and the answer has to do with the arrangement of those vascular bundles. The arrangement in the rhizome resembles vascular arrangement in the stems, not in the roots. Rhizomes also lack root hairs, so they're not absorbing water. The function of the rhizome is to spread the plant vegetatively. Many grasses are rhizomatous, as are many weeds that spread really well, such as Canada thistle. Bamboo is another example. Rhizomatous trees include aspens and poplars. The Pando aspen organism that we mentioned in the roots lecture is also connected through a series of rhizomes.

Another type of horizontal stem that grows on top of the ground, rather than underneath it, is a *stolon*, or runner. These are most familiar in strawberry. Although the stolon can have some rooting from the nodes of the stem, the main function of the stolon is the same as the rhizomes, to allow the plant to spread clones. Other stem modifications are bulbs and corms. Yes, it's true that a bulb is often underground, but tulip bulbs and onions are both botanically stem tissue, rather than root tissue, for the same reason that rhizomes are stems.

It may surprise you to learn that tulips are not actually Dutch. Although Holland exports more tulips than any other country—about 2 billion bulbs in 2014—the tulip itself originated in central Asia and was brought to Holland in the 16th century. You wouldn't know it now, though, since about half of the country is under agricultural production, and all for a modified stem: the tulip bulb.

When we think about the layers of an onion, these are actually leaves—they just happen to be fleshy, underground leaves that have lost their chlorophyll. The outer papery layer is called a tunic, so the onion is a tunicate bulb. But however much you peel away, an onion is just layers of leaves attached to a stem. And did you know that in 2008, botanists created the first onion that doesn't cause tears when you cut it? Scientists in the lab have been genetically modifying onions to see what might be the best way to turn off the gene that creates the enzymes that make our eyes water, though they haven't decided if this new genetically modified onion will be commercialized.

A corm is a bulb that is solid, meaning it is swollen stem tissue that does not have the leaves, and it has a basal plate where the roots will come from. Gladiola and crocus both have corms. A tuber, such as the potato, is another modified, swollen stem that is underground, but it lacks the basal plate that corms and bulbs have. All of these modified stems are for storing sugars, but they can also give rise to new plants via vegetative propagation. Vegetative just means clonal, with the offspring identical to the parent plant. Although we typically don't think of stems as the site of photosynthesis, there are a number of stems that do photosynthesize. Basically, any stem that is green is photosynthesizing. Just because the stem is storage and conductance doesn't mean it can't do a little photosynthesis, too. Stem photosynthesis is especially prevalent in areas where leaf size might need to be reduced. This would be climates that are hot or dry, or maybe very windy.

Cacti, the family Cactaceae, are the major plants that come to mind here. All of the photosynthesis that occurs in the cactus family occurs in the stem. What we think of as cactus needles are botanically called spines, and they're actually the modified leaves, but there is no photosynthesis taking place in the spines. And about 25 other families have member species with stem photosynthesis, including the fabulous Fabaceae, and the Ephedra family, and the Aster family.

Stems also have gas exchange, which means they can exchange their waste, which is oxygen, for the gas they need, which is carbon dioxide. This gas exchange occurs through small, somewhat sort of spongy openings in the stem called *lenticels*. These lenticels are actually visible to the naked eye, especially on branches of woody plants. On trees like birch, with smooth bark, they're pretty obvious, but trees with thicker, rougher bark have lenticels, too. They look like small splits in the wood, but they are there on purpose to exchange gases. If the stem isn't photosynthetic, then these lenticels can take in oxygen, since cells that don't photosynthesize will respire, just like the respiration in our cells.

When we think of stems—from the long stem of a rose to the amazing cork of *Quercus suber*—they all seem to share two functions: gas exchange; and the transport of water up from the ground, and the movement of sugars down to the roots. But there are exceptions. In early spring, the direction of movement is reversed, until leaves can be formed. Try to imagine your veins and arteries trading jobs.

By contrast, in parasitic plants like Dodder, sugars get taken from the vascular system of a host via plant haustoria that combine tissues of the host and parasite. Modified stem types exist that can store sugars, whether in tubers, corms, or bulbs. Stems can be rhizomes, below ground, where they can look like roots. And in cacti, as well as the 25 other plant families, stems can even make sugars by photosynthesis. The stem is literally the vascular core of the plant, and arguably the overlooked heart of what it means to be a plant.

LECTURE 7

THE LEAF AS A BIOCHEMICAL FACTORY

Leaves can be surprising. We know that they are green and that they harvest light. We also know that we can eat some of them. But leaves showcase a myriad of sizes, shapes, and functions. For example, botanist Heather Whitney studies iridescence in the peacock begonia, whose leaves resemble a peacock feather—deep blue and green and shiny. There are even underground leaves. In Brazil, a species called *Philcoxia minenensis* has normal aboveground leaves, but it also has leaves that grow under the soil.

LEAVES

- Leaves from the coca plant are the raw ingredient for cocaine, but a lot of processing occurs to concentrate the cocaine from the coca leaf. Cocaine is an alkaloid, which means that it is a class of small molecules containing nitrogen. These alkaloids are alkaline, or basic.

- For humans, these compounds can act as psychotropic drugs that alter the brain, and this can happen in insect herbivores, too, causing disorientation. However, the alkaloids' main effect on other animals disrupts the digestive system.

- High alkaloid content will cause nausea in many animals, so a plant that produced these chemicals in high quantities would be less likely to be fed on and would therefore be more successful in producing successive generations.

- These compounds might even have a greater physiological effect on human digestion if our mouths and stomachs weren't so acidic. The acidity neutralizes the alkalinity of the alkaloid compounds. Without these defensive chemicals, the plants would be damaged much more by herbivores.

- Other than leaf protection, these chemicals aren't involved in photosynthesis. Because of this, these class of chemicals are called secondary compounds. It could be that these compounds are actually waste products. And unlike animals, plants store their waste products.

- Plant waste isn't so wasteful: The 2 products of photosynthesis are oxygen and glucose. Oxygen isn't stored at all; it's released to the atmosphere. The sugar glucose does get transformed into other products, such as cellulose, and then assimilated into many different storage products, such as wood and cell walls.

- There are chemicals in the leaf that the plant is using, and most of these involve photosynthesis. One of the major keys to photosynthesis is the plant pigment chlorophyll. A pigment is any chemical that can absorb light, and chlorophyll absorbs the wavelengths of light that are not green, so the red and blue. When the chlorophyll goes away in the autumn, we see the other pigments that are in the leaf: the orange and yellow of beta carotenes and the red of anthocyanins.

- How animals react to the red color in leaves is not so apparently advantageous to the plant. Red is often seen in developing leaves, and primates tend to prefer these leaves. Of course, this is not an advantage to the plant because all of its young leaves are eaten.

- But there are many hungry animals that depend on leaf eating. Although we broadly think of such animals as herbivores, an animal that has only a diet of leaves is called a folivore. There's been a lot of research into what kinds of leaves animals prefer, and this research provides clues as to the nutritional value of these leaves.

- Plants, however, don't make leaves to feed animals, or even to make them sick, or to provide brilliant fall colors for us. Leaves are to absorb light and power the photosynthetic machine. The structure of a leaf is tied to its function.

ELEPHANT EAR

Xanthosoma robustum

- Leaves are flat and thin, typically, but some leaves are more succulent than others, and some leaves, such as those of evergreen conifer trees, are needle-shaped and not flat at all.

- The thickness of the leaf depends on the amount of sunlight the leaf is trying to absorb. If the leaf is in full sun, it will likely be smaller and thicker because the leaf is not limited by light. If the leaf is growing in the shade, it is likely to be thinner and have more area, as it's trying to capture every bit of light it can get.

PARTS OF A LEAF

◈ Plants are typically arranged with nodes, where the leaves come out, and the stem space between the leaves, which is called the internode. The arrangement of leaves on the stem is also called phyllotaxy, and it's an important characteristic in identifying plants.

MAPLE
Acer sp.

◈ Leaves can be opposite of each other on the stem, such as on dogwoods or maples; or leaves can be alternate, such as on elms; or they can be arranged in a whorl, which means a circular arrangement of leaves around the stem, such as on bedstraw.

◈ Overall, the leafy part is called the blade, and the stalk that attaches the blade of the leaf to the stem is called the petiole.

ELM
Ulmus sp.

◈ The way botanists determine if something is a leaf or not is by the presence of an axillary bud in the junction where the petiole attaches to the leaf. If there is a small bud present, then the structure is a leaf. A petiole is just like a mini-stem to the leaf. The difference between a stem and a petiole is that the petiole attaches the leaves to the stem.

FRAGRANT BEDSTRAW
Galium triflorum

- If there is no bud, it's what botanists call a leaflet, attached to a vein. These are very similar to the veins of animals. The veins are where the vascular tissue, the xylem and phloem, come into the leaf, or leaflet.

- There are specialized terms that describe venation patterns: parallel, like in grasses; palmate, like a maple leaf; and pinnate, like elms.

- A leaf that is divided into many leaflets is called a compound leaf. Some are palmately compound, such as buckeye. Many are pinnately compound, such as hickory or walnut.

OHIO BUCKEYE
Aesculus glabra

BLACK WALNUT
Juglans nigra

◇ This is how we can tell that the whole frond of a fern, even a large frond of a tree fern, is all one "leaf." Other trees have simple leaves that are partially divided, such as lobed leaves.

◇ In addition to gathering light, leaves have to not blow off the plant and can't overheat. Being a leaflet is a good solution to both of these problems. Smaller leaves, or leaflets, provide less resistance to the wind, so they would be less likely to blow off of the tree, and the wind going by them would cool them down. This is one reason many desert plants have leaflets.

◇ There are hundreds of terms that describe everything about a leaf—from the surface, to the number of hairs, to the overall shape, to the margin (the outside perimeter). The reason for this proliferation of terms is that leaves are often the easiest or most accurate way to identify a plant.

◇ There are even terms for the shapes of the hairs on the leaves. These hairs are called trichomes, and they are cool because they do more than help botanists tell plants apart.

◇ One function of these hairy trichomes is that they can be glandular. This is sort of like a hair follicle in people with a sweat gland at the base. Plants that have a lot of oils on their surfaces, such as basil, have a lot of trichomes where there are exudates of oil.

COMMON MULLEIN
Verbascum thapsus

◇ Trichomes also serve as anti-herbivory structures. Just the presence of many trichomes can make the leaf unpalatable to an animal or gum up the mouthparts of an insect. Trichomes are also responsible for the sting in stinging nettle.

- Another very important function of trichomes is to increase the boundary layer around the leaf. This is the layer of air just next to the leaf. The hairs will help slow the air passing over the leaf, which decreases the amount of evaporation out of the leaf.

- This reduction of evaporation is important because water is lost through small openings in the leaves called stomata. Typically, leaves have more stomata on the underside to reduce water loss because the underside of the leaf is in the shade.

- The stomata are surrounded by 2 semicircle shapes, which are the guard cells. They open and close the stomata based on how drought-stressed the plant is, balanced with the need for carbon dioxide. So, the stomata are the sites where carbon dioxide diffuses into the leaf and where oxygen is released. At the same time, when the stomata are open, water is evaporating, and this is what makes possible evapotranspiration.

- Some transpirational loss is good because this is the process that pulls water up from the roots. But too much transpirational water loss will cause the cells to lose water and become flaccid, and the plant will wilt.

- Sometimes having leaves can be too costly in terms of water loss. Like so-called evergreens dropping their needles, most tropical trees gradually replace their leaves pretty quickly—anywhere from 2 months to 2 years.

- If the leaves don't last as long as the tree lives, what is the mechanism for the tree losing its leaves? Deciduous trees lose their leaves before winter. Some plants drop their leaves before a time of drought, and these plants are called drought-deciduous. These plants are common in deserts and arid lands. These may be the ultimate cues, but what is the mechanism in the plant?

- Plants have hormones that cause many of the developmental changes throughout a plant's life. Hormones tell the plant when to grow, which way to grow, and when to shut off certain parts, such as leaves.

◇ Nineteenth-century German botanist Julius von Sachs first proposed the idea that chemicals produced in one part of the plant would effect changes in another part of the plant. Much of the early knowledge of hormones was derived from similar studies occurring in animals.

◇ There are 5 main types of plant hormones. Of these, auxin was the first hormone to be discovered, and it's one of the first hormones to be expressed. Along with the hormone cytokinin, auxin is required for the plant to live. The ratio of these 2 hormones control many functions in the plant.

◇ Auxins are the longest-studied group of hormones, but there are other growth hormones, and botanists are discovering new chemical messengers and regulators all the time.

READINGS

Coombes, *The Book of Leaves*.

Farmer, *Leaf Defence*.

Lee, *Nature's Palette*.

Vogel, *The Life of a Leaf*.

QUESTIONS

1   How is a leaf botanically determined to be a leaf and not some other part of the plant or a leaflet?

2   Describe a tropism. Think about how differently plants and animals respond to their environments. Have you seen a plant tropism in action? What was it like?

LECTURE 7 TRANSCRIPT
THE LEAF AS A BIOCHEMICAL FACTORY

Leaves can be surprising. We know they are green, we know they harvest light, we know that we can eat some of them, but just when we think we know leaves, something about them surprises us. For example, Heather Whitney, a botanist from the University of Bristol, has been studying iridescence in the peacock begonia. This plant grows in the understory of the Malaysian rainforests, and the leaf really does resemble a peacock feather: deep blue and green, and shiny. And this is a leaf. Why is it blue? And what is a leaf in the shaded understory doing expending light on iridescent shimmer? Dr. Whitney and her colleagues have found that the chloroplasts within this leaf are highly structured, so that they are able to absorb about 5%–10% more red and green light with this arrangement of chloroplasts. The mysterious blue is the light left over, and the shimmer comes from the intricate stacking. Surprising.

There are even underground leaves. In Brazil, a species called *Philcoxia minensis* has normal aboveground leaves, but it also has leaves that grow under the white, sandy soil. Because the soil is white, it reflects light onto the leaves that are growing underground, so these leaves still photosynthesize. But, these same underground leaves also secrete a sticky substance that allows them to trap nematodes, which are microscopic worms that live in the soil. Again, surprising.

Leaves showcase a myriad of sizes, shapes, and functions. As a botanist, I measure the temperature of leaves. I might remove leaves to see how thick they are, which can tell me about adaptations to drought or high light levels. I might dry leaves in a drying oven, which is a regular oven, just found in a lab. I might also analyze the shape of leaves to determine particular adaptations in one population of plants versus another. Sometimes, I actually measure a leaf's rate of carbon dioxide consumption, which lets me calculate its rate of photosynthesis.

But when I think of my most memorable leaf encounter, it involved the leaves of the coca plant. When I was in northern Chile doing research, it was common to stop in and meet with the chief of police to let them know you were going to be in the area. When I first met the chief of police in the small town where I was working, he offered me coca leaf tea, and I was surprised, because I thought coca leaves were illegal in Chile. These leaves from the coca plant are the raw ingredient for cocaine, but of course there's much processing that occurs to concentrate the cocaine from the coca leaf. What's also ironic is that the coca bush is a rather straggly-looking shrub with innocuous-looking, ordinary leaves.

So the police chief tells me that he regularly confiscates coca leaves when truckers try to smuggle tons of leaves out of the country to make cocaine in other parts of the world. Yet he understands that the coca leaf tea is a way that traditional peoples in the Andean Plateau greet each other, and he knows that the tea is meant to ease the altitude sickness that can occur when one is above 10,000 feet in elevation. The concentration of cocaine in a coca leaf is only about 0.5%, so there is no high when you drink coca leaf tea, though I will admit my lips got a bit numb. And coca leaf tea is legal in many parts of the South American countries where the Andes Mountains run.

Cocaine is an alkaloid, which means that it is a small class of molecules containing nitrogen. These alkaloids are alkaline, or basic, because the nitrogen atom takes the place of a hydrogen in the carbon and hydrogen ring structure. Most of the well-known alkaloids end with "ine" to signify their relation to the amine groups, which is the ring with the N. So, coca-ine, or cocaine, caffeine, morphine, and nicotine are all alkaloids, as is theobromine, the active alkaloid in chocolate.

For humans, these compounds can act as psychotropic drugs that alter the brain, and this can happen in insect herbivores, too, causing disorientation. However, the alkaloids' main effect on other animals disrupts the digestive system. High alkaloid content will cause nausea in many animals. So, a plant that produced these chemicals in high quantities would be less likely to be fed on and would therefore be more successful in producing successive generations. These compounds might even have a greater physiological effect on human digestion if our mouths and stomachs weren't so acidic. The acidity neutralizes the alkalinity of the alkaloid compounds. Without these defensive chemicals, the plants would be damaged much more by herbivores.

Other than leaf protection, these chemicals aren't involved in photosynthesis. Because of this, these class of chemicals are called secondary compounds. It could be that these compounds are actually waste products. And unlike animals, plants store their waste products. There is this great cartoon of animal waste and plant waste in Francis Halle's brilliant and eclectic book *In Praise of Plants*. It shows a dog on top of a pile of its waste next to a palm tree with the caption: "Solution to the problem of excretion. An animal that stored its excrement would also be capable of becoming very tall."

We don't often think of this aspect of plants, but it certainly sets them apart from animals. What this really means is that plant waste isn't so wasteful. The two products of photosynthesis are oxygen and glucose. Oxygen isn't stored at all; it's released to the atmosphere. The sugar glucose does get transformed into other products like cellulose, and then assimilated into many different storage products, such as wood and cell walls. Of course, there are chemicals in the leaf that the plant is using, and most of these involve photosynthesis. One of the major keys to photosynthesis in the plant is the pigment chlorophyll. A pigment is any chemical that can absorb light, and chlorophyll absorbs wavelengths of light that are not green, so the red and the blue. When the chlorophyll goes away in the autumn, we see the other pigments that are in the leaf: the orange and yellow of beta-carotenes and the red of anthocyanins.

Or do we? I mean, yes, we see these autumn colors, but are they in the leaf all along, just masked by chlorophyll? How could we test this? There is a process called chromatography that separates the pigments in the leaf. "Chroma" means color and "graphy" means to draw or graph. What we know is that the anthocyanins are produced as the plant is reabsorbing nitrogen. How do we know this?

Let me show you how we know that anthocyanins are not in the green leaves. This is a process called paper chromatography, and you can do it in your own kitchen. I have some filter paper, but you could also use a coffee filter, and I have some leaves that I've collected that are green and leaves that are red from the fall. If you needed to, you could freeze the green leaves until the leaves turn red and are ready for this experiment.

So, what I've done is I've cut the filter paper, and you generally don't want to handle these with your fingers. And then you just sort of take some scissors or something flat and you're going to score the leaf onto the filter paper. And what I'm doing is getting the pigments from the leaf onto the filter paper. And I'm going to do this with my green leaf, and then I'm going to do it with my red leaf. And then, when I've scored the pigments on there, I will put it into a small dish, and then I'm simply just going to pour some acetone—this is just nail polish remover—and I'll pour a little bit into the dish.

And after I let that sit for about 20 minutes, you can start to see the pigments separating on the filter paper. And what I want you to notice is that the one that was red, you can see those red pigments really well. The green ones, you can see the yellow pigments, the beta-carotenes, but there are no anthocyanins, and that's because the leaf develops them in the fall as the temperatures get cooler.

What's so interesting is that if you do this with, say, a maple leaf in midsummer, the carotenoids, or the yellow-orange pigments, will appear, but the red ones, the anthocyanins, will not. It's because the anthocyanins are not there—the leaf doesn't contain red pigments in the midsummer. Through a series of experiments, botanists determined that the anthocyanins are produced in the leaf as autumn approaches. Why would the anthocyanins be produced when the leaf is getting ready to shut down for the winter? In order to understand this, we need to think about what is happening in the leaf as winter approaches, and, of course, this is true for deciduous leaves of perennial plants. The leaf is about to be dropped for the winter—that's what deciduous means. The leaf has a lot of chlorophyll in it, and each chlorophyll molecule contains a ring of four nitrogen atoms. Nitrogen is very valuable to the plant, so the plant wants to reabsorb all the nitrogen in the leaf before the leaf is dropped. Anthocyanins, by contrast, have no nitrogen in their core structure.

However, as the chlorophyll with its valuable nitrogen is being reabsorbed, the photosynthetic machinery in the leaf can be damaged. This can happen through a process called photoinhibition, which is the slowdown of photosynthesis. And photoinhibition can occur when sunlight is very intense, but it's even worse when temperatures are cold, because the enzymes that facilitate photosynthesis don't work as well. So, here we have this leaf, trying to photosynthesize as much as it can before winter, all the while trying to reabsorb its chlorophyll that contains the valuable nitrogen. Enter—dun-dun-dah—the anthocyanins. A new pigment comes to the rescue. Some trees start to produce anthocyanins as autumn approaches because these red or sometimes purple pigments are better at avoiding photoinhibition. Why would this be?

Well, you know that antioxidants are good for you, and antioxidants are good for the leaf. The antioxidants do the same thing in us that they do in the leaf; they absorb those nasty free radicals that cause aging in humans and photoinhibition in plants. All free radicals are unpaired electrons in the atoms, but anthocyanins can neutralize these by donating electrons to the atoms that have the unpaired electrons. There is still work to be done in this area because most of the anthocyanins are stored in the vacuole and many of the free radicals are outside the vacuole in the cell, and how freely these atoms and anthocyanins can cross the vacuole membrane is in question.

The other way that anthocyanins protect the leaf is by absorbing light in the blue and green wavelengths, and this protects the chloroplasts. That blue and green being absorbed is the "cyan" in the name anthocyanin. So, experiments showed that red leaves with anthocyanins had less photoinhibition than green leaves. Experiments also showed that reabsorption of chlorophyll was occurring in these leaves, because more anthocyanin equaled less nitrogen. So, on a cold day, the anthocyanins are acting like a sunscreen for the chloroplasts. They are protecting the leaf from photoinhibition, even though the leaf is reabsorbing its chlorophyll.

How animals react to the red color in leaves is not so apparently advantageous to the plant. Red is often seen in developing leaves, and primates tend to prefer these leaves. Of course, this is not an advantage to the plant, because having your young leaves eaten by hungry monkeys does not seem too advantageous. And there are many hungry animals that depend on leaf eating. Although we broadly think of such animals as herbivores, an animal who has only a diet of leaves is called a *folivore*. There's been a lot of research into what sorts of leaves animals prefer, and this research provides a lot of clues as to the nutritional value of these leaves. Perhaps the most familiar folivore is the koala, eating nothing but eucalyptus leaves. Anyone who has smelled eucalyptus knows it has a strong scent, and this scent comes from a number of chemical compounds in the leaf that are very hard to digest. This is most likely the reason that koalas sleep all day, because it takes a lot of energy to digest those leaves.

Plants, however, don't make leaves to feed animals, or even to make them sick, or to provide brilliant fall colors to us. Leaves are to absorb light and power up the photosynthetic machine. As we've seen many times before, structure is tied to function. Let's look at the structure of a leaf so we can better understand its function.

Leaves are flat and thin, typically, but some leaves are more succulent than others, and some leaves, like those of evergreen conifer trees, are needle-shaped and not flat at all. Like I say, botany isn't rocket science, it's more complicated. But even though there aren't laws in botany, there are still some general guidelines that help us understand some more about leaves. The thickness of the leaf will depend upon the amount of sunlight that the leaf is trying to absorb. If the leaf is in full sun, it will likely be smaller and thicker because the leaf is not limited by light. If the leaf is growing in the shade, it is likely to be thinner and have more area, as it's trying to capture every bit of light it can get.

One such large leaf is called elephant ear because it resembles the shape of an elephant's ear. This plant, which is also called taro, is native to southeastern Asia, and grows commonly in the understory of tropical rain forests. The elephant leaf may be one of the largest, simple, entire leaves, meaning a leaf that is not a compound leaf and has a smooth margin, but the longest leaves are found in the species *Raphia regalis*, which has leaves that have been measured at 80 feet long. This is the same palm that provides the material raffia for crafts, small housewares, and some kinds of clothing. I use it most commonly as wrapping ribbon. The raffia we use is derived from the veins of the leaf, which have parallel venation, like for most of the palms.

The first thing we should consider is what makes a leaf. How is a leaf determined? And to answer this, I'm going to take you back to the stem for just a moment. Plants are typically arranged with nodes where the leaves come out, and the stem space between the leaves is called the internode. The arrangement of the leaves on the stem is also called *phyllotaxy*, and it's an important characteristic in identifying plants.

Leaves can be opposite of each other on the stem, like dogwoods or maples, for example. Or leaves can be alternate, like elms. Or they can be arranged in a whorl, which means a circular arrangement of leaves around the stem, like bedstraw. Overall, the leafy part is called the blade, and the stalk that attaches the blade of the leaf to the stem is called the petiole. Perhaps you didn't know that when you eat celery or rhubarb, you are actually eating the petiole.

But how is this actually determined? One could easily think that the celery stalk is a stem. The way botanists determine if something is a leaf or not is by the presence of an axillary bud in the junction where the petiole attaches to the leaf. If there is a small bud present, then the structure is a leaf. A petiole is just like a mini-stem to the leaf. Celery plants have been selected over time to have even bigger stalks and smaller leaves. The difference between a stem and a petiole is that the petiole attaches the leaves to the stem. If there is no bud, it's what botanists call a leaflet, attached to a vein. That's really the term that botanists use: veins. These are very similar to the veins of animals. The veins are where the vascular tissue—the xylem and phloem—come into the leaf or leaflet. And, yes, there are specialized terms that describe venation patterns. There's parallel, like in grasses; palmate like in a maple leaf; or pinnate like in elms.

A leaf that is divided into many leaflets is called a compound leaf. Some are palmately compound, like buckeye. Many are pinnately compound, like hickory or walnut, or Boston fern. This is how we can tell that the whole frond of a fern, even a large frond of a tree fern, is all one leaf. Other trees have simple leaves that are partially divided, such as lobed leaves, like oaks.

So, what might be the advantage of having leaflets over a single leaf? In addition to gathering light, leaves have to not blow off the plant, and they can't overheat. Being a leaflet is a good solution to both of these problems. Smaller leaves, or leaflets, provide less resistance to the wind, so that they would be less likely to blow off the tree, and the wind going by them would cool them down. This is one reason many desert plants, like paloverde, have leaflets.

Now that we know what constitutes a leaf, we can begin to explore some different types of leaves. There are hundreds of terms that describe literally everything about a leaf, from the surface, to the number of hairs, to the overall shape, to the margin, or the outside perimeter—there is a term for every little thing about the leaf. The reason for this proliferation of terms is that leaves are often the easiest way to identify a plant. I like to think about this way; I like to think about paint in a paint store. You go in and ask for red paint, and there are hundreds of shades. Leaf morphology, or shape, is a lot like this.

There are even terms for the shapes of the hairs on the leaves. These hairs are called *trichomes*, and they're really cool because they do more than help botanists tell the plants apart. One function of these hairy trichomes is that they can be glandular. This is sort of like a hair follicle in people, with a sweat gland at the base. Plants that have a lot of oils on their surfaces, like basil, for example, have a lot of trichomes where there are exudates of oil. Trichomes also serve as antiherbivory structures. Just the presence of many trichomes can make the leaf unpalatable to an animal, or gum up the mouthparts of an insect. Imagine trying to eat the leaves of a very hairy plant like mullein. Trichomes are also responsible for the sting in stinging nettle. Interestingly, in the book *100 Plants That Almost Changed the World*, Chris Beardshaw reports that nettle was used in the Middle Ages to treat baldness. He supports this use by stating that modern science now shows that nettle controls dihydrotestosterone, or DHT, which causes baldness. Less DHT equals more hair.

Another very important function of trichomes is to increase the boundary layer of the leaf. This is the layer of air just next to the leaf, and the hairs will help slow the air passing over the leaf, which decreases the amount of evaporation from the leaf. This reduction of evaporation is important because water is lost through small openings in the leaves called *stomata*. *Stoma* is opening or mouth in Greek, so stomata is "little mouth," and that is sort of what they look like, though you can't really see them with the naked eye. Typically, leaves have more stomata on the underside to reduce water loss because the underside of the leaf is in the shade. One way to see the stomata is the paint the underside of the leaf with nail polish and then just let it dry.

The stomata are surrounded by two semicircular shapes, and these are the guard cells. They open and close based on how drought-stressed the plant is, balanced with the need for carbon dioxide. So, the stomata are the sites where carbon dioxide diffuses into the leaf and where oxygen is released. At the same time, when the stomata are open, water is evaporating, and this is what makes possible evapotranspiration.

As you know, some transpirational loss is good because this is the process that pulls water up from the roots. But too much transpirational water loss will cause the cells to lose water and become flaccid, and the plant will wilt. To see how this might happen, we should think about the cross section of the leaf. The leaf has a waxy cuticle on the top and the bottom to prevent water loss. Just inside of this cuticle is a layer of cells called the epidermis, and that works just like the epidermis in humans. This cell layer protects the inner part of the leaf, which is mostly parenchyma cells. These are cells with thin walls, so that movement of materials into and out of the cells can occur relatively easily. These parenchyma cells have vacuoles with a lot of water both for structure and for photosynthesis.

But sometimes having leaves can be too costly in terms of water loss. Like so-called evergreens dropping their needles, most tropical trees gradually replace their leaves pretty quickly, anywhere from 2 months to 2 years. One reason that palm leaves can get so long in such a short amount of time is that palm trees only grow up and not out. They get taller but not wider through their lives, and this is due to the fact that they don't have a vascular cambium for secondary growth that would increase their width.

So, if the leaves don't last as long as the tree lives, what's the mechanism for the tree losing its leaves? Deciduous trees will lose their leaves before winter. Some plants will drop their leaves before a time of drought, and these plants are called drought deciduous. These plants are common in deserts and arid lands. These may be the ultimate cues. But what is the mechanism in the plant?

Plants have hormones that cause many of the developmental changes throughout a plant's life. Hormones will tell the plant when to grow, which way to grow, and when to shut off certain parts, like leaves. The 19th-century German botanist Julius von Sachs first proposed the idea that chemicals produced in one part of the plant would effect changes in another part of the plant. Much of the early knowledge of hormones was derived from similar studies occurring in animals.

We now know there are five main types of plant hormones. Of these, *auxin* was the first hormone to be discovered, and it's one of the very first hormones to be expressed. Plus, along with the hormone *cytokinin*, auxin is required for the plant to live. The ratio of these two hormones control many of the functions in the plant—no auxin, and the plant will be kaput. Botanists know this because there are no mutant plants that don't express auxin. Wow, folks, we may have found a botanical law: auxin is required for life. As an aside, I think auxin would be a tremendous company name. It sounds like apple, it begins with A, and it promotes growth. OK, so there are already a few companies out there named Auxin.

One thing that many people don't know about Charles Darwin, the British scientist who introduced natural selection and evolution to the world, is that he was a botanist. And one of his classic experiments demonstrates where auxin is produced. The experiment Darwin conducted was actually done with his son Francis. Together, they wanted to investigate plant movement toward light. Plants grow toward the light, and this is called phototropism—"photo" for light and "tropism" is a growth response in plants. That plants grew toward the light was well known, but what caused this growth was not known. The Darwins used *coleoptiles* of canary grass. Coleoptiles are the first leaves to emerge out of a developing seed, and it probably goes without saying that these coleoptiles are very sensitive to light. The Darwins discovered that the signal for growth toward the light was coming from the tip of the coleoptile, the immature leaf. When they covered the tip with a cap that prevented light, the coleoptile did not grow toward the light. They wrote about their experiment in an 1880 book called *The Power of Movement in Plants*. In this book, they also hypothesized that the tip of the plant was producing a substance that was promoting the growth on the dark side of the plant that would bend the plant toward the light.

In 1913, Peter Boysen-Jensen supported the Darwins' hypothesis by showing that there was a substance that was produced in the tip that was causing the growth toward light. He cut off the top of the coleoptile and put a piece of gelatin block between the tip and the rest of the coleoptile. If there were a mobile substance, it should be able to diffuse through the gelatin block and the plant should still exhibit a phototropic response. This is exactly what happened. He also showed that if an impenetrable mica sheet was inserted on the side facing the dark, then no growth response would occur. This supports the hypothesis that the mobile chemical travels from the tip and to the dark side of the plant.

At this point, many botanists tried to isolate the substance responsible for growth. Many of these experiments tried to grind up the tip of the coleoptile to extract the substance, but this proved unfruitful because substances that were normally tied up in the cell inhibited the growth response. In 1926, Frits Warmolt Went came up with an elegant experiment to isolate the substance. Again using the gelatin blocks, he let the mystery substance diffuse into the blocks from excised coleoptile tips. Then, he placed the blocks on one side of the coleoptile that had its tip removed. Sure enough, the side where the gelatin block was placed would then start to grow. The increase of growth on the side with the gelatin would force the plant to bend in the opposite direction, thus demonstrating how phototropism works in plants. Because this substance was responsible for growth in the plant, it was named auxin from the Greek *auxein*, meaning "to increase" or "to grow."

Most of the effects of auxin involve growth and phototropism in the shoot, but auxin also promotes gravitropism, which is growth response to gravity. Auxin's gravitropic responses are positive in the root so that it grows down, and negative in the shoot so that it grows up. Auxin also promotes apical dominance in plants, which is the dominant growth of the uppermost shoots to continue growing up. The auxin produced in these apical shoots will suppress the growth of lateral and axillary buds, and this is why we prune shrubs to get them to fill out, so the lateral buds will no longer be suppressed and they will develop.

Auxin also delays leaf fall. As long as auxin is being expressed in the plant, the leaves won't drop, and this is the same thing for flowers and fruits. Auxins are actually a whole group of hormones, and there are synthetic auxins that are also used as herbicides on lawns to kill all the weeds that aren't other grasses and so-called broad-leaved plants. Like many things in biology, a little auxin is a growth enhancer, but too much and the plant will grow itself to death. Auxins are the longest-studied group of hormones, but there are other growth hormones called gibberellins and cytokinins. There are also senescence hormones; abscisic acid, or ABA; and ethylene. Botanists are discovering new chemical messengers all the time.

Far more than just a pretty shape, each leaf is a complex biochemical factory guided by hormones and other chemical signals. Next time, we'll turn to the most important biochemistry of all: how leaves and all of life are powered by photosynthesis.

LECTURE 8

PHOTOSYNTHESIS EVERYONE SHOULD UNDERSTAND

Almost everything in our world depends on the biochemical process of photosynthesis. It is through this cycle that plants and trees are able to become so gigantic through molecules so small we can't even see them. The main difference between animals and plants is that animals have to consume mass to get energy, whereas plants are able to make their own energy from nothing other than sunlight, gas, and water.

THE ANATOMY OF A LEAF

- Beginning on the top of the leaf, there is usually a layer of wax, called the cuticle. The waxy cuticle prevents water loss through the top tissue of the leaf. This is important, because the top is where most of the sunlight will fall, and this part of the leaf can heat up, and the evaporative loss could be quite great, but the cuticle will prevent most of this loss. The cuticle can also help reflect some sunlight as well, which is important in keeping the leaf from overheating.

- Next is a layer of epidermal cells. Like the epidermal cells in our own skin, these cells help protect the underlayer of cells in the leaf.

- After the epidermis, we get to the action site of the leaf, the parenchyma. This is typically where the majority of photosynthesis takes place. Parenchyma cells have thin walls, and these will be the majority of cells in the center of the leaf. These cells contain organelles called chloroplasts, which is where photosynthesis takes place. An organelle is to a cell what an organ is to a body. They are smaller subcellular structures that enable certain physiological processes.

- The chloroplast is so named because it contains chlorophyll, which is the green pigment that enables the process of photosynthesis.

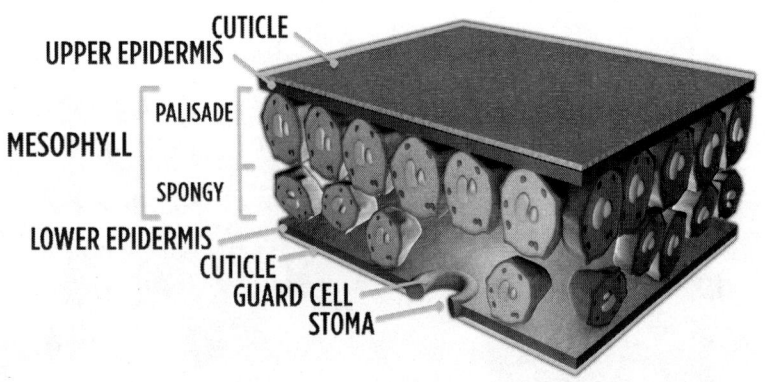

◇ Plants have to be green to photosynthesize. Not all cells of the plant have chloroplasts; they are not in the roots, for example. However, because the leaf has chloroplasts, it has chlorophyll, and that is why the leaves appear green. Chlorophyll appears green because it absorbs most wavelengths of light, except green, which it reflects.

◇ There are other pigments in the leaf, and these are important for absorbing other wavelengths of light. Beta carotenes appear orange-yellow because they are reflecting those wavelengths and absorbing in the green-blue wavelengths. Anthocyanins reflect red and absorb mainly in the green wavelengths, where chlorophyll doesn't absorb at all.

CHLOROPLAST

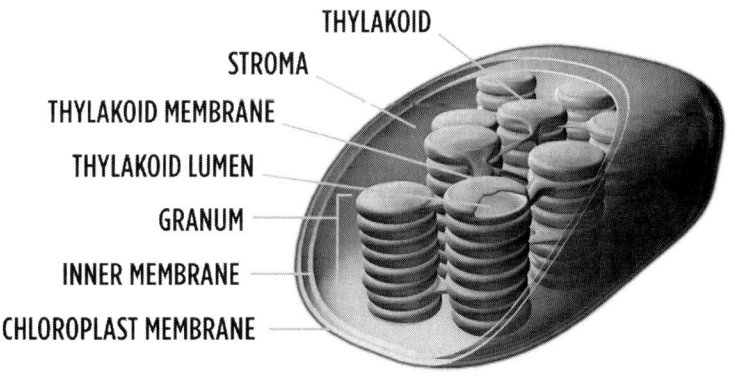

PHOTOSYNTHESIS EVERYONE SHOULD UNDERSTAND

- The chloroplast is shaped like an elongated jellybean. The outer part of the jellybean is called the chloroplast membrane. Inside the chloroplast jellybean are single layers and stacks of pancakes or pita bread, called thylakoids. The purpose of the shape of the pitas is to maximize surface area: The higher the surface-area-to-volume ratio is, the more efficient the structure can be.

- The outer part of thylakoids, the "skin" of the pita, is called the thylakoid membrane. The inner part, or the space inside the pita pancake, is called the thylakoid lumen.

PHOTOSYNTHESIS

- There are 2 steps in photosynthesis: the light-dependent reaction (or light reaction), which occurs within and across the thylakoid membrane; and the light-independent reaction (or dark reaction), which occurs in the stroma, which is inside of the chloroplast but outside of the thylakoids.

- The light-dependent reaction harvests sunlight energy, and the light-independent reaction uses the energy from the light reaction to turn carbon dioxide into sugar. Both of these 2 reactions are necessary for the process of photosynthesis. It's also important to understand that the light-independent, or dark, reaction doesn't occur at night; the phrase "light-independent" means that this reaction isn't actively harvesting light. The light-independent reaction also uses products from the light reaction to turn carbon dioxide into sugar—specifically, a 6-carbon sugar called glucose.

- The light-dependent reaction takes place on and through the surface of the thylakoid membrane, or the skin of the pita pancake. Embedded within the thylakoid membrane are many chlorophyll molecules. Chlorophyll is the green pigment that can absorb light (except green light, which it reflects).

- Many chlorophyll molecules are specially arranged together within the thylakoid membrane to make an antenna that intercepts photons, which are discrete packages of light. All the photons within several chlorophyll molecules get concentrated together, and this group of chlorophyll molecules is called a reaction center.

- All the photon energy within the reaction center then causes one particular chlorophyll molecule to have an electron that gets excited. This excited electron is unstable, meaning that it can be transferred to another molecule.

- In the reaction center of photosynthesis, phaeophytin will pass an electron down to plastoquinone. This is like the biochemical pathway of photon energy, only now the energy from the photons has been changed into the energy carried by one electron, so this electron is moved down the line, called the electron transport chain. This is how energy is transferred from molecule to molecule.

- Interestingly, the electron that gets excited was originally donated from the splitting of water to create an electron, a proton, and oxygen, so this is where and when plants produce oxygen.

- Together, the reaction center and the electron transport chain are called the photosystem, and there are actually 2 photosystems, but the first one to generate the excited electron from the splitting of water is called photosystem II. Photosystem I is where the excited electron goes to generate the products that will be necessary for the light-independent reactions.

- The whole point of photosystem II is to use the light energy to drive one proton, also called a hydrogen ion, across the thylakoid membrane and into the thylakoid lumen. As this process continues, we have a buildup of protons on one side of the thylakoid membrane.

- There are many protons on one side of the membrane and many fewer protons on the other side, and they all want to get through the membrane. The protons want to move down their concentration gradient—from the side with a lot of protons to the side with fewer protons.

◇ However, the only way the protons can get to the other side is through a turnstile called ATP synthase complex. ATP stands for adenosine triphosphate, and it is a molecule that can carry energy around in the cell. ATP synthase is an enzyme—a protein that can make reactions occur faster—that can make ATP, but it can only make ATP if it has the proper starting material, which is ADP, adenosine diphosphate, and the energy from protons moving down their gradient.

◇ In addition to creating the ATP from the ATP synthase, photosystem I also generates a molecule called nicotinamide adenine dinucleotide phosphate (NADPH). This molecule is important because it can carry protons and electrons around in the cell to do work. ATP and NADPH are both used in the light independent reactions.

CALVIN CYCLE

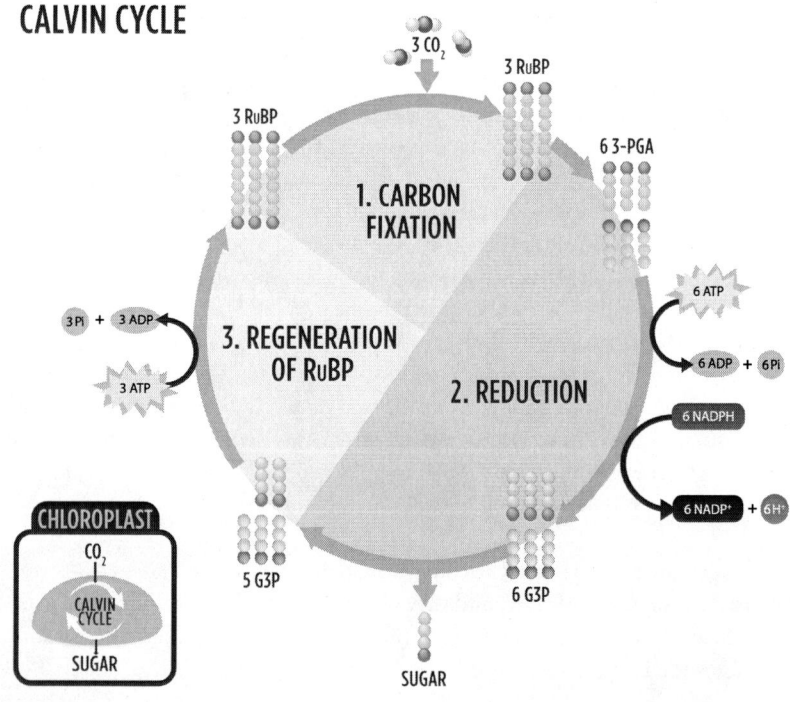

- The ATP and NADPH that were made from the light-dependent reaction now go to the light-independent reaction, which occurs outside of the pita pancake, and this space is the stroma. The light-independent reaction is also called the Calvin cycle, named after Melvin Calvin, who discovered it in 1950. It is also called the C3 cycle, because the cycle starts with 3 carbon dioxide molecules entering the cycle, one at a time.

- The light-independent reaction is a cycle, so we could start anywhere in the cycle and the cycle will just keep going. We're going to start with a 5-carbon sugar by convention, but this is the last product of the cycle. In an overview, this is the part of the cycle where carbon dioxide is taken up to create sugars. When carbon dioxide is added onto other sugars, this is called carbon fixation.

- The carbon dioxide is taken into the leaf through the small pores called stomata, which typically are on the underside of the leaf so as to be in the shade and reduce evaporation of water out of the leaf. These pores are surrounded by guard cells that regulate the opening and closing of the leaf based on the water status of the plant.

- When the plant has plenty of water, the stomata are open and carbon dioxide is diffusing into the leaf, but if the plant is drought-stressed, the guard cells will keep the stomata closed, and then the plant can't get carbon dioxide and can't undergo photosynthesis.

- Just as in the light reaction, the dark reaction is also catalyzed by an important enzyme. The enzyme in the dark reaction is called RuBisCO, which stands for ribulose-1,5-bisphosphate carboxylase/oxygenase. Ribulose bisphosphate is the 5-carbon sugar.

- Through the action of RuBisCO, a carbon atom from a carbon dioxide molecule is added to the 5-carbon sugar ribulose bisphosphate. Now we have a 6-carbon sugar, but it almost instantaneously splits into 2 3-carbon sugars called 3-phosphoglycerate.

- ATP from the light reaction will be used to put an extra phosphate atom onto the 3-carbon sugar, and it becomes 1,3-bisphosphoglycerate. Then, the other product from the light reaction, NADPH, breaks that phosphate off.

- This might seem curious, but it occurs because the phosphate bond to the sugar contains a lot of energy. When it's formed, it requires energy, using the ATP, and when it's broken, it releases energy that allows other reactions to occur. What really needs to occur in this process is that sugar must get made—and the cycle continues.

- The sugar that is formed then leaves the Calvin cycle. This sugar is called glyceraldehyde-3-phosphate (G3P), and it can make many different kinds of carbohydrates, such as cellulose or starch. In fact, 6 of these G3P sugars are made, and one of these molecules leaves the Calvin cycle to become sucrose or starch. The remaining 5 3-carbon sugars then get shuffled around with the help of ATP to form 3 5-carbon sugars. And each one of these 5-carbon sugars is the beginning of the cycle again: ribulose bisphosphate.

- About 95% of the plants on Earth accomplish photosynthesis in the manner that was just described, but there are a few variations on this theme. The first variation is called the C4 pathway (C4). In this pathway, carbon dioxide is picked up by a different enzyme, called PEP carboxylase, and then the first product is a 4-carbon sugar called oxaloacetate.

- The benefit of the C4 pathway is to avoid photorespiration. This can occur when there is light, but photosynthesis isn't happening—respiration is happening instead. The enzyme RuBisCO used in the Calvin cycle has an affinity to pick up carbon dioxide, but it will also pick up oxygen, and its probability of doing so increases with increasing temperatures.

- Plants that use the C4 pathway pick up carbon dioxide using PEP carboxylase and then transfer the carbon as the sugar malate or aspartate to a whole different cell, called the bundle cell, where the normal Calvin cycle will take place. In this way, the carbon dioxide and RuBisCO, which is still present in the Calvin cycle, are spatially segregated so that RuBisCO doesn't encounter oxygen.

- A little less than half of the grasses use the C4 pathway, and this group makes up the majority of plants that use it. Plants growing in warm conditions are most likely to use the C4 pathway. Economically important species that use the C4 pathway are sugarcane and corn.

- Another variation on this theme of photosynthesis is called crassulacean acid metabolism (CAM), which is commonly found in desert plants.

READINGS

Halle, *In Praise of Plants*.

King, *Reaching for the Sun*.

QUESTIONS

1. How does gas become mass in plants?

2. What are some environmental factors that would limit photosynthesis?

3. Farmers and growers aren't that interested in photosynthetic rates because these often don't correspond directly with yield. Why might this be?

LECTURE 8 TRANSCRIPT
PHOTOSYNTHESIS EVERYONE SHOULD UNDERSTAND

When you look at a massive tree—a redwood, for example—have you wondered where all that mass comes from? There are redwoods that are so large, you can drive a car through the base of the tree, and there is still enough trunk around the opening that the tree can still live. Where do the materials come from that create such mass?

In 1995, a group of researchers from Harvard University asked a number of students graduating from their university this very question. Where does the mass come from in such huge trees? Many answered that it was from the water, some said it was from the soil, but what botanists have understood for a long time is that the mass comes from gas—from carbon dioxide, to be specific. I like that phrase: the mass comes from gas.

This is truly amazing when you think about it: a virtually weightless substance is the main ingredient that trees and all plants transform into herbaceous tissues and wood through the process of photosynthesis. This is very different than the process in animals. We have to consume mass to make energy and create more mass, but trees can create and make energy from thin air and sunlight and turn it into incredible mass. Given how we currently operate, everything in our world pretty much depends on this biochemical process of photosynthesis, so it's worth understanding it.

What's funny is that if you take a biochemistry class in college, the focus is almost entirely on glycolysis and the citric acid cycle. Although photosynthesis appears in the textbook, the students want to focus on the processes of their own bodies instead of the process that makes life on Earth possible. In order to understand the process of photosynthesis, we'll first have to look more closely at the anatomy of a leaf. Although leaves come in a huge variety of shapes and sizes, the internal anatomy is generally the same, much like that of humans. One thing that will be true of most leaves is that they are green. In fact, for something to photosynthesize, it has to be green.

When you look at a carrot, which is technically the root of that plant, that part of the plant isn't photosynthesizing. So roots don't photosynthesize; they aren't green. Some stems can photosynthesize because they're green, like cactus stems, so it isn't only the leaves that can undergo photosynthesis, but the leaves are where most of the photosynthetic action will occur. So, if you think about a typical leaf, an oak or a maple, for example, it's flat and thin, and this shape of the leaf has a lot of surface to collect sunlight. If we investigate the anatomy of this leaf, we can look at that thin leaf in cross section.

So, we'll talk about the leaf's anatomy from top, through the middle, to the bottom, beginning on the top of the leaf, where there is usually a layer of wax, and this is called the cuticle. The waxy cuticle prevents water loss through the top tissue of the leaf, and this is so important, since the top is where most of the sunlight will fall, and this part of the leaf can heat up, and the evaporative loss could be quite great, but the cuticle will prevent most of this loss. The cuticle can also help to reflect some of the sunlight as well, which is important in keeping the leaf from overheating.

Next in our cross section of the leaf is a layer of epidermal cells. Like the epidermal cells of our own skin, these cells help protect the underlayer of cells in the leaf. Although our epidermis and leaf epidermis are very different in structure, they are called by the same name because the prefix "epi" means "on top of"; the epidermis is on top of the leaf. After the epidermis, we get to the action site of the leaf: the parenchyma. This is typically where the majority of the photosynthesis takes place. Parenchyma cells have thin walls, and these will be the majority of cells in the center of the leaf. These cells contain organelles called chloroplasts, and this is where photosynthesis takes place. An organelle is to a cell what an organ is to a body. They are smaller subcellular structures that enable certain physiological processes.

The chloroplast is so named because it contains chlorophyll, which is the green pigment that enables the process of photosynthesis. Remember, I said that plants have to be green in order to photosynthesize. Not all cells of the plant have chloroplasts—they are not in the roots, for example. But because the leaf has chloroplasts, it has chlorophyll, and that is why the leaves appear green. Chlorophyll appears green because it absorbs most wavelengths of light, except green, which it reflects. So, let's test your understanding. What would be the worst color of light for growing plants? If you said green, that's correct.

There are other pigments in the leaf, too, and these are important for absorbing other wavelengths of light. Beta-carotenes appear orangish-yellow because they are reflecting those wavelengths and absorbing in the greenish-blue wavelengths. Anthocyanins reflect red and absorb mainly in the green wavelengths, where chlorophyll doesn't absorb at all.

I want to show you how botanists would measure photosynthetic rates in plants. I have here a machine very much like the one I used in my dissertation to measure the photosynthetic rates of plants, and it consists of sort of three pieces here: a cuvette, which I'm going to put on the leaf; a computer, which will record data and can measure environmental variables happening around the leaf; and lastly, this part is called an infrared gas analyzer, or an IRGA, and it's going to measure the amount of CO_2 that the leaf is taking up. Let me show you how it works.

So you take the cuvette and I'm going to put a leaf in here, and I'll close the cuvette down. The cuvette—this container—has a thermocouple and also a light sensor, so I know how much light the leaf is receiving and I know its temperature. I can also control how much air is moving into and out of the cuvette, as well as the relative humidity, by using the controls on this computer. So, I would have my leaf in a steady state. I want to make sure that the humidity isn't changing, the airflow isn't changing, and the temperature is staying relatively constant. So I'm checking all of that on the computer. When I feel like I have a steady measurement, I can press a button over here, and the computer will record the measurement. Then I can release the leaf and I'm ready to move on to the next leaf or the next plant.

Because the chloroplast is so important to understanding how photosynthesis works, we're going to investigate the structure of the chloroplast. Remember, these are special organelles within the cells of the leaves. The chloroplast itself is shaped like an elongated jellybean. The outer part of the jellybean is called the chloroplast membrane. Inside our chloroplast jellybean will be a single layer and single layers of stacks, or pancakes, or pita bread if you've had enough sugary metaphors with our jellybean. These stacks of pita and single pitas are called *thylakoids*. Again, the shape of these pitas are to maximize the surface area. The outer part of the thylakoids, the skin of the pita, is called the thylakoid membrane. The inner part, or the space inside the pita pancake, is called the thylakoid lumen.

OK, so we've got the basic structure of the chloroplast. There are pita pancakes both in single layers and in stacks, and the pita pancakes are called thylakoids, and they are surrounded by a thylakoid membrane, and the space inside the thylakoid is called the thylakoid lumen. And this shape is important because it maximizes the amount of surface area for the membranes. The higher the surface area to volume ratio, the more efficient the structure can be.

There are two steps in photosynthesis, and one is called the light-dependent reaction, which occurs within and across the thylakoid membrane. The other reaction is called the light-independent reaction, and it occurs in the *stroma*, which is inside of the chloroplast but outside of the thylakoids. There are a few different names for these two steps of photosynthesis. The light-dependent reactions are also simply called the light reactions. The light-independent reaction is also called the dark reaction.

What I like about these terms is that they are descriptive of what is happening: the light-dependent reaction harvests sunlight energy, and the light-independent reaction uses energy from the light reaction to turn carbon dioxide into sugar. What is important to understand is that both of these reactions are necessary for the process of photosynthesis. It's also important to understand that the light-independent or dark reaction doesn't occur at night. The phrase light-independent means that this reaction isn't actively harvesting light. The light-independent reaction also uses products from the light reaction to work. Now we have these two reactions that are occurring simultaneously in the leaf, but we'll talk about them sequentially, beginning with the light reaction, and then using the products from the light reaction to run the dark reaction to turn carbon dioxide into sugar, specifically a 6-carbon sugar called glucose.

So let's begin with the light dependent reaction, which is going to take place on and through the surface of the thylakoid membrane. Remember that it is the skin of our pita pancake. So, embedded within the thylakoid membrane are many chlorophyll molecules. Remember that chlorophyll is the green pigment that can absorb light, except green light, which it reflects. Many chlorophyll molecules are specially arranged together within the thylakoid membrane to make a sort of dish antenna, rather like one you would see on the side of houses to intercept satellite television. Instead of television, the chlorophyll molecules are all intercepting photons, which are discrete packages of light. All the photons within several chlorophyll molecules get concentrated together, and this group of chlorophyll molecules is called a reaction center.

All of the photon energy within the reaction center then causes one particular chlorophyll molecule to have an electron that gets excited, and this excited electron is unstable, meaning it can be transferred to another molecule, sort of like a fire line. So, think of a fire in a house, and instead of a hose, people are passing buckets of water from the creek to the fire. Each person would represent a different molecule.

In the reaction center of photosynthesis, phaeophytin will pass an electron down to plastoquinone. This is like the biochemical pathway of photon energy, only now the energy from the photons has been changed into energy carried by one electron, so this electron is moved down our fire line. And this is called the electron transport chain, and this is how energy is transferred from molecule to molecule. Interestingly, the electron that gets excited was originally donated from the splitting of water to create an electron, a proton, and oxygen, so this is where and when plants produce oxygen.

Together, the reaction center and the electron transport chain is called the photosystem, and there are actually two photosystems, but the first one to generate the excited electron from the splitting of water is called photosystem II. It's not that botanists don't know how to count, it's because that photosystem was discovered second. Photosystem I is where the excited electron goes to generate the products that will be necessary for the light-independent reactions. Of course, as it's called photosystem I, it was discovered first. Both of these photosystems together are called the Z-scheme because they form the letter Z on its side. And as we'll see, the whole point of photosystem II is to use the light energy to drive one proton, also called a hydrogen ion, across the thylakoid membrane and into the thylakoid lumen. As this process continues, we have a buildup of protons on one side of the thylakoid membrane.

To understand why this buildup of protons is so important, and to understand the process that occurs next, imagine you are trying to leave the zoo and you have to go through one of those big vertical turnstiles, the kind that you can't jump over but you have to walk through. If you imagine everyone is leaving the zoo because it's closing time, then you can imagine a lot of people on one side of the turnstile waiting to go through. There are many people on the one side concentrated together in the zoo waiting to leave. On the other side of the turnstile, people are walking away and are more diffuse. This is our concentration gradient. There are more people on one side of the turnstile, and this is like the concentration difference across our thylakoid membrane. There are many protons on one side of the membrane and many fewer protons on the other side. They all want to get through the membrane. The protons want to move down their concentration gradient from the side with a lot of protons to the side with fewer protons.

However, like you leaving the zoo, the only way the protons can get to the other side is through the turnstile, only now the turnstile is called ATP synthase complex. ATP stands for adenosine triphosphate, and it's a molecule that can carry energy around in the cell. Whenever you hear the suffix "ase" on the end of a molecule name, that implies an enzyme, which is a protein that can make reactions occur faster. So, ATP synthase is a big enzyme that can make ATP, but it can only make ATP if it has the proper starting material, which is ADP, adenosine diphosphate, and the energy from the protons moving down their gradient.

In addition to creating ATP from ATP synthase, photosystem I also generates a molecule called NADPH. This molecule is important because it can carry protons and electrons around in the cell to do work. When the molecule doesn't have the H, or proton, on the end, it's $NADP^+$, or nicotinamide adenine dinucleotide phosphate—again, abbreviated $NADP^+$—and the plus is on the end to tell us that it has a positive charge. When a hydrogen gets attached to the molecule, it's reduced in charge, which means it loses its positive charge to become NADPH. Again, this molecule is good for donating protons and electrons as needed around the cell. We'll see that ATP and NADPH are both used in the light-independent reactions.

So, the protons want to go from one side of the thylakoid membrane to the other, and they can only travel through the turnstile ATPase. As they travel down their concentration gradient to the other side, they essentially spin the ATPase with their energy, which can then add a phosphate group to the ADP, the adenosine diphosphate, to create the ATP, adenosine triphosphate. And it's the addition and removal of this phosphate that can provide energy to other reactions, particularly the other main reaction of photosynthesis: the dark reaction.

So let's review the light reaction. Essentially, packets of light—photons— are intercepted by antennas of chlorophyll in the thylakoid membranes of the chloroplast. The energy from the photons excites an electron, which is transported from molecule to molecule in an electron transport chain to drive a proton against its concentration gradient on the other side of the thylakoid membrane and to create NADPH. Now the proton wants to go with its concentration gradient to go back into the thylakoid lumen. As the proton moves down its concentration gradient through the ATPase, the energy it creates allows the ATPase to make adenosine triphosphate from adenosine diphosphate. This adenosine triphosphate will be used in the dark reaction.

The ATP and NADPH that was just made from the light-dependent reaction now goes to the light-independent reaction, which occurs outside of our pita pancake, and this space is the stroma. Recall that the light-independent reaction doesn't involve light specifically, but it can only occur with the products of the light reaction, namely the ATP and NADPH. The light-independent reaction is also called the Calvin cycle, named after Melvin Calvin from the University of California at Berkeley, who discovered it in 1950. It is also called the C3 cycle, because the cycle starts with three carbon dioxide molecules entering the cycle one at a time.

As I've just alluded to, the light-independent reaction is a cycle, so understand that we could start anywhere in the cycle, and the cycle is just going to keep going. We're going to start with a 5-carbon sugar by convention, but understand that this is the last product of the cycle. In an overview, this is the part of the cycle where the carbon dioxide, or CO_2, is taken up to create sugars. When CO_2 is added onto the other sugars, this is called carbon fixation.

So, the CO_2 is taken into the leaf through small pores called *stomata*, which is Greek for "little mouth." Typically, the stomata are on the underside of the leaf so as to be in the shade and reduce evaporation of water out of the leaf. These pores are surrounded by guard cells that regulate the opening and closing of the leaf based on the water status of the plant. When the plant has plenty of water, the stomata are open and CO_2 is diffusing into the leaf. But if the plant is drought-stressed, the guard cells will keep the stomata closed, and the plant can't get CO_2 and it can't undergo photosynthesis.

Just as we saw in the light reaction, the dark reaction is also catalyzed by an important enzyme, and recall that an enzyme is a protein that allows reactions to occur faster. The enzyme in the dark reaction is called RuBisCO, which stands for ribulose-1,5-bisphosphate carboxylase/oxygenase. That name sounds long, but it actually describes the structure of the enzyme and offers clues to its function. Ribulose-1,5 is a sugar with five carbons. Aha! Remember, I said we would start with a 5-carbon sugar. Ribulose bisphosphate is that sugar, and bisphosphate means that it has two inorganic phosphate groups stuck onto it as well. The phosphate group is a phosphate atom with four oxygen atoms.

Sugars are carbohydrates, and as that name suggests, those are molecules with carbon and oxygen and hydrogen: carbohydrate. So, from the name ribulose, we have a clue that this enzyme will have something to do with a 5-carbon sugar. So, we have a 5-carbon sugar, ribulose bisphosphate, that meets up with an enzyme, RuBisCO, or 5-ribulose bisphosphate carboxylase/oxygenase. The latter part of this enzyme's name describes the action that the enzyme will catalyze. Carboxylase means to add carbons, and oxygenase means to add oxygen. Because this enzyme, RuBisCO, is found in every chloroplast in every leaf, it is often surmised to be the most abundant protein on Earth.

As I said, we'll begin discussing the Calvin cycle where the CO_2 molecule is incorporated, or fixed. Through the action of our enzyme RuBisCO, a carbon atom from a carbon dioxide molecule is added to the 5-carbon sugar we talked about, the ribulose bisphosphate. Now, if you're following the accounting here, we took a 5-carbon sugar, and we added a carbon atom from carbon dioxide, so now we have we should have a 6-carbon sugar, and we do, but it almost instantaneously splits into two 3-carbon sugars called 3-phosphoglycerate. Recall that the "phospho" part of the name means we still have a phosphate atom, only now we just have one phosphate with each 3-carbon sugar. This is also why the Calvin cycle is sometimes referred to as C3 photosynthesis.

Now, with our 3-carbon, 1-phosphate sugar, we will use some energy from the light reaction. Remember the beginning of this lecture? In the light-dependent reactions, the plant used light energy to split water to gain an electron to eventually create that ATP molecule that carries energy. At this stage in the Calvin cycle, ATP will be used to put an extra phosphate atom onto our 3-carbon sugar, so it becomes 1,3-bisphosphoglycerate. The "biphospho" tells us that there are two phosphate atoms.

At the next stage, the other product from the light reaction, NADPH, breaks that phosphate off. Now, it's curious isn't it that in one step ATP is used to add a phosphate, and in the very next step NADPH is used to remove that phosphate? This occurs because the phosphate bond to the sugar contains a lot of energy. When it's formed, it requires energy, using the ATP, and when it's broken, it releases energy that allows other reactions to occur. What really needs to occur in this process is that the sugar must get made, and the cycle continues. Let's review what's happened so far.

An enzyme called RuBisCO catalyzes a carbon joining a 5-carbon sugar, ribulose bisphosphate, to become a 6-carbon sugar, which immediately breaks into two 3-carbon sugars. Now, ATP is added, which adds a phosphate, and then it is broken off in the next step, which is all done so that a sugar can be formed that then leaves the Calvin cycle. The sugar that leaves is called glyceraldehyde-3-phosphate, which is called G3P. If you recall that photosynthesis makes glucose, or $C_6H_{12}O_6$, you may be wondering why we're left with G3P. The reality is that G3P can make many different kinds of carbohydrates, like cellulose, essentially the stuff celery is made of, or starch, the material found in potatoes.

We're still not finished with the Calvin cycle, because although the sugar was produced, the reaction to create the beginning of the product still needs to occur. Remember that? It was the 5-carbon sugar: ribulose bisphosphate. So, let's determine how the cycle moves from a 3-carbon sugar—G3P, glyceraldehyde-3-phosphate—to ribulose bisphosphate. At this point, we left off with our G3P sugar from the Calvin cycle. Actually, there are six of these molecules made, and, as I mentioned, one of the G3P sugars leaves the Calvin cycle to become sucrose or starch. The remaining five 3-carbon sugars, with three carbons each, get shuffled around with the help of ATP to form three 5-carbon sugars. And, yes, each one of these 5-carbon sugars is the beginning of the cycle again: ribulose bisphosphate.

In summary, if we look at the whole system, light will enter the chloroplast from the top of the leaves, along with water from the stems. First, in the thylakoid membrane, the photosystems create ATP and NADPH, which are then used in the light-independent cycle. In the light-independent cycle, a carbon from carbon dioxide is added to the sugar ribulose bisphosphate. This splits into two 3-carbon sugars, and that's why it's called C3 photosynthesis, or the Calvin cycle. The Calvin cycle eventually produces G3P, which can be exported to be made into starch, amino acids, or sucrose.

Now, about 95% of plants on Earth accomplish photosynthesis in the manner I just described, but there are two variations on this theme, and they're worth mentioning here. The first variation is called the C4 pathway, or simply C4. If C3 was so called because the first product was a 3-carbon sugar, I'll bet you can guess the first product of the C4 pathway. If you think it's a 4-carbon sugar, you're right. In this pathway, CO2 is picked up by a different enzyme called PEP carboxylase, and then the first product is a 4-carbon sugar called oxaloacetate.

The benefit of the C4 pathway is to avoid photorespiration. This can occur when there is light, but photosynthesis isn't happening; respiration is happening instead. You see, the enzyme RuBisCO used in the Calvin cycle has an affinity to pick up carbon dioxide, but it will also pick up oxygen, and you can tell that by the name. RuBisCO stands for ribulose-1,5-bisphosphate carboxylase/oxygenase. That oxygenase means it will sometimes pick up oxygen, and its probability of doing so will increase with increasing temperatures.

Plants that use the C4 pathway to pick up CO_2 using PEP carboxylase, and then transfer that carbon as the sugar malate or aspartate to a whole different cell called the bundle cell, where the normal Calvin cycle will take place. And so, in this way, the CO_2 and the RuBisCO, which is still present in the Calvin cycle, are spatially segregated so that the RuBisCO doesn't ever encounter oxygen. The bundle sheath cells are visible in a cross section of the leaf. A little less than half of the grasses use the C4 pathway, and this group makes up the majority of plants that use it, though there are a few others. Plants growing in warm conditions are most likely to use the C4 pathway. Economically important species that use the C4 pathway are sugarcane and corn.

There is another variation on this theme of photosynthesis, and it's called the crassulacean acid metabolism, or CAM. We'll learn more about CAM when we talk about desert plants, because this photosynthetic pathway is commonly found in desert plants. All of photosynthesis—the light-dependent and light-independent reactions—takes place in the chloroplasts, the organelles that give leaves their green color. Within parenchyma cells there are roughly 100 chloroplasts, and within each leaf there are thousands of cells. It is through this cycle that plants and trees are able to get so gigantic through molecules so small we can't even see them. The main difference between animals and plants is that animals have to get their energy from energy, whereas plants are able to make their own energy from nothing other than sunlight, gas, and water. Imagine just being able to go outside and get energy and food instead of having to go to the grocery store.

LECTURE 9

DAYS AND YEARS IN THE LIVES OF PLANTS

How does a plant know when to flower? And is climate change causing plants to flower earlier? This lecture will consider these questions by taking a 21st-century approach that will involve genes, which are recipes for proteins, and gene regulation, which determines which genes get activated. This determines which proteins get made, and those proteins, in turn, cause changes in the plant.

HOW DOES A PLANT KNOW WHEN TO FLOWER?

- Flowering is promoted by internal cues of hormones and the status of buds. If the bud is at the tip of a branch or plant, it's called the apical bud. This bud is also called the terminal bud, because it's at the terminal end of the branch or plant.

- Those buds that promote growth to the sides are called the lateral buds. Flowering can also be promoted by environmental cues. Two important cues are light and temperature. The light cue is called photoperiod, or the amount of light a plant sees during a certain time frame. With regard to the temperature cue, the amount of cold a plant experiences is just as important as the amount of heat it experiences.

- Phenology is the study of the timing of biological events. This timing can be seasonal or daily. A circadian rhythm is a pattern that exists over 24 hours, or 1 day.

- Plants are classified according to the length of photoperiod they need to flower. Short-day plants need a day that is less than 12 hours to flower, which would mean a short day, so they would flower in early spring or late fall. An example of a short-day plant is a Christmas cactus, which flowers in the late fall.

CHRISTMAS CACTUS
Schlumbergera sp.

- A long-day plant would need a day longer than about 12 hours to flower, so these are plants that flower in the midsummer. Many of our midsummer vegetables are long-day plants, such as potatoes and spinach. In both cases, there is a critical length that must be met so that flowering occurs.

- Some plants are day-neutral plants, which don't flower in response to day length and instead flower in response to a different environmental cue. An example of a day-neutral plant is a tomato, which typically flowers in response to temperature. Many other day-neutral plants also respond to temperature, and this is why we see a proliferation of flowers after a particularly warm spring weekend.

- Although photoperiod is important for all plants, the degree of its importance is different for different species; that is, some plants use photoperiod as a cue, but they also respond to temperature—usually soil temperature, but also air temperature.

- How are plants sensing the length of the daylight, or the photoperiod? Plants sense the length of the night instead of the length of the day. Plants that flower early in the spring flower after long nights. Plants that flower in the middle of summer flower after short nights. And this effect is reversible because of a unique chemical in their leaves called phytochrome.

- If you interrupt a long night with light, then the plant "sees" a short night, and if it's a short-day plant, it won't flower. Likewise, if you break a long night up with a shot of light in the middle, the plant "sees" a short night and will flower, if it's a long-day, and thus short-night, plant. This interruption of flowering occurs inadvertently because of nighttime lighting.

- In the 1930s, Russian scientist Mikhail Chailakhyan conducted grafting experiments to determine how flowering was induced. A graft is when you take part of one plant—for example, the tip, which is also called the scion—and attach it to the base, or rootstalk, of another plant. When a plant is grafted, one plant can be connected to another by small pieces of tape.

- By conducting these experiments, Chailakhyan showed that the leaves were communicating the readiness of flowering. If Chailakhyan took a plant that was ready to flower and grafted it onto a plant that was not ready to flower, the plant would flower. And if a leaf was taken from a plant that was not ready to flower but was placed on a rootstalk that was ready to flower, the plant would not flower.

◇ So, it's been known for a long time that there is a substance produced in the leaves that signals to the rest of the plant that it is time to flower. Scientists also knew that this substance must travel through the phloem to get to the part of the plant where the flower buds would develop. This substance was named florigen by Chailakhyan. However, proving its existence was elusive.

◇ A paper by Jan Zeevaart in 2006 outlined the pathway of trying to find florigen over 70 years. This paper includes Chailakhyan's work. After this work, there wasn't much done in locating florigen and determining its exact role until the 1980s and the advent of molecular genetics.

◇ Most of these studies involved looking at mutant varieties of a rock cress plant called *Arabidopsis thaliana*, which is very useful in the lab because it has a very fast generation time from seed to mature plants with seeds—about 6 weeks. It also has a small genome, meaning the entire collection of its DNA, and it is fully mapped out, so scientists know where all the genes are on the chromosomes.

◇ Because it's such a good model organism, scientists were able to create various mutants of *Arabidopsis*. Some of these mutants flowered later than plants in nature, so scientists tried to determine the gene and/or proteins that these plants were missing.

THALE CRESS
Arabidopsis thaliana

- These experiments determined that 2 genes were important in flowering: CONSTANS (CO) and FLOWERING LOCUS T (FT). The way these genes interacted and the exact proteins they produced remained inconclusive until 2005. During that year, Tao Huang, and his colleagues in Sweden and France, conducted experiments to unify the CONSTANS gene and the FLOWERING LOCUS T gene.

- They knew that the day-length signal was phytochrome but then found that the plant's response to the day-length signal would switch on the FLOWERING LOCUS T gene. They knew that turning on just one of these genes in the leaf was enough to induce flowering. They also determined that it was the messenger RNA of this gene that traveled to the shoot apex, where it would induce flowering. Thus, the messenger RNA of this gene becomes the mystery florigen.

- Whether the same mechanism is true in all plants remains to be seen.

- Another environmental cue that many plants need to flower is a period of cold, called vernalization. To flower, many seeds must be given a cold treatment. Flowering can be promoted by a cold treatment given to a seed or a mature plant.

EFFECTS OF CLIMATE CHANGE

- Some of the sunlight that comes to Earth reflects back into space from the atmosphere, and some passes through our atmosphere and hits the Earth. As the light makes its way through the atmosphere and hits the Earth, it loses energy and slows down to become heat. This heat can't pass back out of our atmosphere.

- This is a natural phenomenon, and it's a good thing, because without our atmosphere trapping heat, we would all freeze to death. The problem comes from releasing carbon dioxide, methane, and other greenhouse gasses into the atmosphere. These gases trap the heat even more, thus warming the planet.

◇ Some plants are responding to climate change by flowering earlier. For example, Seth Munson and Anna Sher used herbarium records to determine how much earlier plants have been flowering in Colorado. Herbaria are collections of dried plants that have been collected in the field. The plants also contain information about the collection.

◇ Using herbarium records for 12 different rare plants in Colorado, they showed that the flowering date was on average 42 days earlier than in the late 1800s. That breaks down to 3.1 days per decade. On average, their flowering was about 1.5 days earlier per decade.

◇ There is a network called Nature's Notebook that is devoted to taking data about all kinds of plants in various areas of the world. On the organization's website, amateur naturalists can report the flowering of different plants in their area, and the data are available for anyone to analyze.

◇ In addition to species flowering earlier, another well-documented change is the movement of whole plant communities up mountainsides. A study done by Anne Kelly and Michael Goulden in 2008 documented observations of plants across a 2314-meter elevation gradient in 1977 and again in 2006–2007 in the Santa Rosa Mountains of southern California. They showed that the elevation of plant communities shifted about 65 meters upward.

READING

Glover, *Understanding Flowers and Flowering*.

QUESTIONS

1 What is phytochrome, and what does it sense?

2 How do you think plants would respond to 24-hour sunlight?

LECTURE 9 TRANSCRIPT
DAYS AND YEARS IN THE LIVES OF PLANTS

Have you ever been surprised to see a plant flower in the spring? And even though I'm a botanist, I feel like every spring, flowers sort of sneak up on me. There I am walking around campus in March and thinking about midterms, and bam, the magnolia tree is in full bloom. This is remarkable to me because: a) magnolias are not native to Denver—they're better known to the southern climates of the U.S. where they're native; b) this is a large tree that is taller than our three-story science building; and c) I know winter isn't over. In fact, in Denver, March is our snowiest month. Still, there it is on St. Patrick's Day this year, March 17, full of bright pink blossoms. Nothing else on our campus looks remotely close to flowering, and yet here is this precocious magnolia. Even as a botanist, I wonder, how does that tree know when to flower? And I, along with everyone else who occasionally notices nature, wonder: is climate change causing plants to flower earlier?

In this lecture, we'll consider both of these questions. The answer to the first question, how does a plant know when to flower, is considerably more complex than when I learned basic plant physiology last century. Here, we'll take a thoroughly 21st-century approach to the question, and this will involve our good friends genes and gene regulation. Recall that gene regulation determines which genes get activated, and this determines which proteins get made. Those proteins, in turn, cause changes in the plant.

One thing we know for sure is that flowering is promoted by internal cues of hormones and the status of buds. If the bud is at the tip, or the top, of the plant, it's called the apical bud. This bud is also called the terminal bud, since it's at the terminal end of the branch or the plant. When you look at the very tip of the branch or the very top of the plant, particularly a woody plant, you might be able to see the apical bud.

Those buds that promote growth to the sides are called the lateral buds. Flowering can also be promoted by environmental cues, and two important cues are light and temperature. The light cue is called photoperiod, or the amount of light a plant sees during a certain time frame. The temperature cue sounds obvious enough, but we'll actually see that the amount of cold a plant experiences is just as important as the amount of heat.

What we're actually interested in here is called phenology, the study of the timing of biological events. This timing can be seasonal or daily. The time of year birds migrate is also an example of phenology. We use the term circadian rhythm for plants the same way it would be used for humans. A circadian rhythm is a pattern that exists over 24 hours, or one day. *Circa* comes from the Latin for "about" and *diem* for "day."

Photoperiod is another term in botany, and it simply means the amount of light in a given period. So, if I said a plant flowers when it receives a 15-hour photoperiod, you would know that this plant won't flower until midsummer, when the sun shines for about 15 hours. Photoperiod is so important that plants are actually classified according to the length of photoperiod they need to flower. So short-day plants, or SDPs, need a day that is less than 12 hours to flower, which would mean a short day, so they would flower in early spring or late fall. A great example of a short-day plant is a Christmas cactus, which flowers in the late fall, when the days get short. A long-day plant, or LDP, would need a day longer than 12 hours to flower, so these are plants that flower in the midsummer. Many of our midsummer vegetables are long-day plants, like potatoes and spinach. In both cases, there is a critical length that must be met so that flowering occurs.

But there are some plants that are day-neutral. These plants don't flower in response to day length; they flower in response to a different environmental cue. An example of a day-neutral plant is tomato. Tomatoes typically flower in response to temperature. Many other day-neutral plants will also respond to temperature, and this is how we can see a proliferation of flowers after a particularly warm spring weekend. So, although photoperiod is important for all plants, the degree of its importance is different for different species. That is, some plants are using photoperiod as a cue, but they're also responding to temperature, usually soil temperature, and also air temperature as well. In fact, scientists have known since the 1930s that plants had to be sensing the amount of light with their leaves. They believed in a mystery substance that was then transported from the leaves to the other organs of the plant.

We've come a long way from the mystery substance, as we'll see, but first let's ask, how are plants sensing the length of the daylight, or the photoperiod? It turns out that it's not so much about the length of the daylight as it is about the length of the night. Plants have another unique chemical in their leaves called *phytochrome*. *Phyto* is Greek for "plant" and *chrome* is Greek for "color." This chemical is really another pigment, and pigments are substances that absorb light. We learned about the pigment chlorophyll in the previous lecture. Do you remember which types of light it absorbs? If you think about the kind of light it reflects—green—you'll remember chlorophyll absorbs in the red and blue parts of the spectrum.

The discovery of phytochrome took a long time. According to the U.S. Department of Agriculture, the search began around 1936 at the U.S. Department of Agriculture Research Center in Beltsville, Maryland. America was going through a green revolution of agriculture, and understanding plant physiology was becoming very important to crop science. Farmers wanted to know what was controlling flowering in plants. For example, soybeans, despite being planted at different times, all flowered at once, which made them less valuable as a crop. If the soybeans could stagger their maturation times, then the crop could be harvested at different times, which would be more efficient and ensure less waste during harvest.

There were a number of experiments to determine cues necessary for flowering. In 1936, one of the first experiments with flowering in plants confirmed that it was the period of darkness that controlled flowering. This study indicated that in long-day plants, a brief period of light during the dark period encouraged flowering. In short-day plants, a brief period of light during the dark period inhibited flowering. Thus, the stage was set. There was something in the leaves that was sensing light and dark and sending messages elsewhere in the plant about flowering.

The botanists at the USDA then teamed up with a chemist to try and characterize the mechanism for this light sensing in plant leaves. They went from overall amounts of light to amounts of light at specific wavelengths. By using a prism, these scientists used different wavelengths of light to determine which were most effective at promoting or inhibiting flowering. They found that red light strongly inhibited flowering. Later experiments at another lab found a similar reaction with germination. Red light inhibited germination, but far-red light promoted germination. Far-red is a bit longer wavelength—730 nanometers—than regular red.

In order to understand this relationship, we need to understand some key facts about red light versus far-red light. Suppose we have a rainbow, like the kind you learned about in science. You might remember the acronym, ROYGBIV, which stands for red, orange, yellow, green, blue, indigo, and violet. That rainbow is a tiny portion of the overall electromagnetic radiation, from high-energy gamma rays to low-energy radio waves, with visible light somewhere in the middle.

The wavelengths of light get shorter as one progresses through the spectrum. Red light is the longest wavelength of visible light, around 660 nanometers, and violet at the shortest, around 400 nanometers. You've probably heard of ultraviolet light, and it is even shorter than violet light, so anything shorter than 400 nanometers down to 10 nanometers is ultraviolet light. Ultraviolet light is also called ultraviolet radiation, since we can't see ultraviolet as visible light. You've probably heard that this is bad for your skin because ultraviolet radiation causes sunburns and cancer. This is why we wear sunscreen. The point here is for you to recognize that shorter wavelengths have more energy. So, which type of light has more energy: red or blue? The blue light has a shorter wavelength, so it has more energy.

What these early experiments also revealed was that this relationship was reversible. That is, whatever wavelength of light the plant was last exposed to would be the wavelength to which it would respond. So, if a plant was last exposed to red light, even if it had been exposed to far-red light earlier, flowering would not occur. This is important in our understanding of phytochrome. If we think about red light at 660 nanometers, and far-red light at 730 nanometers, which has more energy? If you answered red light, you are correct. Now, knowing that red light has more energy, think of red light as a biological activator and far-red light as an inactivator. The more far-red light a plant senses, the more the plant sees shade.

As we mentioned, the molecule is reversible, which means that it comes in two forms, which are phytochrome red, abbreviated Pr, and phytochrome far-red, or Pfr. Phytochrome red absorbs red light and converts to phytochrome far-red, which is the active form of the pigment and will promote biologic functions like seed germination or flowering. Phytochrome far-red absorbs far-red light to become phytochrome red, or the inactive form. It's fully reversible. This is a bit confusing, so let's go over it again. Far-red light at 730 nanometers has less energy than red light at 660 nanometers. OK, so far, so good. But far-red light makes the phytochrome become phytochrome red, which is the inactive form. And, red light converts phytochrome to phytochrome far-red, which is the active form.

Since sunlight contains both far-red and red light, the presence of light will cause an equilibrium of sorts of phytochrome red and phytochrome far-red. But, in the dark, the phytochrome far-red will decrease steadily as long as the plant is kept in the dark. But one little flash of red light will cause an instantaneous conversion of phytochrome red to phytochrome far-red, and flowering will be inhibited. All of this is to say that plants sense the length of the night instead of the length of the day. As we all know, some plants flower after long nights—those that flower early in the spring. Other plants flower after short nights, and those are plants that flower in the middle of summer. And we know that this effect is reversible because of phytochrome. So, if you interrupt a long night with light, then the plant sees a short night, and it's a short-day plant it won't flower. Likewise, if you break up a long night with a shot of light in the middle, the plant sees a short night and it will flower, if it's a long-day and thus a short-night plant.

This interruption of flowering is actually occurring inadvertently because of nighttime lighting. A 2015 study compared LED white lights to sodium orange-glow lights and found that the orange-glow lights inhibited flowering in *Lotus pedunculatus*, which is a member of the pea family. The researchers speculated that this could be important, since insects might depend on this flower at a particular time.

So, now that we've established phytochrome as the way plants sense the photoperiod, how does that translate into actual flowering? Also, if plants are only using photoperiod as a cue for flowering, then how might longer-term changes in climate cause plants to flower earlier or later in summer? Much work has been done to answer this question. In fact, in the 1930s, a Russian scientist named Mikhail Chailakhyan conducted grafting experiments to determine how flowering was induced. A graft is when you take part of one plant, the tip for example, which is also called the scion, and attach it to the base, or rootstalk, of another plant. When a plant is grafted, one plant can be connected to another by small pieces of tape. They sort of look like plants that have nicked themselves shaving where the petiole of the leaf meets the stem and have bits of tissue to stop the bleeding.

So I have here two plants, and let's suppose that I want to put the leaf from one onto the rootstalk of another, or I just want to take a branch from this one and put it onto this one. So, I can use a procedure called grafting, and I want to show you how that works. So, the first thing I need is a razor blade. I'm using a box cutter, but a razor blade would be just fine, and I've got some tape here that I'm going to use to tape my two plants together.

So, first I want to cut off the branch that I want to use, and then I'm going to make a small incision in the other plant where I want to graft this material onto. And then, when I'm finished with that, I put some tape—when I've got my scion inserted into my rootstalk—I put some tape, I tape it together, and now I've grafted.

By conducting these experiments, Chailakhyan was able to show that the leaves were communicating the readiness of flowering. What this means is that if Chailakhyan took a plant that was ready to flower and grafted it onto a plant that was not ready to flower, the plant would flower. And if the leaf was taken from a plant that was not ready to flower, but placed on a rootstalk that was ready to flower, the plant would not flower. Chailakhyan's results were intriguing. Notice the induced leaves on an uninduced plant—the leaves win out every time.

So, it's been known for a long time that there is a substance produced in the leaves that signals the rest of the plant that it's time to flower. Scientists also knew that this substance must travel through the phloem to get to the part of the plant where the flower buds would develop. This substance was named *florigen* by our Russian scientist Chailakhyan: *flor* for flower and *gen* for generate—a catchy name. However, proving its existence was elusive.

A paper by Jan Zeevaart in 2006 outlined the pathway of trying to find florigen over 70 years. This paper includes Chailakhyan's work from the 1930s. After this work, there wasn't much done in locating florigen and determining its exact role until the 1980s and the advent of molecular genetics. Most of these studies involved looking at mutant varieties of a rockcress plant called *Arabidopsis thaliana*, which is really the white mouse of plant science work. In nature, this small member of the mustard family is a little herb with white flowers. It's so useful in the lab because it has a very fast generation time from seed to mature plants with seeds in about 6 weeks. It also has a small genome, meaning the entire collection of its DNA. This plant's genome is also fully mapped out, so scientists know where all the genes are on the chromosomes. Because it's such a good model organism, scientists were able to create various mutants of *Arabidopsis*. Some of these mutants flowered later than plants in nature, so scientists tried to determine the gene and/or proteins that these plants were missing.

These experiments determined that two genes were important in flowering. Remember that genes are the recipes for proteins that then make things happen in the plant. The two genes that were discovered in *Arabidopsis* were CONSTANS (CO) and FLOWERING LOCUS T (FT). The way these genes interacted and the exact proteins they produced remained inconclusive until 2005. During that year, Tao Huang and his colleagues in Sweden and France conducted experiments to unify the CONSTANS gene and the FLOWERING LOCUS T gene.

They knew the day-length signal was phytochrome, but then found that the plant's response to the day-length signal would switch on the FLOWERING LOCUS T gene. They knew that turning on just one of these genes in the leaf was enough to induce flowering. They also determined that it was messenger RNA of this gene that travelled to the shoot apex, where it would induce flowering. Thus, messenger RNA of this gene becomes the mystery florigen. This was pretty interesting stuff because no one suspected that a messenger RNA would be the mystery flowering signal florigen.

Of course, whether or not the same mechanism is true in all plants remains to be seen. In trees, for example, there is usually a long juvenile period where flowering doesn't occur. Let's compare the juvenile time for a number of woody plants. A rose bush might mature after about 30 days, but an orange or lemon tree, or any species of the genus *Citrus*, might take five years to produce fruit. At the longest end of this spectrum is the European beech tree, *Fagus sylvatica*, which can take up to 40 years to end the juvenile stage and begin flowering in the adult stage. Research is supporting that the trigger to end this juvenile period involves the CONSTANS and FLOWERING LOCUS T genes, so it may be that this is a universal system in plants, and not just limited to herbs like the *Arabidopsis*.

Another environmental cue that many plants need to flower is a period of cold. This is called *vernalization*. *Vernalis* is Latin for "of the spring." In order to flower, many seeds must be given a cold treatment. Flowering can be promoted by a cold treatment given to the seed or a mature plant. Let's compare two images of the same kind of plant. First is our old friend *Arabidopsis thaliana* that has not been given a cold treatment. There are many leaves but only one tiny, little flower. On the other photo there of the plant has been given a cold treatment. There are about nine leaves and a flower, plus about four other flower buds.

So now we have an idea of what causes plants to flower. But what about those cherry trees? If they were cued only to photoperiod, the flowering times wouldn't change. But the flowering time is changing. In a 2001 paper from the journal *Biodiversity and Conservation*, scientists looked at 100 different species around the Washington, DC area to see if their flowering times had changed in the past 30 years. They looked at these species because there were at least 19 different records of flowering times for each species. What they found was that, on average, these 100 species were flowering 2.4 days earlier than 30 years ago. Even more, if they removed the 11 species that actually flowered later, the average flowering time for the remaining 89 species was 4.5 days earlier than

30 years ago. Over that same time period, the average minimum temperature from December to May has increased 1.2° Celsius, which is about a 2.7° Fahrenheit increase. This temperature increase is likely due to climate change.

Let's take an overview of possible effects of climate change. The sunlight that reaches the Earth contains many wavelengths. Recall that ultraviolet light has more energy than visible light. Also remember that red light had more energy than far-red light. If the wavelengths get even longer, and thus have less energy than far-red, they become infrared, which is heat. The Latin prefix *infra* means "below," so infrared is below the red, with a longer wavelength. Infrared radiation is basically heat radiation.

So, sunlight comes in, and it might be represented by very squiggly lines to denote short wavelengths of this intense light. Some of the sunlight reflects back into space from our atmosphere, and some passes through the atmosphere and hits the Earth. As the light makes its way through the atmosphere and hits the Earth, it loses energy and becomes heat, and begins to slow down. This heat can't pass back out of our atmosphere. Now, this is a natural phenomenon, and it's a good thing, because without our atmosphere trapping heat, we would all freeze to death. The problem comes from releasing carbon dioxide, methane, and other greenhouse gases into the atmosphere. These gases trap the heat even more, thus warming the planet.

Here is an analogy for the greenhouse effect. Imagine that you've parked your car in a sunny parking lot on a sunny but cool day. When you get to your car, is it warmer or cooler than the outside air? It's warmer, right? And the analogy is that the light comes through your car windows, but then it slows down to become heat. Once it's heat inside the car, it doesn't have the energy to escape back out of the window. In our analogy, the car window is the Earth's atmosphere. So, by adding greenhouse gases, we are in effect thickening the windows of the car.

So, we heard that Washington, DC plants are responding to climate change by flowering earlier. Similar evidence has been found in my home state of Colorado. Interestingly, the results are even stronger than those found for the Washington, DC area. Seth Munson and Anna Sher used herbarium records to determine how much earlier plants have been flowering in Colorado. They specifically wanted to look at rare plants, but they also needed to find rare plants with good records. They found 12 species of rare plants with herbarium records going back to 1862. Herbaria are collections of dried plants that have been collected in the field. The plants also contain information about the collection. Most herbaria are digitized with photos of the herbarium sheets in the collection. In the paper, Munson and Sher present photos that show how much earlier each species is flowering in the more recent collection.

Getting plants into a herbarium is a bit like putting a book in a library. I want to show you my plant press and how I would press flowers if I were in the field. So this is my plant press, and this strap just holds it together. There is a wood panel here to give it some structure, and then the cardboard is here to blot out the moisture from the plants, because the plants are going to be wet. And so here you can see I have a specimen from a former collecting trip, but let's say that I want press this flower.

So I've collected this flower in the field, and I want to make sure that I get a lot of the stem, so I'm going to sort of break this stem. And then I also want to make sure that you can see the parts of the flower. So here you can see the stigma and the pistil, the style; you can see the anthers. I want to be able to see all that. Sometimes, I might even take a petal and sort of put it over here so that I can get a nice press of what that looks like.

And then I'm just going to take another piece of blotter paper, and this is really all I'm going to do. I'm just going to come right along and I'm going to press that down and flatten it, make sure my stem stays in there. And then I'm just going to put the cardboard back on top. And I might have several—if I were in the field, I'd have several plants—and then I'd put my wood back on top, and then the strap just sort of presses it all together. And in about a week or so I'll have a dried specimen that I can mount on a herbarium sheet with the information of where I've collected the plant. I'd want to record the elevation, the latitude, the longitude, and also the flower color, because sometimes when the plants dry they lose a bit of their flower color. And then my specimen could go to any herbarium in the world.

So, using these collections for 12 different rare plants in Colorado, they were able to show that the flowering date was on average 42 days earlier than in the late 1800s. That breaks down to 3.1 days per decade, which is an even greater response than the Washington, DC plants showed. On average, their flowering was about 1.5 days earlier per decade. The reason for the greater difference in Colorado may be attributed to the fact that mountain systems, particularly alpine regions, are warming faster than other parts of the globe. This faster warming may have to do with less snowfall. The snow reflects light, and with less snow, the ground absorbs more heat.

What's also so interesting about this earlier flowering is that there is a whole citizen network devoted to taking data about all kinds of plants in various areas of the world. The network is called *Nature's Notebook*. On their website, amateur naturalists can report flowering of different plants in their area, and the data are available for anyone to analyze, including students from kindergarten through college.

In addition to species flowering earlier, another well-documented change is the movement of whole plant communities up mountainsides. Here, I'm showing a cartoon of a mountain slope with various types of plants on it as one moves up in elevation. By the way, ecologists generally use the word elevation for changes on the Earth. The word altitude is reserved for levels in the atmosphere. In any case, the base of our mountain shows a beautiful picture of the Flatirons, which are famous rock-face mountains that begin in the Rocky Mountains just to the west of Boulder. In the foreground, we see a grassland ecosystem. Moving up in elevation, there are conifer trees like Ponderosa pine, or *Pinus ponderosa*, which are well adapted to drier climates than the trees further up the slope. After moving up in elevation again, there is a mixed conifer forest. And finally, at the very top, is the land above the trees, the alpine region.

Here, we see how these plant communities have moved up a mountainside. This figure was published in 2008 by David D. Breshears and colleagues to describe a study done by Anne Kelly and Michael Goulden in 2008. Their study documents observations of plants across a 2314-meter elevation gradient in 1977, and again in 2006 and 2007, in the Santa Rosa Mountains of Southern California. They showed that elevation of plant communities shifted upward by about 65 meters.

So, plants know when to flower by the length of the night by using a pigment called phytochrome. The chemical signal florigen, which is messenger RNA, then triggers flowering in the plants. But plants can also change their flowering times due to temperature. Indeed, as annual temperatures rise, as they have at various times in Earth's long history, many plants are flowering earlier in the year, moving to high latitudes, and moving to higher elevations.

LECTURE 10

ADVENT OF SEEDS: CYCADS AND GINKGOES

Plants that have vascular tissue but lack flowers are called gymnosperms. Instead of flowers or fruits, these plants produce seeds. In flowering plants, the seed develops inside a fruit, but in gymnosperms, there are no fruits. The word "gymnosperm" is a morphologic rather than a taxonomic description. The gymnosperms include 4 major phyla. This lecture will address 3 of those phyla: Cycadophyta, Gnetophyta, and Ginkgophyta (with the fourth being Coniferophyta).

GYMNOSPERMS

◇ In the ferns and mosses, gametes were produced by a gametophyte generation. The gametes fused and gave rise to a sporophyte, which produced spores. Most of those plants just made one size of spore, so they were homosporous. However, some of the seedless vascular plants were heterosporous, meaning they had megaspores, which evolve into structures that produce the ovule, and microspores, which evolve into structures that produce the pollen grains. The whole gametophyte generation is reduced to the formation of the ovule or the spore.

◇ Gymnosperms have separate sexes. Male and female parts can be on the same plant or on different plants. If the sexes occur on different plants, the species is called dioecious; if there are male and female parts on the same plant, the species is monoecious.

◇ The seed was a big evolutionary move, because its success is much longer lived than a spore's. Even though spores are much smaller and can travel farther distances via wind, seeds have several advantages. Although spores are lightweight, they don't carry anything extra. The seed is heavier, but it has some reserves. The seed is basically an ovule that contains an embryo, so it's an embryonic plant with embryonic roots, stems, and leaves all ready to go in a package with food. A spore was one cell.

◇ Different seeds have different amounts of reserves. But all seeds have some, so that they can lie dormant for a while until conditions are right not only for germination, but also for the highest chance of survival. Seeds can remain in what is called a seed bank until conditions are best. Some seeds will only remain viable for a few months; others remain viable for years.

◇ The advantages of a seed plant can also be found in the pollen. Ferns and mosses are dependent on water for sperm to swim to the egg in the gametophyte generation; in seed plants, sperm and egg come together differently.

- The entire male gametophyte has been reduced to the pollen grain, which in seed plants contains a tube cell and a generative cell. When the pollen grain lands on an ovule via wind dispersal, the tube cell will grow a pollen tube through a special hole in the seed coat called the micropyle. The generative cell will eventually divide into 2 sperm cells, one of which will then fertilize the ovule. Only the cycads have retained a motile sperm.

- The seed is now ready to be dispersed. Seed dispersal can be done a variety of different ways, but pollination in the gymnosperms is entirely by wind, except for the cycads. Some members of the cycads are pollinated by beetles.

CYCADS

- Cycads flourished during the Triassic Period about 250 million years ago. Fossils of cycads have been found on every continent, suggesting a warmer world at that time. Now, there are only about 300 species in 11 different genera.

SAGO PALM
Cycas revoluta

- Cycads superficially look like palm trees, but palm trees are flowering plants, and cycads have cones. Some of the cycad cones get really big, too. Cycads and palm trees tend to occupy the same kind of habitat, but cycads actually prefer subtropical habitats—those just north or south of the true tropics. The cycads that do occur in the tropics grow at higher elevations, which have lower temperatures and lower humidity, according to Loren Whitelock, author of *The Cycads*.

- Just to confuse nonbotanists, there is a plant called a sago palm, which is actually a cycad, *Cycas revoluta*. Even more confusing is that there is also an actual palm called a sago palm, where the pith is eaten and has many uses. Both are food sources. Although cycad ingestion is known from a number of cultures, these days most botanists understand that cycads are poisonous.

- Despite their poisonous toxins, cycads are heavily sought after and are in decline around the world. Part of the reason for their decline is that they poison livestock, so ranchers have removed cycads. Interestingly, only introduced animals seem to be effected. If animals are native to a region that has cycads, they either avoid the leaves or have developed some sort of immunity to the toxin.

- Another threat to cycads is the clearing of their habitat to make way for ranching and other kinds of settlement and development. Native forests don't provide the immediate economic benefit that ranching and farming do, so cycads need some kind of conservation strategy put into place if a cycad habitat is to survive.

- Cycads are also under threat from poachers. Due to their exotic good looks, cycads are removed from the native habitat to be placed as ornamentals in lawns and gardens. In fact, even though the country of South Africa has tried to establish preserves to protect cycads, there is still an active black market of cycads in that country. Like many black markets, when the product can be bought legally for a comparable price, the illicit market disappears, so governments must support nurseries and gardens that cultivate cycads for legal sale.

GINKGOES

- Ginkgophyta is another gymnosperm phylum that probably evolved around the same time as the cycads, around 200 million years ago. Although there is only one extant, or still-living, member of this phylum, it's an important member with much to set it apart.

GINKGO
Ginkgo biloba

- *Ginkgo biloba* is an oddity among this group of vascular plants with seeds but no flowers. All of the other gymnosperm plants have evergreen leaves. Even the conifers that shed their needles, such as the larch, have needles and not broad leaves.

- *Ginkgo biloba* is named for its characteristic leaf. It looks like a fan with a notch in the apex, thus giving the leaf 2 distinct lobes—biloba. Ginkgo leaves attach directly to the main branch with very little peripheral branching.

- Another curious feature of ginkgoes is that they all seem to lose their leaves overnight. No one knows why the synchronous leaf drop occurs, and the mechanism behind it is utterly befuddling.

- Many of the gymnosperms have male and female plants. Female ginkgoes produce seeds. The seeds of the female ginkgo have a peculiar scent—described as akin to vomit—that is due to butyric acid, which is found in the fleshy seed coat. Because of this, only male trees are planted these days.

- Ginkgo trees are native to Asia. Many Asian cultures eat ginkgo seeds in a variety of ways. They are purported to taste like chestnuts. Ginkgo also has some medicinal purposes. A chemical compound found in the leaves increases circulation and is said to reduce many of the effects of aging.

- The only way for reproduction to occur in the gymnosperms is for the pollen grains to land on the ovule within the strobilus, or cone. There are a few gymnosperms that are pollinated by insects, but for the most part, the gymnosperms have to rely on the wind. They must synchronize pollen formation with ovule ripeness.

- To increase the chance that the pollen will make it to the ovule, many gymnosperms have ovules that will secrete a special drop of fluid to hold the pollen grain above the opening to the ovule, which is called the micropyle. This bit of fluid is called a pollination drop. In ginkgoes, from formation to pollination to disappearance of the pollination drop takes about 10 days. As soon as the pollination drop receives pollen, it can completely disappear in 5 hours.

GNETOPHYTES

- Like the cycads and ginkgoes, the gnetophytes have some unique characteristics. For example, they have vascular tissue but lack fruits.

- Some molecular analyses place Gnetophyta as the closest gymnosperm group to the flowering plants. There are some similarities between the gnetophytes and flowering plants. There are flower-like structures on the strobili of Welwitschia, and many gnetophytes produce nectar on the tip of the cone. Also, the gnetophytes have special xylem cells called vessel elements that look like those found in the flowering plants.

- Gnetophyta is an even smaller phylum than the cycads, though not as small as the lonely ginkgo. There are 3 families that each contain 1 genus apiece: Welwitschiaceae, Gnetaceae, and Ephedraceae.

- Welwitschia is the only member of its genus and of its whole family: the Welwitschiaceae. It has a bizarre growth pattern: It only produces 2 leaves, but these leaves get really long and continue to grow throughout the life of the plant. Welwitschia also has giant strobili.

WELWITSCHIA
Welwitschia Mirabilis

- The genus *Gnetum* has about 40 species. Most of these are vines that are restricted to rain forest habitats, and this may be due to the fact that this group is insect pollinated.

Gnetum Macrostachyum

- The last genus of Gnetophyta is *Ephedra*, which contains about 67 plants, mostly shrubs, that are native to Asia, Europe, and North and South America. *Ephedra sinica*, which is native to China, gained some rather infamous notoriety during the 1990s as a popular weight-loss drug, because it is a stimulant. The plant's potent alkaloid, ephedrine, falls into the same category as caffeine and nicotine. However, with weight loss came heart problems and strokes, so the U.S. Food and Drug Administration banned the sale of dietary supplements containing ephedrine in 2004.

CHINESE EPHEDRA
Ephedra sinica

READINGS

Crane and von Knorring, *Ginkgo*.

Moore, *Medicinal Plants of the Mountain West*.

Whitelock, *The Cycads*.

QUESTIONS

1   What is the difference between a seed and a spore? What are the advantages and disadvantages of each?

2   Why is the ginkgo so unique?

LECTURE 10 TRANSCRIPT
ADVENT OF SEEDS: CYCADS AND GINKGOES

As a field botanist, I value lab work, but my main research is going to the field and seeing how plants make a living, so to speak. I want to learn the story of a plant in its native habitat. As such, I've been travelling the world for 25 years meeting fascinating plants in their native habitats. I've been to every continent except Antarctica, which does have two species of vascular plants I haven't seen—Antarctic hair grass, and a pillowy-looking little flower plant called Antarctic pearlwort. In all this time, I've seen many, many cool plants, but I've also been waiting to see one plant that I think would top any botanist's top 10 list of weird and interesting plants. The plant is called welwitschia, and it grows in the Namib Desert of Namibia to the north and west of the country of South Africa. I've even been to South Africa, and why I didn't make it up to Namibia is a mystery to me, but I will get there, and I will see this plant.

There are several features that make this plant so exciting. First, it only grows in the Namib Desert, so you have to be a bit of adventurer just to get to it. Second, it's the only member of its genus and of its whole family, the Welwitschiaceae. This means that there is pretty much nothing like it in the whole world. Third, it has a truly bizarre growth pattern. It only produces two leaves, but these leaves get really long, and continue to grow throughout the life of the plant. Remember how, in the leaf lecture, I said that small leaves were an adaptation to dry and hot places? But now we have a desert plant with enormous leaves, and this is why botany is even harder than rocket science.

Those leaves are a big reason the main way this plant gets its water is by dew condensation on the leaves. That's weird, too, because most plants get their water from their roots. Welwitschia has roots, but it also uses condensation on its large leaves. Lastly, welwitschia has giant *strobili*, which are basically cones where the seeds are produced. Cones are strobili that are technically reserved for the members of the Coniferophyta, but everyone knows what a cone is, and most people think that a strobili is a delicious Italian sandwich.

Seeds? Did someone say seeds? A major innovation from the ferns and mosses, the plants we'll discuss starting in this lecture, is that they produce seeds. These welwitschia plants don't produce flowers or fruits, but they do have seeds. The seed is only produced after fertilization, or pollination, which we'll devote a whole lecture to later, but because seeds are found in this group, we'll begin talking about them here.

Vascular plants with seeds but no flowers—that's the group of plants we'll discuss in this lecture and the next. Collectively, all seed plants that have vascular tissue but lack flowers are called gymnosperms. The word gymnosperm is a morphologic rather than a taxonomic description. The gymnosperms include four major phyla. We'll discuss three of those phyla in this lecture: Cycadophyta—the cycads; Gnetophyta, which includes the welwitschia; and the Ginkgophyta, which includes one species, the ginkgo tree. Then, we'll devote an entire lecture to the Coniferophyta, since this conifer phylum of gymnosperms has far more species and dominates more of the globe than the other three combined.

Gymnosperm, from the Greek *gymnos*, means "naked." Sort of makes me think a little differently about a gymnasium. Anyway, the *sperm* part is Greek for "seed," hence "naked seed," a testament to the fact that these plants don't have flowers or fruits. As we'll see in flowering plants, the seed will develop inside a fruit. Here in the gymnosperms, there are no fruits. Still, the evolution of the seed was a major deal.

In the ferns and mosses, gametes were produced by a gametophyte generation. The gametes fused and gave rise to a sporophyte, which produced spores. Most of those plants just made one size of spore, and so they were *homosporous*. However, some of the seedless vascular plants were *heterosporous*, meaning they had megaspores and microspores. Seeds have just carried this differentiation of spores to a larger degree. Basically, the megaspores evolve into structures that produce the ovule, and the microspores evolve into structures that produce the pollen grains. The whole gametophyte generation is now reduced to the formation of the ovule or the spore.

This is interesting because all the gymnosperms have separate sexes. Male and female parts can be on the same plant or they can be in different plants. If the sexes occur on different plants, the species is called *dioecious*, meaning "two houses" in Greek. If there are male and female parts on the same plant, the species is *monoecious*, Greek for "one house."

The seed was a big evolutionary move, because its success is much longer-lived than a spore's. Even though spores are much smaller and can travel farther distances via wind, seeds have several advantages. Although spores are lightweight, they don't carry anything extra. The seed is heavier, but it has some reserves. The seed is basically an ovule that contains an embryo, so it's an embryonic plant with embryonic roots, stems, and leaves all ready to go in a package with food. A spore was one cell.

Different seeds have different amounts of reserves, but all seeds have some, so that they can lie dormant for a while till conditions are not only right for germination but also right for the highest chance of survival. For example, a seed from a low elevation might need four months of chilling time before its dormancy will be broken, but a seed from high elevation might only need a month or so. Although conditions might be right for the seed from low elevation to germinate in February because of a warm day, this would not be advantageous to long-term success. The alpine seed would never see a day warm enough until it actually is warm enough to germinate.

Thus, seeds can remain in what is called a seed bank until conditions are right for germination and survival. Some seeds will only remain viable for a few months; other seeds remain viable for years. Svetlana Yashina of the Russian Academy of Science reported in 2012 that they had found a seed buried in the Siberian permafrost and were still able to germinate that seed. Using radiocarbon dating, they estimated the seed was 31,800 years old, so there is a distinct advantage for seeds to stay dormant, yet viable, for long periods of time.

The advantages of a seed plant can also be found in the pollen. Remember that ferns and mosses were dependent on water for the sperm to swim to the egg in the gametophyte generation. In seed plants, sperm and egg come together differently. The entire male gametophyte has been reduced to the pollen grain, which in seed plants contains a tube cell and a generative cell. When the pollen grain lands on an ovule via wind dispersal, the tube cell will grow a pollen tube through a special hole in the seed coat called the micropyle. The generative cell will eventually divide into two sperm cells, one of which will then fertilize the ovule. Only the cycads have retained a motile sperm.

Our seed is now ready to be dispersed. Seed dispersal can be done in a variety of different ways, but pollination in the gymnosperms is entirely by wind, except for the cycads. Some members of the cycads are pollinated by beetles. At this point, we've learned two weird things about cycads: they have motile sperm, and they can be pollinated in a way different from the rest of the gymnosperms. The cycads are pretty different altogether. Cycads flourished during the Triassic, about 250 million years ago. Fossils of cycads have been found on every continent, suggesting a warmer world at that time. As you can imagine, they were probably a major dinosaur food source. Now, there are only about 300 species in 11 different genera. Despite not being speciose, or even terribly economically important, there is a true following of die-hard cycad lovers. There's even a group called the Cycad Society dedicated to the growth and conservation of these plants.

Cycads superficially look a lot like a palm tree, but they're not even close to palm trees, because palm trees are flowering plants and cycads have cones. Some of the cycad cones get really big, too. Cycads and palm trees tend to occupy the same sort of habitat, but cycads actually prefer subtropical habitats, those just north or south of the true tropics. The cycads that do occur in the tropics grow at higher elevation, which have lower temperatures and lower humidity, according to Loran Whitelock, author of *The Cycads*, a definitive book on all things cycad.

Of course, just to confuse non-botanists, there is also a plant called a sago palm, which is actually a cycad, *Cycas revoluta*. Even more confusing is that there is also an actual palm called a sago palm, where the pith is eaten and has many uses. Both are food sources. Loran Whitelock reported that the Japanese used to make miso out of the *Cycas revoluta* seeds. Miso is normally made from fermented soybeans, but in times when the soybean crop was exhausted, the cycad starch was used. The starch came from the stems or seeds.

Even in Florida, cycad starch was milled into a flour between 1845 and 1920. The mills essentially almost wiped out the native populations of *Zamia integrifolia* in the region. This was the cycad that was used for the flour. Although cycad ingestion is known from a number of cultures, these days most botanists understand that cycads are really poisonous. In fact, the cycad sago palm, *Cycas revoluta*, is the most popularly cultivated cycad in the southern United States, and more than one dog has been poisoned by chewing on its leaves, so be careful where you plant your cycads.

I first learned of cycad's mysterious poisons from the famous ethnobotanist Paul Cox. I had the good fortune to meet Dr. Cox on a two-week fellowship at the Coconut Grove, Florida location of the National Tropical Botanical Garden. There were many cycads there, and one of the fellowship directors, Dr. Barry Tomlinson from Harvard University, would demonstrate the prodigious amount of pollen produced by the male cycad. When he blew on the male cones to release a cloud of pollen, Dr. Cox would cringe and move away.

Dr. Cox was cringing for a reason, as he explained that during his research with the Chamorro people on Guam, he suspected a toxin in the cycad plants called BMAA, beta-Methylamino-L-alanine, which is a non-protein amino acid. The Chamorro people had a high incidence of neurodegenerative diseases like Parkinson's, Alzheimer's, and ALS, which stands for amyotrophic lateral sclerosis, also known as Lou Gehrig's disease. The normal incidence of ALS is about 1 in 50,000 people. Within the Chamorro villages, the incidence of ALS was about 200 times higher in the middle of the 20th century, or 1 in 250 people. In more recent years, the rate has dropped, but stayed about 10 times higher than the normal incidence. The BMAA produced by the cycad plants was a potential cause of these diseases.

Even though the Chamorro people eat the seeds of cycads regularly, clinical trials suggested they couldn't possibly be consuming enough of the BMAA to cause disease. Then, Dr. Cox recognized that these people also ate flying foxes, which are a type of bat that lives on Guam. These animals eat the seeds of the cycad as a staple of their diet, which meant that these animals were storing BMAA in their tissues. This was a classic case of biomagnification, where the amounts of a compound that an animal eats are magnified as consumption moves up the food chain. Biomagnification also happens in the ocean with large fish, which explains why pregnant women shouldn't consume too much fish due to mercury. Fish store mercury in their tissues, which magnifies when large fish such as tuna eat smaller fish. So, eating too much tuna can increase mercury in our bloodstream.

The cycad in this case, though, has a symbiotic bacteria that lives in its roots that is actually the source of the BMAA. This bacteria is confusingly called blue-green algae or cyanobacteria. I say confusingly because it's not an algae, it's a bacteria. Blue-green algae is also called spirulina and is touted as a superfood on the web. Of course, it's a jump to say that BMAA on blue-green algae is responsible for neurodegenerative diseases, but it seems that more research on the subject might be warranted. In the meantime, I'm not eating any cycads.

Despite their poisonous toxins, cycads are heavily sought after and are in decline around the world. Cycad conservation is serious business. Part of the reason for their decline is that they poison livestock, and so ranchers have removed cycads. Interestingly, only introduced animals seem to be effected. If animals are native to a region that has cycads, they either avoid the leaves or have developed some sort of immunity to the toxin. Cattle and sheep seem to be the most affected commonly. Certainly, these animals are not native to the areas that have native cycad populations. Symptoms of poisoning are paralysis similar to the paralysis found in ALS patients. Another threat to cycads is the clearing of their habitat to make way for ranching and other kinds of settlement and development. Unless some sort of conservation strategy is put into place, native forests don't provide the immediate economic benefit that ranching and farming do, so cycads need some sort of conservation strategy put in place if the cycad habitat is to survive.

Lastly, cycads are under threat from poachers. Due to their exotic good looks, cycads are removed from the native habitat to be placed as ornamentals in lawns and gardens. In fact, even though the country of South Africa has tried to establish preserves to protect cycads, there is still an active black market of cycads in that country. Like many black markets, when the product can be bought legally for a comparable price, the illicit market disappears, so governments must support nurseries and gardens that cultivate cycads for legal sale.

I mentioned at the beginning of this lecture that we would talk about three of the gymnosperm phyla in this lecture. We've discussed Cycadophyta, the cycads, so next in this discussion of the taxonomy of seeded vascular plants, I want to devote some time to the Ginkgophyta, another gymnosperm phyla. Although there is only one extant, or still living, member of this whole phylum, it's an important member with much to set it apart and make it worthy of our attention.

Ginkgo biloba, or just ginkgo, for there is only one amongst the genus in the family, is an oddity amongst this group of vascular plants with seeds but no flowers. We can start with the leaves. All the other gymnosperm plants have evergreen leaves—the welwitschia and its members, all of the cycads, and most of the members of the Coniferophyta have evergreen leaves. The conifers that shed their needles, like the larch, have needles and not broad leaves that look like a relative of the maple, maybe.

Of course, *Ginkgo biloba* is named for its characteristic leaf. It looks like a fan with a notch in the apex, thus giving the leaf two distinct lobes: biloba. Ginkgo leaves also attach directly to the main branch with very little peripheral branching, and this appearance is a bit like a minimalist Christmas tree. The leaves turn a spectacular golden hue in the fall. Dr. David Lee of Florida International University notes that the ginkgo leaves contain a molecule called 6-hydroxykynurenic acid, which Dr. Lee likens is to a laundry whitener. The chemical is produced in the leaves before they drop, and it intensifies their color.

In 2014, Oliver Sacks, the famous and now deceased neurologist, wrote a short piece in *The New Yorker* about another curious feature of ginkgoes: they all seem to lose their leaves overnight. Apparently, there are many ginkgoes in New York City, and this phenomenon has not gone unnoticed, and yet, Dr. Sacks writes, "No one knows what lies behind this synchronicity, but it is surely related to the antiquity of the ginkgo, which has evolved along a very different path from that of more modern trees." It's true that the ginkgo probably evolved around 200 million years ago, around the time as the cycads. It's also true, as we mentioned, that the ginkgo is unique from other trees, but that no one knows why the synchronous leaf drop occurs or the mechanism behind it. It's utterly befuddling.

In 2007, Denise Corkery wrote in the *Chicago Tribune* that when trees prepare to lose their leaves, the tissue that holds the petiole, the stalk that attaches the leaf to the branch, begins to form a protective layer in preparation for being exposed to the elements. She says that other trees drag this process out, which is why leaf drop takes place over several weeks. A hard frost will eventually cause all the petioles on these trees to harden. Ginkgoes, on the other hand, all form this protective barrier at once, and then the first hard frost will cause all the leaves to be dropped.

Surprisingly, I could not find evidence supporting this theory in the scientific literature. That doesn't mean it isn't there, but it does mean that it's elusive if it is there. I do wonder if the morphology of the tree branches—the way the petioles are attached directly to primary leaves—has anything to do with the cue of petiole hardening. Perhaps, due to the arrangement of the leaves on the branches, all the leaves are feeling the same temperature at the same time. Perhaps petiole hardening is cued by light levels received by the leaves. In any case, this is one of the things I love to share with my students: there are many unanswered questions in botany. I also like to point out that curiosity and creativity are two extremely important traits in a scientist.

As we've learned, many of the gymnosperms have male and female plants. Although New York has many ginkgoes, I'm willing to bet that a good number of them are male. See, there is a bit of botanical sexism occurring with the ginkgo. It's a great city tree because it is very tolerant to pollution, and it's a low VOC emitter. VOCs are volatile organic compounds. When mixed with sunlight and car exhaust, VOCs enable the formation of ozone, or smog. Since ginkgoes don't emit much VOCs, this is another reason they are ideal as a city tree. One thing female ginkgoes do though is produce seeds. If you've ever seen the seed of a ginkgo, you may be tempted to think of it as a fruit, but it is not a fruit. Repeat after me: it's a ginkgo seed, not a ginkgo fruit.

Looking at it, one might think, what's the difference? Remember, the difference is huge. The pollen lands on the ovule and the seed develops in the open—in the naked, if you will. Remember, gymnosperm means naked seed. Although the ginkgo seed may look fleshy, that outer covering is what botanists call a *sarcotesta*, an enlarged seed coat. It falls off the ginkgo seed and doesn't provide nutrients to the developing embryo.

The seeds of the female ginkgo have a peculiar scent. Many have described it akin to vomit. I might add that it sort of smells like rancid butter because of the butyric acid which is found in the fleshy seed coat. When city planners first started planting ginkgo trees, it was impossible to tell which sex they had planted. It takes about 25 years for ginkgo to mature to produce seeds. These days, however, we can tell the difference, and only male trees are planted, but a few female trees remain here and there.

Even though there are numerous ginkgo trees in New York City, they are native to Asia. Peter Crane has written a wonderful book called *Ginkgo: The Tree that Time Forgot*. In it, he describes how there are sanctuaries dedicated to these trees, and how revered they are in the Asian culture. He even reports that some ginkgo trees survived the atomic bomb of Hiroshima in 1945. Apparently, those trees leafed out the following spring. Many Asian cultures also eat the ginkgo seeds in a variety of ways. You can find them in Asian food markets without their smelly, fleshy seed coat. They are purported to taste a bit like chestnuts. Ginkgo also has some medicinal properties, too. What was it good for? Oh yes, memory. That's another one I like to tell my students. The chemical compound ginkgo flavonglycosides within the leaves increases circulation, and so it is said to reduce many of the effects of aging.

Another really interesting thing that can be seen really well in the ginkgoes is the formation of the pollination drop. Think about this: the only way for reproduction to occur in the gymnosperms is for the pollen grains to land on the ovule within the strobilus, or the cone. There are a few gymnosperms that are pollinated by insects, but for the most part, the gymnosperms have to rely on the wind. They must synchronize pollen formation with ovule ripeness. In order to increase the chance that the pollen will make it to the ovule, many gymnosperms have ovules that will secrete a special drop of fluid to hold the pollen grain above the opening to the ovule, which is called the micropyle. This bit of fluid is called a pollination drop. In ginkgoes, from formation to pollination to disappearance of the pollination drop takes about 10 days. As soon as the pollination drop received pollen, it can completely disappear in five hours.

So far, in the gymnosperms, we've talked about cycads and ginkgoes, but I began the lecture with the unique welwitschia plant from the Namib Desert. This plant is the phylum Gnetophyta, and I like to say that gnetophytes are neat because they, like the cycads and the ginkgo, do have some unique characteristics. There are some similarities between the gnetophytes and flowering plants. There are flower-like structures on the strobili of the welwitschia, and many gnetophytes produce nectar on the tip of the cone. Also, the gnetophytes have special xylem cells called vessel elements that look a lot like those found in the flowering plants.

The Gnetophyta are an even smaller phylum than the cycads, though not as small as the lonely ginkgo. There are three families that each contain one genus apiece. We've already met the genus *Welwitschia*, and there are two other genera: *Gnetum* and *Ephedra*. The genus *Gnetum* has about 40 species. Most of these are vines that are restricted to rainforest habitats, and this may be due to the fact that the group is pollinated by insects. The strobili in this group actually have a scent like a flower.

The last genus of the Gnetophyta is *Ephedra*. The name for this genus actually comes from the famous book *Naturalis Historia* by Pliny the Elder. In this book, the name ephedra was actually given to a type of horsetail, which makes sense because this plant looks a bit like a horsetail. However, horsetails were fern allies and reproduced by spores; ephedra produces by seed. Ephedra are also shrubs with woody bases, while horsetails have no wood at all. The genus *Ephedra* contains about 67 plants, mostly shrubs, that are native to Asia, Europe, and North and South America, but none of them rival the fame of *Ephedra sinica*, which is native to China. The specific epithet *sinica* means Chinese.

This plant gained some rather infamous notoriety during the 1990s as a popular weight loss drug, and people did lose some weight with the drug, because it is a stimulant. The plant's potent alkaloid, ephedrine, falls into the same category as the other alkaloids we're so fond of, like caffeine, nicotine, and theobromine. However, with weight loss came heart problems and strokes, so the Food and Drug Administration banned the sale of dietary supplements containing ephedrine in 2004. The plant *Ephedra sinica* has been used in China for thousands of years to treat nasal congestion, and the synthetic drug pseudoephedrine is still found over the counter in decongestants. It all depends on dose.

The 10 *Ephedra* species in the American West are collectively known as Mormon Tea. Although none of the plants seems to have the same amount of ephedrine that the Chinese variety has, they're all purported to have some sort of stimulant properties, and were no doubt helpful on that long trip across the Intermountain West. Too bad the American desert doesn't have a welwitschia plant, too.

LECTURE 11

WHY CONIFERS ARE HOLIDAY PLANTS

Although there are many species of trees that are used as Christmas trees, they are all conifers, which simply means cone-bearing. Although the cycads and Welwitschia also have cones, these are called strobili, and only the cones found in the conifers are true cones. This group used to be called Coniferophyta, but there isn't actually a conifer family, so now the name is Pinophyta, after the largest family found in the conifers, the pine family, or Pinaceae.

CONIFERS

- All Christmas trees have cones, and they all have a unique shape—a sort of conical shape, small and narrow at the top and expanding to the base. Not all conifers have this shape, though. For example, Monterey pine trees, which grow on the coast of California, look more like a rectangle.

- Although conifers grow all over the world, the ones people use as Christmas trees tend to be from places that have cold and snowy winters. A very popular Christmas tree species is the Norway spruce, from Norway. The shape of the Norway spruce, and all conical conifers, does 3 important things.

  - Having short branches at the top of the tree prevents snow loading on the top of the tree, which could damage the top or cause it to break off completely.

  - The tips being smaller reduces the wind loading on the tips of the trees, which could rip the tip off.

  - The conical shape allows greater light penetration to the bottom branches and leaves.

- Many spruces also have branches that tend to droop. This, too, is to prevent snow loading. Because the Monterey pines don't face a lot of snow, their shape isn't the same.

- Although we tend to think of conifers as growing in cold places, that isn't the only type of place they grow. Douglas fir and Ponderosa pine trees live near the West Coast of the United States, from California up through Washington. These trees are also found scattered throughout the intermountain West. There are also conifer forests in the Southeast. So, their conifers must be adapted to something other than just harsh winters to flourish in these different regions.

PONDEROSA PINE
Pinus ponderosa

- Conifers are well adapted to a few different climates. They're adapted to the northern boreal forest because of their ability to thrive in snow and wind. This forest occurs at about 50° latitude, and it is the largest forest on Earth, covering ⅓ of Earth's forested land.

- The conifers' leaves, popularly known as needles, are also well adapted to not freezing in the wintertime. Many conifers have antifreeze proteins that they store in their apoplasts (area surrounding the cell) so that the tissue doesn't freeze. Sometimes, the apoplast will freeze, but this protects the cells inside from freezing. Also, photosynthesis slows down and will stop on very cold days, so the stomata are closed, and water is not moving through the leaf.

- Conifers are also well adapted to fires, and this may be one reason that they dominate in the Southeast, which has a lot of lightning. There are several different adaptations that make pines fire-resistant. Of course, no tree is going to be fire-proof, but many trees, especially the conifers, can survive mild to moderate fires.

- A moderate fire typically doesn't get into the canopy of the forest, so it burns along the understory, which is the layer of vegetation under the main canopy. Conifers that are adapted to these understory fires usually have thick bark around the trunks. The bark may burn a bit, but then the fire will pass by, and the tree will survive. Fires like these will clear out the understory and ensure that there isn't a great buildup of dead wood and shrubs, which could then serve as fuel for another fire.

- Ponderosa pine has really thick bark that forms plates on the tree. This pine is adapted for small fires, and frequent small fires maintain a ponderosa pine forest. When we prevent these types of small fires from happening, the fuel in the understory can burn up and cause severe fires. Even fire-adapted conifers can't survive severe forest fires.

- Ponderosa pines also have an adaptation to fire by not having branches close to the ground. This is not true for spruces and other conifers that don't grow in fire-prone areas.

- Some conifers have cones called serotinous cones that will only open when exposed to the heat of fires. The degree of serotiny seems to depend on the fire regime of a particular area.

- Conifers are well adapted to soils that don't have a lot of nutrients, so we find conifer trees where flowering trees wouldn't thrive. Conifers have a lesser-developed root system than many other kinds of trees. That conifers don't have to put out a new bunch of needles every year is probably advantageous in a soil with fewer nutrients.

PINES, SPRUCES, AND FIRS

- Conifers include pines, spruces, firs, and other plants. Pines are the easiest group to identify. There are other families in the Coniferophyta, but Pinaceae, the pine family, is the largest both in terms of species and numbers of trees on the planet. The pine family contains pines, spruces, larches, cedars, firs, and hemlocks.

- All the pines, except for one, have the needles attached together before they are attached to the branch of the tree. This group of needles is called a fascicle. The number of needles within the fascicle is also a distinguishing trait to determine a particular species of pine tree.

- There are a little more than 100 species of pines, and they probably radiated out of Mexico. *Pinus* is the largest genus in the Pinaceae family, and there are native pines all over the Northern Hemisphere, but they are not native to the Southern Hemisphere. There are conifers in the Southern Hemisphere, but they are not members of the genus *Pinus*.

MONTEREY PINE
Pinus radiata

- The *Pinus* genus also gives us the oldest individual trees in the world. Generally, the Bristlecone pine, *Pinus longaeva*, which grows in the White Mountains of California, is thought to be the longest-lived tree on Earth.

- Pines are some of the fastest-growing and slowest-growing trees. Their adaptations to fire have shaped much of our more recent management policies as we understand that they are accustomed to mild or moderate frequent ground fires. Pines are one of the heaviest of all the conifer trees so they are used for furniture and construction.

- We often use the word "pine nut" to describe that tasty morsel that goes on salads or is used to make pesto, but a pine nut is not a nut—it's a seed. Pine seeds have been fertilized, so there is an embryonic plant inside. In pines, these seeds are released by the female pinecone. The seeds of pinecones are an important food source for many animals.

NORWAY SPRUCE
Picea abies

- In addition to the pine family, there are many other families that are conifers. The spruces and firs are probably the next most well known in this group.

- Telling a spruce from a fir is almost as easy as distinguishing a pine. A fir is friendly—its needles are flat and not so prickly. A spruce does not have flat needles; it is decidedly unfriendly and prickly. A spruce also has the old needle scars apparent on the older parts of the branches. These will appear as bumps on the twigs.

EUROPEAN SILVER FIR
Abies alba

- Also in the pine family are the hemlocks. Their needles are also somewhat flattened but not like a fir. They are not as pointy as a spruce's needles. The firs and hemlocks make friendlier Christmas trees.

NORTHERN JAPANESE HEMLOCK
Tsuga diversifolia

WHY CONIFERS ARE HOLIDAY PLANTS

- There is also another tree in the pine family called the Douglas fir (*Pseudotsuga menzesii*), which is not actually a fir. It isn't a hemlock either. It's sort of a cross between a fir and a spruce; its needles are not as stiff and pointy as a spruce's needles and are not as flat as a fir's needles.

DOUGLAS FIR
Pseudotsuga menzesii

- Douglas firs have extremely recognizable cones. Note that they are not called pinecones because Douglas firs are not pines. Their needles are not in fascicles. Douglas firs have a characteristic bract protruding between the scales of the cone.

- Douglas firs, along with all of the other conifers up to this point, have recognizable needles. The needles are the modified leaves of conifers. They are small and impregnated with a lot of lignin, which gives them their stiff texture. The small size means that they don't heat up too much in the summer, but also that there isn't a huge amount of water in the cells to freeze in the wintertime. The other families in the conifer group have needles that look much the same as the needles in the Pinaceae family.

- In addition to claiming the oldest tree with the Bristlecone pine, conifers also claim the tallest tree, with the giant sequoia, which commonly grows to more than 300 feet tall.

- The redwood, *Sequoia sempervirens*, is more susceptible to drought than other conifers. Also, its roots tend to spread wide rather than go very deep. Redwoods are also very disease resistant. This is one reason why their lumber is so sought after, especially for outdoor construction. Like other conifers, they are also resistant to fire. They also have an astonishingly fast growth rate, with some reports of up to 6 feet per year.

- One member of the conifer group is responsible for one of the most-prescribed cancer drugs on the market, Taxol. The Pacific yew tree is in the Taxaceae family and grows only on the Northwest coasts of the United States and Canada.

- Another family in the conifer group, Araucariaceae, does not exist in the Northern Hemisphere. This family contains a species called the Norfolk Island pine, which is a common house plant because it can tolerate low light levels as a juvenile. It's in the family Araucariaceae, so it's not actually a pine.

- One other interesting tree in this family is the monkey-puzzle pine, which is not a pine, but a very interesting-looking plant native to Chile, where it is the national tree, and Argentina. It's been successfully started in other places with mild climates and good rain, but it grows slowly and is somewhat endangered.

NORFOLK ISLAND PINE
Araucaria heterophylla

MONKEY-PUZZLE PINE
Araucaria araucana

READINGS

Farjon, *A Natural History of Conifers*.

Richardson, *Ecology and Biogeography of* Pinus.

QUESTIONS

1. In what kinds of habitats do conifers thrive? What are some advantages of conifers over deciduous trees?

2. What is a cone, in botanical terms?

LECTURE 11 TRANSCRIPT
WHY CONIFERS ARE HOLIDAY PLANTS

It's a sort of strange ritual isn't it, that around the end of each calendar year, we who celebrate Christmas choose to cut down a perfectly happy and healthy evergreen tree and put it inside our houses. Of course, we have to keep the bottom of its stem in water so it doesn't dry out and shed all of its needles, though it will shed some needles regardless. Then, we put lights on it, and all manner of decorations, too. And we love them. Christmas trees are grown in all 50 states, including Alaska and Hawaii. Peoples long before the advent of Christianity have attributed special significance to evergreens, especially in the Northern Hemisphere around the time of the winter solstice. These trees are still green even in the dead of winter, so people would put them in their homes to remind them of the coming spring. Legend has it that Martin Luther, the Protestant reformer, was the first to wire candles on a tree and bring it inside his house. He wanted to mirror the stars in the night sky.

Although there are many species that are used as Christmas trees, they are all conifers, which simply means cone-bearing. Although the cycads and welwitschia also have cones, these are called strobili, and only those cones found in the conifers are true cones. We used to call this group Coniferophyta, but there isn't actually a conifer family, so now the name is Pinophyta, after the largest family found in the conifers, the pine family, or Pinaceae.

So, all Christmas trees have cones, and they all have a unique shape, a sort of conical shape: small and narrow at the top, and expanding to the base. Ha, conifers are conical. And I'm comical. Not all conifers have this shape. When we think about a Monterey pine growing on the coast of California, it doesn't look pyramidal or conical at all. In fact, it looks more like a flat top, more like a rectangle. Why would our favorite Christmas trees have this characteristic shape? Although conifers grow all over the world, the ones we like for Christmas tend to be ones from places that have cold and snowy winters. I can't say exactly why we prefer the conical shape to our Christmas trees, but it probably has something to do with Christmas trees starting in Germany, which meant they would have been from a cold place, in the mountains of Germany. A very popular Christmas tree species is the Norway spruce from, you guessed it, Norway.

The shape of the Norway spruce, and all conical conifers, does three important things. First, having short branches at the top of the tree prevents snow loading on the top of the tree, which could damage the top or cause it to break off completely. Second, the tips being smaller reduces the wind loading on the tips of the tree, which again could rip the tip right off. Third, the conical shape allows greater light penetration to the bottom branches and leaves. Many spruces will also have branches that tend to droop, for lack of a botanical word to describe this feature. This, too, is to prevent snow loading. Since the Monterey pines don't face a lot of snow, the shape isn't the same.

Although we tend to think of conifers as growing in cold places, that isn't the only place they grow. Think of the Monterey pine on the coast of Northern California. Douglas fir and Ponderosa pine also live near the west coast from California up through Washington. These trees are found scattered throughout the Intermountain West. Also, there are conifer forests in the Southeast: the loblolly pines of Mississippi, Georgia, South and North Carolina, and northern Florida. So, the conifers must be adapted to something else other than just harsh winters to flourish in these different regions.

Conifers are well adapted to a couple of different climates. Certainly, they're adapted to the northern boreal forest. This is because of their ability to thrive in snow and wind. This forest occurs about 50° latitude, and it is the largest forest on Earth, covering 1/3 of the Earth's forested land. The conifers leaves, popularly known as needles, are also well adapted to not freezing in the wintertime. Many conifers have antifreeze proteins that they store in their *apoplasts*, which is the area surrounding the cell, so that the tissue doesn't freeze. Sometimes, the apoplast will freeze, but this protects the cells inside from freezing. Also, photosynthesis slows down and will stop on very cold days, so the stomata are closed and water is not moving through the leaf.

Conifers are also well adapted to fires, and this may be one reason that they dominate in the Southeast, which has a lot of lightning. There are several different adaptations that make pines fire-resistant. Of course, no tree is going to be fireproof, but many trees, especially in the conifers, can survive mild to moderate fires. You may wonder, what is a moderate fire? Something burns or it doesn't burn, right? A moderate fire typically doesn't get into the canopy of the forest, so it burns along the understory, which is the layer of vegetation under the main canopy. Conifers that are adapted to these understory fires usually have thick bark around the trunks. The bark may burn a bit, but then the fire will pass by and the tree will survive. Fires like these will clear out the understory and ensure that there isn't a great build up of deadwood and shrubs, which could then serve as fuel for another fire.

Ponderosa pine is a great example of this kind of tree. It has really thick bark that form these sort of plates on the tree. Interestingly, this bark also has a strong smell of butterscotch or vanilla, though no one seems to know if this smell serves any sort of purpose for the tree. So, the Ponderosa pine is adapted for small fires, and frequent small fires will maintain a ponderosa pine forest, that is actually called a pine savanna by ecologists because the trees are far apart with grasses growing in between them. When we prevent these sort of small fires from happening, the fuel in the understory can burn up and cause large or severe fires. Even the fire-adapted conifers can't survive severe forest fires.

Ponderosa pines also have an adaptation to fire by not having branches close to the ground. The branches on mature trees don't seem to start till higher up the stem. This is not true for spruces and other conifers that don't grow in fire-prone areas. These fires aren't all bad, however, as some conifers have cones that will only open when exposed to the heat of the fire. These cones are called *serotinous*. The degree of serotiny, from completely serotinous to not at all, seems to depend on the fire regime of a particular area. For example, if trees are from a region that has seen frequent fire, these cones might be more likely to be serotinous. If the region has not seen much fire, cones that opened because of other cues may have been just as likely to survive and reproduce. Lastly, conifers are well adapted to soils that don't have a lot of nutrients, so we find conifer trees where flowering trees wouldn't thrive. Ironically, conifers have a lesser developed root system than many other kinds of trees. That conifers don't have to put out a bunch of new needles every year is probably advantageous in a soil with fewer nutrients.

As I've mentioned, conifers includes pines, spruces, and firs, and other plants that we'll meet in this lecture. But, do you know the difference between a pine and a spruce and a fir? This is one of my favorite things to teach in all of botany because it's so approachable that even the struggling students feel good after this lab, because they just can't get it wrong. Pines are the easiest group to identify. All of the plants we've talked about so far are actually in the Pinaceae, the pine family. There are other families in the Coniferophyta, but the pine family is the largest in terms of species and in terms of numbers of trees on the planet. The pine family, Pinaceae, contains pines, spruces, larches, cedars, firs, and hemlocks. The ending "aceae" tells us this is a plant family, and this is the level to which most botanists can identify a plant by sight.

So, a pine tree is technically in the genus *Pinus*. It's a good thing that *Pinus* takes the Latin pronunciation, because the Greek pronunciation of penis might be too much for a botanist to bear. All the pines, except for one— remember, there are always exceptions in botany—have the needles attached together before they are attached to the branch of the tree. This group of needles is called a *fascicle*. The number of needles within the fascicle is also a distinguishing trait to determine a particular species of pine tree. In biology, generally a fascicle is any sort of bundle, so a bundle of nerve tissues would also be a fascicle.

What's sort of cool about the fascicle in the pine tree is that they're not born with it, so to speak. The developing pine tree begins with the needles attached directly to the branch. These young needles are arranged in a spiral around the stem. After one or two years of growth, the needles begin to develop the fascicles, and usually they are wrapped at the base by a modified leaf. There you have it—needles in bundles attached to the stem: it's a pine. There are a little over 100 species of pines, and they probably radiated out of Mexico. This is the largest genus in the Pinaceae and there are native pines all over the Northern Hemisphere, but they are not native to the Southern Hemisphere.

There are conifers in the Southern Hemisphere, but they are not members of the genus *Pinus*. Though, ironically, the Monterey pine, native to the coast of California, is the most common plantation tree in the Southern Hemisphere.

The Monterey pine, *Pinus radiata*, is a fast-growing pine species that is fire-resistant and useful as timber for construction. Plantations started in the 1960s in Chile, and New Zealand, and Australia. Now, however, some of these trees have escaped the plantation and are seen as weedy. This species of pine does has very long roots, and so it can tolerate dry landscapes well. The *Pinus* genus also gives us the oldest individual trees in the world. Like most people, botanists love records. Just as baseball fans like to memorize the records of the greatest players, botanists can't help but be enthralled by the oldest, largest, tallest, heaviest, et cetera. There can probably be some debate about this, but generally the Bristlecone pine is thought to be the longest-lived tree on Earth. We're talking an individual tree; we're not talking about a clone.

The Bristlecone is near and dear to my heart because of my dissertation research, which was in a special place called the White Mountains of California. It's special because it has Bristlecone pines, and because it's east of the Sierra Nevada mountains. These mountains are on the border of Nevada, and they're much drier than the Sierra. Within these mountains live the Bristlecone pine, *Pinus longaeva*. Of course, we know that *Pinus* places them in the pine genus, and that the last name gives us a good clue about their long lives.

On the road to my research site in the Whites, as we would call them, was the Schulman Grove. This grove was named after Dr. Edward Schulman, who in 1957 discovered that many of these trees were over 4000 years old. One over 4000 years was especially old, and he called this tree Methuselah, from the biblical figure in Genesis who lived 960 years. Schulman was able to accurately age these trees by counting the annual rings, where the darker ring represents the winter, and the lighter rings represent spring and summer. Radiocarbon dating has since confirmed these findings.

There is another story that goes with the Bristlecone pine. This story is famous among graduate students of field botany. I was very interested to see this story on the Great Basin National Park website. I guess it wasn't a myth after all. The story goes that a young graduate student from back East was working in the area and found a tree he suspected was older than any other. He tried to no avail to core the tree, but he was unable to get a good record of it. After the Forest Service decided that this tree was not any sort of important landmark, they granted permission for it to be felled. Of course, it turned out to be 5062 years old. Sure enough, it was the oldest tree ever recorded, and now it was dead.

I guess it's a common story among graduate students, because no matter how badly you messed something up, at least you didn't kill the oldest living tree on Earth. In 2012, another old tree was found in the same grove, and it was dated to 5065 years, a whole 3 years older. That graduate student did go on to have a fruitful academic career, but let's be clear; he was looking at the trees to make inferences about past climates through their tree rings. He was a geographer and not a botanist. Ha! How's that for plant blindness?

So, pines are some of the fastest growing and slowest growing trees around. Their adaptations to fire have shaped much of our more recent management policies as we understand that they are accustomed to mild or moderate frequent ground fires. Pines are one of the heaviest of all conifer trees, so they're used for furniture and construction. Pines also have some special stories with animals that involve their seeds. We often use the word pine nut to describe that tasty morsel that goes on salads or is used to make pesto, but a pine nut is not a nut at all. Botanically, a nut is a fruit, which means it is a fertilized embryo, the seed inside a swollen ovary. You can imagine their faces when you tell people that they're eating swollen ovaries as they bite into that apple. But in this case, we know all gymnosperms, which pines are, have naked seeds, so these seeds aren't inside of anything. So, botanically, they are pine seeds and not pine nuts.

The pine seeds have been fertilized, so there is an embryonic plant inside. In pines, these seeds are released by the female pinecone. Interestingly, the female pinecone is a great example of the sequence of Fibonacci numbers. Look at the spiral arrangement of the scales on the bottom of the cone. If you look at the set of spirals going one way, there are 8; if you count the spirals going the other way, there are 13. These two numbers are in the Fibonacci series, which was discovered by an Italian mathematician in the 13th century. The Fibonacci sequence is a number followed by the sum of the two preceding numbers, so 1, 1, 2; 1, 2, 3; 2, 3, 5; 3, 5, 8, and so on. Pinecones don't actually know math. The Fibonacci sequence is a very efficient design, so it shows up a number of times in botany.

In any case, the seeds of the pinecones are actually an important food source for many animals. In fact, grizzly bears use the seeds of whitebark pine as a pre-hibernation snack. These particular seeds have more calories ounce per ounce than chocolate. While most pinecones will open to release the seeds when they're mature, the seeds of the whitebark pine don't open. These pinecones have to be opened by animals. The animals that open them are Clark's nutcracker—a type of bird—and red squirrels. The birds cache the seeds and then feed off of them all winter long. So, while the birds depend on the seeds, the tree depends on the bird for dispersal.

The whitebark pine is found at high-elevation tree line from California to Canada to the middle of Wyoming. However, this tree is being decimated by a fungal pathogen, the white pine blister rust. On top of all that, it can't stand the heat of even a slight warming in the climate, so there is a bit of debate among conservation biologists if it should be relocated, so to speak. The trees can survive further north than is their current range, but the animals it depends on will have to move, too, if it is to survive in the long run.

We've spent a lot of time on the pine family, Pinaceae. The ending "aceae" tells you it's the family level, but there are many other families that are in the conifers. The spruces and firs are probably the next most well known in this group. Telling a spruce from a fir is almost as easy as distinguishing a pine.

So, we call these plants the conifers because they all have cones, but do you know the difference between a pine, a spruce, and a fir? I have here a pine, a spruce, and a fir, and I want to show you how you can easily tell the difference between these three types of plants.

Here, we have the pine, and what you can see with the pine is that the needles are in fascicles. So, the needles form a bunch before they attach to the stem. So they form a little grouping, and then they attach to the stem. Here, I've pulled the fascicle off so you can see what it looks like. But if you look closely, you can see that these needles are actually joined together before they come on to the stem.

When we look at a spruce, when you just touch the spruce, it's very prickly. It's sort of a spiky spruce, I like to say. And then notice how the needles are connected directly to the branches. So they're not in groups before they attach to the branch. And then one last trait of spruces is that you can see these bumps, these little tiny bumps, and so the older twigs of the spruce where the needles have fallen off will leave behind these needles scars that you can see, and that's a good trait to identify a spruce.

And then we come to the friendly fir. And the friendly fir is so called because it's not as spiky as the spruce, and also the needles are flat. So, if you look really carefully at the needles of the fir, you can see that they're flattened, and that's what makes it so they're not so spiky. So, friendly fir, spiky spruce, pines have their needles in bunches before attaching to the stem. Now you can tell what kind of Christmas tree you're going to buy.

Also in the pine family are the hemlocks. Their needles are also somewhat flattened, but not like a fir. They are certainly not as pointy as a spruce. Needless to say, the firs and hemlocks make friendlier Christmas trees. There is also another tree in the pine family called the Douglas fir, which is not a fir at all. The scientific name is *Pseudotsuga menziesii.* The first part, *Pseudotsuga*, actually means "false *tsuga*," and *tsuga* is the genus name for hemlock, so it isn't a hemlock and it isn't in the *Abies* genus, which is a fir, so it's not a fir. So what is it? It's sort of a cross between a fir and a spruce. The needles are not as stiff and not as pointy as a spruce, but not as flat as a fir.

The best thing about Douglas fir is that it has extremely recognizable cones. Note that I'm not calling it a pinecone because Douglas fir is not a pine; its needles are not in fascicles. Douglas fir has characteristic bract protruding between the scales of the cone. I first learned it by imagining it looked like the tree was giving you the middle finger, but I later learned a myth about mice surviving a fire inside the Douglas fir, and that the cones have those protruding bracts to represent mice feet. One thing Douglas fir and the other conifers up to this point have is recognizable needles. The needles are the modified leaves of conifers. They are small and impregnated with a lot of lignin. It's this lignin that gives them their stiff texture. The small size means that they don't heat up too much in the summer, but also that there isn't a huge amount of water in the cells to freeze in the wintertime.

The other families in the conifer group have needles that look much the same as the needles we've looked at so far in the Pinaceae. We already know that the conifers claim the oldest tree with the Bristlecone pine, and they can also claim the tallest tree with the giant sequoia. The Douglas fir is the second tallest conifer tree reaching high heights, and it's common in this group. The giant sequoia, or *Sequoia sempervirens*, commonly grows to over 300 feet tall. The tallest of them all was measured at 379.7 feet and it's named Hyperion.

In 1998, Dr. Todd Dawson, now at Berkeley, used stable isotopes to determine that redwoods were actually getting as much as 34% of their water from fog. This was a unique finding because the redwood is more susceptible to drought than other conifers. Also, its root tends to spread wide rather than go very deep, so being able to use fog water, especially during the summer when there is a lot less rain, is a huge advantage. Redwoods are also very disease resistant. This is one reason why their lumber is so sought after, especially for outdoor construction. This disease resistance is partly due to a high concentration of tannins in their bark and needles. Yes, these are the same tannins that provide that dry feeling in your mouth when you drink wine, as tannins are found in grape skins. Aside from the seedling diseases, nothing infects adult redwoods. Like other conifers, they are also resistant to fire. Add to this an astonishingly fast growth rate—some reports of up to 6 feet a year—and it's easy to see why this is the tallest tree on Earth.

Another member of the conifer group is responsible for one of the most prescribed cancer drugs on the market. The Pacific yew tree is in the Taxaceae family. In the 1960s, the United States National Cancer Institute ran assays on thousands of natural plant chemicals against cancer. The Pacific yew tree, which grows only on the Northwest coasts of the U.S. and Canada, was found to have anti-cancer properties. Scientists discovered a particular chemical component of the bark was especially good at preventing cancer cells from dividing. This substance was named Taxol, after the scientific name of the yew tree, *Taxus brevifolia*. Interestingly, this drug cannot be synthesized in a lab, but scientists have found ways to produce it by starting with similar chemicals found in a tree from the same genus, *Taxus baccata*.

This tree, *Taxus baccata*, is native to Europe, where it is often associated with churchyards. One reason why this might be is that all of the plant is fairly toxic, so it was associated with death. Another more accepted reason is that all evergreens were thought to be sacred in pre-Christian times, so people worshipped them and had sacred sites near the trees. When Christianity arrived, churches were built near these sacred sites.

Also interesting about this plant, and others in the genus *Taxus*, is that they look like they form fruits. They produce bright red balls on the branches that look like berries, but berries are fruits, and these are seeds. These seeds sit in an open, red, fleshy cup called an *aril*. The aril develops from the immature cone, which starts out as green. This aril is the only part of the plant that isn't toxic, so that the seed inside the aril can be dispersed by an animal after eating the aril. *Taxus baccata* is mainly dispersed by small mammals.

There are other families in this conifer group, and one more family that is worth mentioning doesn't exist in the Northern Hemisphere at all. This family is the Araucariaceae, and it contains a species that you may know, even though it doesn't live natively here. That tree is the Norfolk Island pine, a common houseplant because it can tolerate low levels of light as a juvenile. Is it a pine? Think about the leaves attached to the branch. Well, I just told you it was in the Araucariaceae, so it can't be in the Pinaceae. So, no, it's not a pine. Of course, as a kid—OK, probably up until college—I thought this plant was native to an island off the coast of Virginia, because I knew there was a city in Virginia on the coast called Norfolk, and I assumed that there was an island off the coast from whence these plants came. Norfolk Island is actually an Australian territory off the east coast of Australia and the northwest of New Zealand.

The other really interesting tree in this family, the Araucariaceae, is the monkey puzzle pine, which again is not a pine but a super interesting looking plant native to Chile, where it is the national tree, and Argentina. It's been successfully started in other places with mild climates and good rain, but it grows slowly and is somewhat endangered. According to the Kew Botanic

Garden in London, England, the monkey puzzle was given its name by an observer who thought that monkeys wouldn't be able to climb the spiky branches.

That's if the branches are still there. Another unusual trait is that, as the tree matures, it drops whole branches from the lower trunk, which is also a way of dropping leaves, since each branch is entirely covered with spiky leaves. Male trees have cones the size of baseballs, while female trees have cones the size of softballs. It's just a very unique looking tree.

We opened this lecture talking about Christmas trees, and what would Christmas be without mistletoe? Although mistletoe is a flowering plant and it infects somewhere around 200 different species of other plants, it's very noticeable in the conifer forests of the Southwestern U.S. Yes, mistletoe is a parasitic plant—it has no roots into the soil. Instead, it has structures that grow into the conifer tree to obtain the nutrients it needs. Talk about a lover. Maybe that's why we kiss underneath it?

Although it was thought to be merely a pest, ecologists have come to recognize that mistletoe is very important as a source of food. Its berries are eaten by birds, especially the silky flycatcher, and squirrels. Also, the way it forms a gnarled mass in the tree, sometimes called a witch's broom, happens to be great real estate for all of the sorts of nest-building animals. So maybe that's why we kiss under mistletoe. See what we have to learn from botany? So, although our kiss-inspiring parasite may seem like a pest, they're actually crucial for survival. Is that a description of mistletoe or a significant other?

LECTURE 12

SECRETS OF FLOWER POWER

The first flower from the fossil record appeared 125 million years ago. Then, a mere 40 million years later, flowers were everywhere. This is a relatively short evolutionary time. Now, flowering plants are the dominant plant form on the planet in terms of species. Depending on how you count, there are anywhere from 330,000 to 400,000 species of flowering plants. How could natural selection, operating on a long timescale, explain the rapid evolution of flowering plants?

THE RISE OF FLOWERS

- We now have a pretty good idea of where flowers came from and how that evolution could have happened so quickly. We know about mechanisms for the creation of new species, and we also have a pretty good picture of the fossil record.

- It could be that flowers just don't fossilize very well, which they don't. It may also be that early flowers were just small. Additionally, the early flowers may not have been woody, and this would have limited their ability to form fossils.

- Interestingly, molecular evidence puts the advent of the flowering plants back to about 290 million years ago, in the Late Carboniferous Period. Why the fossil record and the molecular evidence don't match up probably has to do with early flowering plants not being the best at forming fossils.

- Whatever the exact origin of the flower, we do have some good evidence of what very early flowers may have looked like and how they might have evolved from conifers.

- Botanists also tend to agree that the flowers form a good group, also called a monophyletic group—that is, a group with one descendant and all of the offspring. Yet other botanists suggest that the flower may have evolved more than once.

- Still, conifers and ferns and mosses have been around much longer than flowering plants. How were there so many different species of flowering plants in such a short time?

- The answer may lie in the fact that many flowering plants are both pollinated and dispersed by animals. An animal moving pollen between plants could become more specialized and get pollen from only one type of plant. Now, the plants pollinated by that animal are reproductively isolated from other plants. This is the key to speciation, or the formation of new species.

- The other way animals can be key in speciation is through dispersal. If an animal disperses a seed to a new location where it can no longer breed with its old population, it becomes reproductively isolated and speciation occurs.

- Generally, field botanists, those interested in identifying plants, focus on the family level. Families are named after the type genus within that family, which is the genus that typifies the traits of the family. There are about 600 families of plants, and some of that number depends on the different ways molecular evidence might be used to define some of the families.

- It used to be thought that the earliest flowering plants were those like the Magnolia family—big, showy flowers, and you can almost imagine the center part of the flower looking like a cone. Next came a hypothesis that the earliest fossil flowers were quite small and looked more like those in the Piper family, or Piperaceae.

- Molecular evidence provides a bit of an alternative. In 2013, the entire genome of a small, weedy shrub native to New Caledonia, an island in the South Pacific, had been sequenced.

- For the decade before this full genome sequence was completed, botanists had used a variety of molecular techniques to determine that this shrub, *Amborella trichopoda*, was the oldest extant flowering plant. This shrub is the only member of Amborellaceae, and it probably had other relatives that were alive sometime in the past.

- It turns out that this shrub is the sister group to all the flowering plants, indicating that, indeed, it is a common ancestor to all flowering plants.

- This shrub gives us some clues about what the earliest flowering plants might have looked like. Because it is a shrub, botanists can confirm the hypothesis that the earliest flowering plants were woody. However, due to its weedy growth in the understory, likely the first flowering plants were not towering trees. Like many of the gymnosperms, this plant is dioecious—meaning that they have male plants and female plants—which is another clue about the early flowering plants.

Amborella trichopoda

⬦ There are several models for how a flower, even a rudimentary flower, might have developed. The best genetic techniques indicate that the same genes that are responsible for flower development in *Amborella* are the same genes that are responsible for the formation of the male cones of the gymnosperms. So, it seems that the flower developed from the male cone.

FLOWER PARTS

⬦ All the flowering plants have an outer ring of male parts with an inner ring of female parts. In botany, when something forms a ring, it is called a whorl. So, the male parts of the flower whorl around the female parts. That this is the case for all flowers supports the idea that the flower only arose once in evolutionary history.

- Male parts are called stamens, which has an anther with pollen and a stalk called the filament.

- The female parts are a bit more complicated. There is a top part called the stigma, which is where the pollen will land. The stigma then becomes the style, which is usually long and thin and connects to the ovary at the base. The ovary is where the ovule is housed. Collectively, the female parts are called the carpel or the pistil.

- At a minimum, flowers have either male or female parts. The typical flower has both male and female parts, and they are called perfect flowers, but some flowers can be all male, called staminate flowers, and others can be all female, called pistillate or carpellate flowers. The petals and sepals are the perianth, and this part is optional: Some flowers have the perianth, which is the outer whorl of petals, which are usually colored, but other flowers do not have the perianth.

- Flowers don't have to have sepals, which are typically green. Sometimes, the petals and sepals can't be distinguished, so they are called tepals.

- The arrangement and number of stamens and carpels, and petals and sepals, varies with various families.

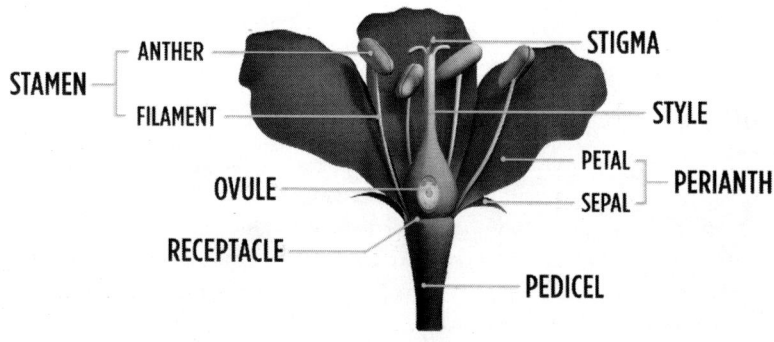

CLASSES OF FLOWERING PLANTS

- Flowering plants are called Anthophyta, or angiosperms. Instead of cones, which have open seeds, these plants will enclose the seed in an ovary that then swells to become the fruits.

- The phylum of flowering plants, the Anthophyta, is traditionally divided into 2 classes: dicots and monocots. However, the term "dicot" isn't used anymore because it doesn't represent a true evolutionary group. Instead of dicots, botanists use the term "eudicots."

- The term "dicot" refers to dicotyledon. The cotyledon is the seed leaf, or the first leaf that emerges in the developing plant. In dicots, there are 2 cotyledons that emerge at the same time. In monocots, there is only 1 seed leaf. The difference between these 2 groups of plants can also be seen in several different physical aspects of the flowers, the veins in the leaf, and the arrangement of the vascular tissue in cross-section.

- Monocots have flower parts in groups of threes. They can have 3, 6, or even 9 stamens and 3 or 6 carpels. Monocots will also have 3 petals and 3 sepals or 6 petals and 6 sepals. Dicots have flower parts in fours or fives. The sepals, petals, stamens, and carpels are all in multiples of fours or fives.

- In monocots, the leaf veins—veins running through the leaf of xylem and phloem—are parallel. In dicots, the leaf veins are netted, or like a feather or a palm.

- The dicots have a tissue arrangement that is reminiscent of that which exists in woody plants. In fact, most trees are eudicots or belong to one of the groups that was removed from the dicots; all together those removed groups are the basal angiosperms.

- The eudicot vascular tissue arrangement has rings around the outside of the stem, with phloem on the outside and xylem on the inside, just like the woody stem, except there is no cork cambium and no bark. The monocots have vascular tissue arranged in bundles throughout the plant, and they consist of 2 large columns of xylem and phloem cells.

- There are a number of flowering plant families that don't really fit into either of these 2 categories, so botanists lumped them together into a group called the basal angiosperms, which make up less than 5% of all the flowering plants. Basal angiosperms often show combinations of the following traits: numerous flattened (laminar) stamens with wide filaments; numerous tepals; many separate carpels; and alternate, spirally arranged leaves.

- The old dicot name used to include many of the basal angiosperms. Once these plants were taken out of the dicots, the new name became the eudicots, or true dicots.

FLOWER IDENTIFICATION

- Flowers can be radially symmetrical or bilaterally symmetrical. There are a few descriptors for the shapes of flowers, such as tube-shaped and bell-shaped flowers, but other terms are reserved for specific flowers, and botanists don't use them broadly, probably because there are so many other ways to determine the family to which a plant belongs by looking at the flower.

- Most wildflower identification books use color; others use inflorescence types. An inflorescence describes how the flowers are arranged on the stem.

- There are quite a few descriptors of inflorescences. A single flower on a stem is called a solitary inflorescence, and flowers that alternate up the stems are an inflorescence called a raceme. If those flowers are tightly compressed on the tip of the stem, it is called a spike. An inflorescence that has numerous flowers forming a sort of umbrella shape is called an umbel.

- But the most influential factor in the determination of the species is the number and arrangement of the male and female parts.

THE EVOLUTIONARY TREE OF THE FLOWERING PLANTS

The oldest flower from the fossil record, *Archaefructus*, was a simple shrubby plant with small flowers. Fast-forward about 125 million years, and we have the species *Amborella trichopoda*, which is the living sister group to all the remaining flowering plants. From there, the basal angiosperms were likely next on the scene, and from a common ancestor to the basal angiosperms, the monocots and eudicots appeared.

In terms of diversity, only about 5% of flowering plants are basal angiosperms. The monocots comprise about 25%, and all the rest are eudicots.

READINGS

Buchmann, *The Reason for Flowers*.

Essig, *Plant Life*.

Howell, *Flora Mirabilis*.

QUESTIONS

1. Why were flowers considered an abominable mystery for Darwin?

2. Do you think that flowers evolved more than once in evolutionary history? Why or why not?

TITAN ARUM, CORPSE FLOWER
Amorphophallus titanum

PIMPERNEL
Anagallis arvensis

ST. BERNARD'S LILY
Anthericum liliago

BUTTERFLY MILKWEED
Asclepias tuberosa

HILL MUSTARD
Bunias orientalis

TALL LARKSPUR

Delphinium barbeyi

MOUNTAIN AVENS

Dryas octopetala

MEDITERRANEAN STORK'S BILL

Erodium malacoides

PURPLE DEAD-NETTLE

Lamium purpureum

LANTANA

Lantana camara

SECRETS OF FLOWER POWER

TRUMPET HONEYSUCKLE

Lonicera sempervirens

COMMON MALLOW

Malva sylvestris

OLEANDER

Nerium oleander

WHITE WATER LILY

Nymphaea alba

BERMUDA BUTTERCUP

Oxalis pes-caprae

MATICO

Piper aduncum

CORPSE FLOWER
Rafflesia arnoldii

CREEPING BUTTERCUP
Ranunculus repens

SPANISH OYSTER THISTLE
Scolymus hispanicus

WOOD LILY
Trillium grandiflorum

WAVY-LEAVED MULLEIN
Verbascum sinuatum

LECTURE 12 TRANSCRIPT
SECRETS OF FLOWER POWER

The next time you're admiring a plant, consider that you and every other animal are meant to admire a flower's color, shape, and smell. Flowering plants may have arrived late in the Earth's history, but they evolved fast, and they displaced many other kinds of plants. We often think of flowers as very delicate plants, and that's not exactly wrong, but they are also world-class competitors who took over the planet. What is the secret of flower power?

Charles Darwin was also confounded by this question. In 1879, Darwin read an essay by another scientist, John Ball, and that essay was titled "On the origin of the flora of the European Alps." The essay was given at a meeting of the Royal Geographical Society in June of 1879. In Ball's essay, he notes that the fossil evidence of flowers appears very rapidly. He then goes on to wonder how the flowers appeared so rapidly and diversely if evolution by natural selection were really responsible for their appearance.

In other words, according to Bell and the fossil record, flowers appeared in the Late Cretaceous, about 145 million years ago, and then there were so many different types seemingly all of a sudden. John Ball questioned how natural selection could account for flowers' sudden appearance. Where did their ancestors come from? And how could natural selection, operating on a long timescale by Darwin's definition, provide so many different kinds of flowers so quickly.

So, later that same year, 1879, Darwin wrote a letter to his friend and fellow scientist, Joseph Hooker. In it, Darwin writes the following sentence: "The rapid development as far as we can judge of all the higher plants within recent geological times is an abominable mystery." The phrase abominable mystery becomes one of the most oft-quoted Darwin phrases of all time, and probably one of the most popular quotes about flower evolution in general. What Darwin meant was that Ball really had him on this one. He didn't know how natural selection could explain the rapid evolution of flowering plants. It went against one of Darwin's central posits about evolution via selection. He said, "Natura non facit saltum," or "Nature never makes leaps."

Consider this. The first flower from the fossil record appeared 125 million years ago. This was *Archaefructus*, which means "old flower." This plant is extinct now, but there it appeared in the fossil record. The flower may be indiscernible because it's small and inconspicuous, but it is present, and that's what makes this plant unique. It's a short, shrubby plant with small flowers, but the flowers are there. Then, a mere 40 million years later, there's flowers everywhere.

This is a relatively short evolutionary time. Consider that the first land plants arose about 420 million years ago in the Silurian, and it took another 293 million years to get to the first flower. And then, 40 million years later, there are flowers everywhere. Even more puzzling, flowering plants are the dominant form of plants on the planet. Depending how you count, there are anywhere from 330,000 to 400,000 species of flowering plants. You can see why this would be a dilemma. Also, at the time, there was nothing that looked even remotely like an intermediate. There were cones from the conifers, and there were flowers. Cones don't look too much like flowers, so how did flowers arise?

In 1879, though, the fossil record wasn't nearly as clear as it is now. We have a much better understanding where flowers came from and how that evolution could have happened much quicker than Darwin suspected. We now know about more mechanisms for the creation of new species than were known during Darwin's time. For example, we now understand polyploidy, which is where the chromosomes can double or triple during reproduction, and we also have a much better picture of the fossil record.

In fact, in a paper from 2013, Peter A. Hochuli and Susanne Feist-Burkhardt found pollen from an extinct flowering plant called *Afropollis* in northern Switzerland that they estimate to be between 252 and 247 million years ago, or even earlier. That would be 100 million years earlier than the date Darwin found so abominable. They also suggest that the pollen is of a type that is usually insect-pollinated, most likely from beetles, given that beetles were also present during this time. Such a plant could have been like present-day gymnosperms. That extra 100 million years certainly helps Darwin out a bit.

Still, pollen isn't a flower. It could be that flowers just don't fossilize very well, which they don't. Some botanists have suggested we're just not looking in the right place. It may also be that early flowers were just small. The conifers and cycads of the Jurassic were so dominant that flowers were just waiting in the understory, so to speak. This scenario was true of early mammals, which were also small, but got bigger after the demise of the dinosaurs. Additionally, the flowers may not have been woody, and, again, this would have limited their ability to form fossils. Interestingly, molecular evidence puts the advent of the flowering plants back to about 290 million years ago, way before the Cretaceous. How would molecular evidence even work in trying to determine the age when something from that long ago evolved?

The main molecular technique used is that of the molecular clock. This basically looks at DNA within an organism that would change on a regular basis—that is, an organism that has a background rate of mutation that wouldn't be associated with reproduction or natural selection. Generally, with plants, botanists are looking at similarities or differences within a particular piece of chloroplast DNA. This technique assumes that the DNA within the chloroplast changes at a regular rate, and that the more different it is from current DNA, the longer ago it must have come on the scene.

Why the fossil record and the molecular evidence don't match up probably has to do with those ideas we mentioned about early flowering plants not being the best at forming fossils. Whatever the exact origin of the flowers, we do have some good evidence of what very early flowers may have looked like and how they might have evolved from conifers. Botanists also tend to agree that the flowers form a good group, also called a monophyletic group—that is, a group with one descendant and all of the offspring. Yet, other botanists suggest that the flower may have evolved more than once.

Still, conifers and ferns and mosses have been around much longer than flowering plants. So how were there so many different species of flowering plants in such a short time? The answer may lie in the fact that many flowering plants are both pollinated and dispersed by animals. An animal moving pollen between plants could become more specialized and get pollen from only one type of plant. Now, the plants pollinated by that animal are reproductively isolated from other plants. This is the key to speciation, or the formation of new species.

The other way animals can be key in speciation is through dispersal. If an animal disperses a seed to a new environment, it can be in a new island, figuratively speaking. That is, the animal has dispersed the seed to a new location where it can no longer breed with its old population, so it becomes reproductively isolated, and speciation occurs.

Now would be a good time to remind ourselves about the taxonomy that botanists use to describe groupings of plants. Do you remember our mnemonic device? King Philip Came Over From Good Spain. King stands for kingdom, the easiest—we're only doing one kingdom in the whole course, and that's plants. Philip stands for phyla. We've discussed a few phyla. We started with mosses, the Bryophyta; we talked about ferns, Pteridophyta; and then we learned all about Pinophyta, the largest phylum of the conifers. Class is the next grouping of different orders. There are essentially are two classes of flowering plants.

The next level down, the order, is helpful in thinking about evolutionary relations of plants. Like class, it's named for the type family that best represents the order. There are dozens of orders, and so one example is the Lamiales. All orders end in the suffix "ales," so there's a sort of rule for you in botany. It's named after the family Lamiaceae, which is the mint family, but it also includes about 20 other families, like the Verbenaceae, which include the verbena herbs, and the Oleaceae, or the olive family. The family level in botany, such as the Lamiaceae that we just met, always ends in the suffix "aceae." The largest plant families in terms of numbers of species are the Orchidaceae, the orchids; the Asters, which are the sunflowers; and the Fabaceae, or the peas.

And then the last two groupings are genus and species. A genus is a particular grouping within the plant family, and the species name is actually comprised of both the genus and the specific epithet. In a sense, this is like having a last name and a first name, only what we think of as a last name comes first, and this is akin to the Chinese way of naming. So, for example, when we discussed *Pinus ponderosa*, the *Pinus* part is the pine genus, and the specific epithet, *ponderosa*, describes which pine we're talking about: the ponderous or heavy one. Together, the genus and the specific epithet make up the scientific name, and we call this a Latin binomial. Because it's Latin and a scientific name, it should always be written in a certain way, with the genus capitalized and the specific epithet lowercase, and the whole thing should be italicized or underlined.

In fact, this is the way we would write our own scientific name, *Homo sapiens*. The binomial system of biological nomenclature was developed by Swedish botanist Carl Linnaeus, but it's also used for animals, protists, and fungus, too. Bacteria and viruses are harder to classify to the species level because it's difficult to know where one species stops and another species begins.

But generally, field botanists—those interested in identifying plants in the field—focus on the family level, and families are named after the type genus within that family, which is the genus that typifies the traits of the family. For example, the genus *Pinus* is the type genus for which the family is named. Another example would be the sunflower family. The type genus of sunflowers is *Aster*, so the family name is Asteraceae. There are about 600 families of plants, and some of that number depends on the different ways molecular evidence might be used to define some of the families. Some botanists might lump two families together, where another will split them.

The grouping of living things like this is called taxonomy. Carl Linnaeus lived in the 1700s, so he predated Darwin. Now, Linnaeus was a Swedish doctor, and at this time, in the mid-1700s, all doctors were gardeners and botanists, since most medicines were derived from plants or fungus. Linnaeus organized living things into a system that is still the basis of the system we use today. All of his arrangements of plants were based on reproductive structures. Using evolutionary relations within such groupings is called systematics, so there's a slight difference between these two. Taxonomy simply says knife, fork, spoon, spork, chopsticks. Systematics says which came first: knife, as the simplest, and as a necessity to hunt, then fork? Chopsticks clearly didn't evolve from fork, so that must be a separate group. I hope you can appreciate the difference. Botanists don't really do simple taxonomy without evolutionary relations, but knowing a few families, even if you can't remember the evolutionary difference between them, helps a lot when trying to identify plants in the field.

As each new scientific advance came along, the sequence of these groupings were changed a bit, but none changed the thinking nearly as much as molecular evidence. As with many things in botany, molecular evidence has changed the way we once thought about the evolutionary progression of flowering plants. Now, as I mentioned, botanists focus on the family. It used to be thought that the earliest flowering plants were those like in the magnolia family—big, showy flowers, and you can almost imagine the center part of the flower looking a bit like a cone. But then next came a hypothesis that the earliest fossil flowers were quite small and looked more like those in the piper family, or the Piperaceae.

Once again, molecular evidence provides a bit of an alternative. In 2013, the entire genome of a small, weedy shrub native to New Caledonia had been sequenced. Interestingly, Darwin actually suspected that the center of flowering plant evolution had been a distant place that science had not yet discovered, and New Caledonia is an island in the South Pacific. For the decade before this full genome sequence was completed, botanists had used a variety of molecular techniques to determine that this shrub, *Amborella trichopoda*, was the oldest extant flowering plant. This shrub is the only member of the Amborellaceae, and probably had other relatives that were alive sometime in the past. It turns out that this shrub is the sister group to all of the flowering plants, indicating that, indeed, it is a common ancestor to all flowering plants.

This shrub gives us some clues about what the earliest flowering plants might have looked like. Because it's a shrub, botanists can confirm the hypothesis that the earliest flowering plants were woody. However, due to its weedy growth in the understory, likely the first flowering plants were not towering trees like many of the gymnosperms. This plant is dioceious, which is another clue about the early flowering plants.

How might a flower, even a rudimentary flower, have developed? There are several models for how this might have occurred. The best genetic techniques indicate that the same genes that are responsible for flower development in *Amborella* are the same genes that are responsible for the formation of the male cones of the gymnosperms. So, it seems the flower developed from the male cone.

Before we talk about the diversity of flowering plants, and there are anywhere between 300,000 and 400,000 of them depending on how you count, we should assure ourselves that we understand all the different parts of a flower. There are very few rules in botany, but here's one I challenge you to check for yourself. All the flowering plants have an outer ring of male parts with an inner ring of female parts. In botany, when something forms a ring, we call it a whorl. So, the male parts of the flower whorl around the female parts in all flowers. That is the case for all flowers supports the idea that the flower only arose once in evolutionary history.

OK, so what are those male and female parts of the flower? Male parts are called stamens, and that's pretty easy to remember because they are "men." Stamen has an anther with a pollen and a stalk called the filament. The female part or parts are a bit more complicated. There is a top part called the stigma. I like to say the sticky stigma, because this is where the pollen will land. The stigma then becomes the style. I like to say the stylin' style, because it's usually long and thin and connects to the ovary at the base. The ovary is where the ovule is housed. And, collectively, the female part is called the carpel or the pistil.

At a minimum, flowers have either male or female parts. The typical flower has both male and female parts, and they are called perfect flowers, but some flowers can be all male, called staminate flowers, and others can be all female, called pistillate or carpellate flowers. The petals and sepals are the perianth, and this part is optional—some flowers have them and other flowers don't. But flowers, such as black pepper, *Piper nigrum*, do not have the perianth. And flowers don't have to have sepals, which are the typically green leaves under the petals. Sometimes, the petals and sepals can't be distinguished, so they're called tepals. The flowers of magnolia trees have tepals, for example.

The arrangement and the number of stamens and carpels, petals, and sepals varies with various families, and this was the distinguishing factor that Linnaeus used to categorize his families of plants. Linnaeus famously described stamens and carpels in terms of brides and grooms. So, in describing a flower that had five stamens and one carpel, he would describe it as one bride in a bedchamber with five groomsmen. He did so to appear modest, but there was quite a bit of outrage about this seemingly perverse nature.

Flowering plants are called Anthophyta or the angiosperms, which means closed vessel. Recall that gymnosperm meant "naked seed," and so angiosperm comes from the Greek *angeion*, "vessel," and refers to the seed. So, instead of the cones, which have open seeds, these plants will enclose the seed in an ovary that then swells to become the fruits. The phylum of flowering plants, the Anthophyta, is traditionally divided into two classes, going back to a taxonomy published in the 17th century by botanist John Ray, which he called dicots and monocots. However, the term dicot isn't used anymore because it doesn't represent a true evolutionary group. Instead of dicots, botanists use the term eudicots.

So, why weren't these good groups? First, let's look at the difference between these two classes, which are still used, just with eudicot instead of dicot, and then look at why the name dicot had to be changed a bit. As you were probably taught at some point, the term dicot refers to dicotyledon. The cotyledon is the seed leaf, or the first leaf that emerges in the developing plant. In dicots, there are two cotyledons that emerge at the same time. In monocots, there's only one seed leaf. The difference between these two groups of plants can also be seen in several different physical aspects of the flowers, the veins in the leaf, and the arrangement of the vascular tissue in cross section.

So, let's begin with the flowers. Monocots have flower parts in groups of threes. That means monocots can have three, six, or even nine stamens, and three or six carpels. Monocots will also have three petals and three sepals, like *Trillium*, or six petals and six sepals, like most lilies. Dicots have flower parts in fours or fives. Notice how the sepals, petals, stamens, and carpels are all in multiples of fours or fives. What's funny about using botanical keys is that usually after 8 or 10, the number just becomes many, or even infinity, and I like to joke that botanists don't really like to count past 10.

Let's look at leaf venation, and this is just what it sounds like, the veins running through the leaf of xylem and phloem. In monocots, the leaf veins are parallel, like the veins in a grass blade, because a grass is a monocot. In dicots, the leaf veins are netted, like a feather or like a palm—pretty much anything but parallel. The dicots have a tissue arrangement that is reminiscent of that which we saw in woody plants. In fact, most trees are eudicots or belong to one of the groups that was removed from the dicots. All together, these removed groups are called the basal angiosperms.

The eudicot vascular tissue arrangement has rings around the outside of the stem, with phloem on the outside and xylem on the inside, just like the woody stem, only now there is no cork cambium and no bark. So, the dicots can have secondary growth, which is outward growth and wood development, but they don't have to do this. The monocots don't have secondary growth at all, so the monocots don't have true wood. The monocots have vascular tissue arranged in what I fondly learned as monkey faces. Essentially, these bundles are arranged throughout the plant, and they consist of two large columns of xylem, which look like eyes, and phloem cells, which look like a forehead. This vascular bundle also has a small space called a lacunae, which looks a little like a mouth. I don't know, maybe it's not a great analogy, but if you squint, you can just make it out.

So, why the new word eudicots? These seem like nice divisions. But there are a number of flowering plant families that don't fit very nicely in either of these two categories. For example, the water lily family, Nymphaeaceae, have a stem arrangement that looks like a monocot. They have six sepals, and they have what looks to be one cotyledon, but they have netted leaf venation. The question then becomes where to put this family? As it turns out, there are a few, maybe a dozen or so, families in this predicament, and so many botanists lump them together in a group called the basal angiosperms. The term basal implies that they are the base of the phylogenetic tree of the flowering plants. This is not a very large group, as they make up less than 5% of all of the flowering plants.

Basal angiosperms often show combinations of the following traits: numerous flattened, or laminar, stamens with wide filaments; numerous tepals, which are indistinguishable; petals and sepals; many separate carpels; and alternate, spirally arranged leaves. Thus, the old dicot name used to include many of the basal angiosperms. Once these plants were taken out of the dicots, the new name became eudicots, or true dicots. I mean, who knows what those basal angiosperms are? Indeed, they're not a cohesive group, but they fit better on their own than they do in the monocots or the old dicots.

One plant family in this group that students usually remember as a basal angiosperms is the magnolia family, the Magnoliaceae. The ending "aceae" implies a family. The magnolias have always been placed at the bottom of the flowering plants probably because the carpels in this group superficially resemble cones. While they're not cones, this group is still in the basal angiosperms mainly due to the numerous tepals and many separate carpels, which can be seen in the cone-like structure of the carpels. Each one of those divisions is a carpel that houses an ovule, and that is where the bright red seeds of the magnolia are born.

So, thinking about the evolutionary tree of the flowering plants, it might go something like this. Our oldest fossil flower is *Archaefructus*. It was a simple, shrubby plant, but clearly a flower. Fast-forward about 127 million years, and we have the species *Amborella trichopoda*. This is the living sister group to all the remaining flowering plants. From here, the basal angiosperms were likely next on the scene, and from a common ancestor to the basal angiosperms, the monocots and eudicots appeared. In terms of diversity, as we mentioned, only about 5% of flowering plants are basal angiosperms. The monocots comprise about 25%, and all the rest are eudicots.

As I mentioned, there are probably between 300,000 and 400,000 species of flowering plants, but botanists don't seem to have as many words to describe the shapes of flowers as the shapes of leaves. When I think of flower shapes, I generally think of symmetry. Flowers can be radially symmetrical or bilaterally symmetrical. There are a few descriptors like tube-shaped and bell-shaped flowers, but other terms would be reserved for specific flowers, and botanists don't use them broadly. I think this is because there are so many other ways to determine the family to which a plant belongs by looking at the flower. Most wildflower ID books will use color; others will use inflorescence types. An inflorescence describes how the flowers are arranged on the stem.

There are quite a few descriptors of inflorescences. A single flower on a stem is called a solitary inflorescence, and flowers that alternate up the stem is called an inflorescence that's called a raceme. If those flowers are tightly compressed on the tip of the stem, it's called a spike. An inflorescence that has numerous flowers forming a sort of umbrella shape is called an umbel, from the Latin *umbella*, meaning "sunshade," which also gives us the word umbrella. But the most deciding factor in the determination of the species is the number and arrangement of the male and female parts. For example, members of the hibiscus family, the Malvaceae, have the stamen forming a tube around the pistil, or coming off directly from the pistil. This is called a *monodelphous* stamen, and it's a key giveaway for this family.

Of course, most children are taught that flowers are showy and smell nice to attract pollinator animals that will enable the plant to exchange pollen. However, there are other adaptations of flowers that are not reproductive. A 2006 review chapter in the book *Ecology and Evolution of Flowers*, Sharon Strauss and Justen Whittall review a number of published studies that show a correlation between pink and purple flowers and drought or heat stress.

Remember that flowers of the same species can have different color flowers, the same as humans can have different hair colors. Anthocyanins are pigments not only in the leaves but also in the flowers, where they cause flowers to be any hue of red or pink. Anthocyanins are also the most common pigments in flowering plants, perhaps because red and pink are also associated with stress. So, in the Strauss and Whittall review, the plant that had pink or purple flowers had greater levels of anthocyanins. These anthocyanins were also in the leaf, where they helped the plant cope with higher levels of heat or drought. So here is an example where flower color isn't related necessarily to a pollinator preference but to a physiologic response.

Additionally, pollinators may prefer large flowers, but large flowers are energetically and hydrologically expensive from the point of view of the plant. There's a tradeoff between attracting pollinators and making flowers small enough that there's enough energy to make a fruit for later dispersal of the seed, and that keeps the flower size modest in most plants. The world's largest single flower is the *Rafflesia arnoldii*, from the rainforests of Indonesia. What do you think: eudicot or monocot? It has five perigone lobes, which are essentially giant, hardened petals. It's a eudicot. The largest record of the individual flower was measured at over 1 meter across—that's about 3 feet 4 inches. This single flower weighed 24 pounds. How about that for a corsage? It's actually a parasitic plant with no visible leaves, roots, or stem, which may be why it's able to shunt all of its energy into its massive flower size. Interestingly, the common name is corpse flower because it smells like rotting flesh, and this is also the common name of the plant with the largest inflorescence. Remember that the inflorescence is a collection of individual flowers on the stem.

Now, the largest inflorescence goes to another corpse flower, but this one is called *Amorphophallus titanum*, or the titan arum. This is the corpse flower that botanic gardens will often put on special display when it flowers, because it doesn't flower very often. This inflorescence can grow to a height of 3 meters or 10 feet. The leaf of this plant can grow 6 meters, or 20 feet tall. It generally takes about 7–10 years for the first flower to bloom, and then it may bloom more frequently, like every 2 or 3 years—or it might not. No rules in botany.

But there are patterns. Why might both of these super-sized flowers smell like carrion? The answer is to attract flies. The titan arum is a monocot, and the *Rafflesia* is a eudicot, but they both live in Sumatra, so something about Sumatran rainforest has pollinators that are attracted to the smell of rotting flesh. These pollinators turn out to be carrion-eating beetles and flesh flies. Both of these animals seek out rotting meat, and these corpse flowers trick them into thinking they've found it, but not before the animals have done the plant's bidding and moved pollen from one plant to another. Certainly, these beetles and flies must think corpse flowers an abominable mystery.

So the next time you are admiring a flower—its color, its smell, or its shape—remember that you and every other animal are meant to admire it. You're doing exactly what the plant wants.

LECTURE 13

THE COEVOLUTION OF WHO POLLINATES WHOM

Pollen comes from plants—primarily from trees and grasses. Pollen delivers the male sperm to the top of the female part of the flower, the stigma, and this is the process of pollination. Of course, if you suffer from allergies, you're wondering, why there is so much pollen. Is it really necessary? It turns out that it is.

POLLEN

- Many plants are pollen limited, which means they could produce more fruits with seeds inside of them if there were more pollen. The pollen we see and breathe in the air is typically from wind-pollinated plants, especially grasses, trees, and weeds. Only wind-pollinated plants can produce so much pollen. Plants pollinated by animals can be more judicious in their pollen production because the delivery of that pollen is more direct. Wind pollination is called anemophily.

- Wind-pollinated flowers—such as those on birch trees, cedar, or alder—are usually inconspicuous, meaning they lack showy parts, because they don't need to advertise to any pollinators. In addition, showy petals might get in the way of the wind reaching the pollen, or the pollen reaching the stigma.

- Wind-pollinated flowers are also small, because they only need to attract wind. And flowers that are too big may get blown off, especially because wind-pollinated plants tend to live where it's windy.

- Wind-pollinated flowers typically lack any smell, because there are no pollinators to attract with fancy perfumes. The same goes for nectar. The wind needs no reward.

- A common inflorescence for wind-pollinated plants is a catkin, which is an arrangement of small flowers, usually unisexual, that hangs down in pendulous, small branchlets. Usually, the catkins on any one plant are of one sex because many of the plants that have this kind of pollination are dioecious, meaning that they have male plants and female plants. If wind-pollinated plants are bisexual, they are monoecious, meaning that they have separate male and female flowers on the same tree.

- The vast majority of flowering plants, though, have bisexual flowers with both male and female parts within them. Because they have both sexes, they can pollinate themselves, and this is called selfing. One advantage to selfing is that there is most definitely a source of pollen and an ovule. Also, because the plant formed gametes, the sperm and the egg, this is sexual reproduction, so a reordering of the genes will occur.

- The disadvantage is that the genes carried by that sperm and that egg are going to be pretty similar. They won't be identical, because during the formation of the sperm and the egg, only certain alleles, which are flavors of the gene, are passed on to the gametes.

- Even though selfing is still sexual reproduction, the genetic difference between the gametes within a plant wouldn't be as great as the genetic difference with another plant.

- In addition, many plants simply can't self. For those plants, when self-produced pollen lands on the stigma, the pollen tube simply won't grow. The pollen tube is that part of the pollen that will grow down into the ovule. The beginning of the growth of the pollen is called pollen germination because it begins by the pollen absorbing water and swelling, just like a seed germinating.

- This leads to an interesting dilemma: The stigma has to be able to recognize pollen that is self or not self and pollen that isn't even of the same species. There is a debate in botany as to what constitutes a species. With animals, the biological species concept is fairly well established. If 2 animals can mate and produce viable offspring, then they are the same species. If they can't, then they're not. Plants are a bit more promiscuous in terms of reproduction, which could account for so much variety and eventually speciation in plants.

- Some plants have developmental delays to prevent self-fertilization. For example, the pollen will mature and get blown away before the stigma is ready to accept pollen. Other plants will spatially segregate the pollen from the stigma. If the stigma protrudes far outside of the flower but the anthers are deeper inside the petals, then there is less chance that the pollen from the same flower will land on the stigma.

- Still, sometimes pollen from the same individual will land on the stigma. A group of genes is responsible for recognizing pollen as self or not self. When plants cannot self-fertilize, they are called self-incompatible, and about 50% of eudicot species fall into this category.

- The genes then provide a biochemical mechanism to prevent fertilization by pollen from the same individual. There are a number of different mechanisms to prevent self-fertilization.

- There are also mechanisms to promote outcrossing, or fertilization with non-self individuals. This is what pollination is all about.

- The majority of flowering plants are pollinated by insects. Insects get nectar or pollen, or both, and in exchange, the pollen they get on their body gets transferred to another flower when they visit it for more rewards.

- Flowers and insects can be so specialized as to have a monogamous relationship, in which one species of insect only visits one type of flower and no other flowers, and the flower, in turn, is only visited by one species of insect.

- In this relation, the flower should have to produce less nectar because the pollinator will only be visiting that flower. However, the flower runs the risk of that particular insect becoming scarce and not pollinating a significant amount of the plant's population; likewise, the inset will be out of luck if something bad happens to that particular species of flower.

- Alternatively, flowers can be pollinated by a variety of pollinators. The same is true for insects. A plant pollinated by many pollinators can increase its likelihood of successful pollination, but the stigma can get clogged up with foreign pollen. Also, the plant has to ensure a prodigious supply of nectar for any insect that comes to call, even if that insect next travels to an entirely unrelated species.

- Plants and insects can also choose an intermediate lifestyle, which is what most do. In other words, plants are pollinated by a group of insect types, such as beetles or flies. Likewise, insects tend to prefer plants that are suited to their needs, skills, preferences, and body types.

POLLINATORS

- Beetles are the oldest and most diverse group of insects. From the fossil record, we know that beetles were present long before flowering plants were, and we also know that some of the gymnosperms, particularly the cycads, are beetle pollinated.

- Most beetles are not great flyers because they are so heavy. Because they don't fly very well, a flower should provide them with a steady landing site. Also, beetles don't have very long tongues, so the nectar of a beetle-pollinated flower can't be buried too deep inside the flower. Furthermore, beetles don't have great vision, but they do have a good sense of smell, so we would expect a beetle-pollinated flower to be cream or white in color, because this would contrast best against a canopy of green. We would also expect beetle-pollinated flowers to have an attractive smell.

- Typical large flowers pollinated by beetles are magnolia and pond lilies, but smaller flowers that are clustered together are also pollinated by beetles, so the beetle doesn't have to travel too far, such as goldenrods.

POLLEN BEETLE

Meligethes aeneus

- Although beetles may be the oldest pollinators, bees are some of the most important pollinators, especially because a good number of foods we eat are bee pollinated, including almonds, avocados, blackberries, blueberries, peaches, pears, cantaloupes, and watermelon.

- Bees, like beetles, are also fairly heavy insects. Additionally, because bees feed their young, they need a greater amount of nectar than some other pollinators. Bees also collect pollen as a protein source.

- Bees have long tongues, so bee-pollinated flowers can have spurs or nectar pockets. So, flowers pollinated by bees should still be large with a decent place for the bee to be while gathering nectar and pollen. But, unlike beetle-pollinated flowers, those pollinated by bees will be more colorful, but not as heavily scented.

- Flies are another common insect pollinator, but they are smaller and lighter in general than either bees or beetles. They also do not feed their young, so flowers pollinated by flies need not produce as much nectar or pollen as those pollinated by bees.

- Although flies can see a bit better than beetles, the flowers they pollinate still tend to be white, cream, or yellow and typically smaller than those pollinated by bees or beetles.

FLY POLLINATING AN UMBEL PLANT

- Flies are important pollinators to plants that may be flowering at times of the year when there are not many bees or beetles. There tend to be more flies throughout the year, so if a flower is particularly early or late in its blooming, it may be fly pollinated.

- As such, we expect fly-pollinated flowers to be small, without much scent, and of a cream-colored hue. Members of the Apiaceae family are typically fly pollinated.

THE COEVOLUTION OF WHO POLLINATES WHOM

MALLOW SKIPPER
Carcharodus alceae

◈ The lepidopterans are insects that are moths and butterflies, and even though these 2 insects are in the same order and they look superficially alike, they pollinate different types of flowers.

◈ Butterflies are typically active during the day and hold their wings up when at rest, and their antennae are thinner and have a small club at the end. Butterflies also have good color vision, so they tend to prefer flowers that are yellow, red, or orange. Because butterflies don't have much sense of smell, flowers they pollinate don't have much scent.

◈ Moths are generally active at night, hold their wings out or folded on their backs, and have thicker antennae that lack a club at the end. Because moths are active at night, they tend to pollinate white or cream-colored flowers. However, moths have a good sense of smell, and a strong scent can help them find flowers at night, so moth-pollinated flowers have a strong scent, usually at night.

- Both moths and butterflies have long proboscises, which are the tubular structures through which they suck up nectar. As such, flowers that are pollinated by moths and butterflies have long nectar spurs, which are elongated parts of the flower that have nectar in the bottom. Moths have even longer proboscises than butterflies.

- Because butterflies are smaller than moths, they have lower energy requirements, so there is generally less nectar produced in butterfly-pollinated flowers. Despite their larger body size, moths prefer to hover while gathering nectar, which means that moth-pollinated flowers must produce even more nectar. Because of the hovering, moth pollinated flowers tend to be bilateral in shape and have a corolla and nectar spur that are easily accessed by a hovering moth.

- Insects aren't the only animals to pollinate flowers. Like butterflies, birds also have good vision but lack a well-developed sense of smell. Bird-pollinated flowers are larger than butterfly-pollinated flowers, and because birds are larger and have more energy requirements than butterflies, bird-pollinated flowers produce a lot of nectar. The shape of the flower depends on the type of bird performing the pollination.

- Bats, like moths, are major nocturnal pollinators. Vision is not very important to bat pollinators, but like many mammals, bats have a well-developed sense of smell. So, like moths, bats pollinate white or cream-colored flowers that have a strong scent.

- There are also nonflying mammal pollinators. About 50 other species of mammal, usually rodent-type animals, pollinate flowers. Typically, flowers pollinated by nonflying mammals have flowers close to the ground with a strong odor. These flowers also have to be sturdy to hold up to the mammals.

- Many times, the animal receives either nectar or pollen as a reward for its services. Sometimes, however, there is no reward. This is called deceit pollination.

- We use pollination generally for many crops. A huge variety of fruits depend on pollinators to reproduce.

READINGS

Fenner, *Seeds*.

Harder and Barrett, *Ecology and Evolution of Flowers*.

QUESTIONS

1. What, botanically, is pollen, and why do plants produce so much of it?

2. Can you draw a typical pea flower?

3. Think about your favorite flower. Describe its color, scent, and shape, and then postulate what kind of animal pollinates it.

LECTURE 13 TRANSCRIPT
THE COEVOLUTION OF WHO POLLINATES WHOM

What do you do when pollen is in the air? Are you looking at the weather websites to see what the pollen count will be? If you have seasonal allergies, are you secretly thinking of moving to a desert where you hope the pollen counts are low? My allergies get really bad in mid-March and by the end of April, they're pretty much gone. I'm allergic to tree pollen, but grass pollen doesn't bother me—probably because I had allergy shots for that when I was a kid.

We all know pollen comes from plants—primarily from trees and grasses. Pollen delivers the male sperm to the top of the female part of the flower, the sticky stigma, and this is the process of pollination—mating season for plants. Of course, if you suffer from allergies, you're wondering, why so much pollen? Is that really necessary?

It turns out that it is. Many plants are actually pollen limited, which means they could produce more fruits with seeds inside of them if there were more pollen. The pollen we see and breathe in the air is typically from wind pollinated plants. Especially the usual suspects, grasses, trees, and weeds, like ragweed. Only wind pollinated plants would produce so much pollen. Plants pollinated by animals could be more judicious in their pollen production because the delivery of that pollen would be more direct.

Wind pollination is called *anemophily*. The prefix *anemos* is Greek for wind. That would be another great business or a band name. The suffix *phily* means to love, which you Philadelphians already know. Wind-pollinated flowers are usually inconspicuous, meaning they lack showy parts because they don't need to advertise to any pollinators. These are flowers like on birch trees, cedar, or alder. Besides, showy petals might get in the way of the wind reaching the pollen, or the pollen reaching the stigma. Wind-pollinated flowers are also very small. Again, there's no reason to be big if you only need to attract the wind. And flowers that are too big may get blown off, especially because wind-pollinated plants tend to live where it's, well, windy.

Another trait we expect with wind-pollinated flowers is the lack of any smell since there are no pollinators to attract with fancy perfumes. The same goes for nectar. The wind needs no reward. A common inflorescence for wind pollinated plants is a catkin. This is an arrangement of small flowers, usually unisexual, that hang down in pendulous small branchlets. Usually, the catkins on any one plant are of one sex because many of the plants that have this sort of pollination are *dioecious*, meaning, they have male plants and female plants. This is like aspen trees, which are also wind-pollinated. If wind-pollinated plants are bisexual, they are *monoecious*, meaning that they have separate male and female flowers on the same tree, like oaks or birches.

The vast majority of flowering plants, though, have bisexual flowers with both male and female parts in them. And this is a curious state of affairs because if they have both sexes, can't they pollinate themselves? They can, of course, and this is called selfing. The advantages to selfing are there is most definitely a source of pollen and an ovule. Also, because the plant formed gametes, the sperm and the egg, this is sexual reproduction and so there is a reordering of the genes that will occur. The disadvantage is that the genes carried by that sperm and that egg are going to be pretty similar. They won't be identical because, during formation of the sperm and the egg, there are only certain alleles, which are flavors of that gene, that are passed on to the gametes.

So, if selfing is so great, why not always self? Even though it's still sexual reproduction, the genetic difference between the gametes within the plant wouldn't be as great as the genetic difference with another plant. So, selfing is a bit like reproducing with a cousin. They did it in the old days, and sometimes it was fine, but if there was any sort of genetic disease in the family, not such good odds.

In addition, many plants simply can't self. For those plants, when self-produced pollen lands on the sticky stigma, the pollen tube simply won't grow. The pollen tube is that part of the pollen, which will grow down into the ovule. The beginning of the growth of the pollen is actually called pollen germination because it begins by the pollen absorbing water and swelling, just like a seed germinating.

So, this leads to an interesting dilemma, the stigma has to be able to recognize pollen that is self or not self and pollen that isn't even of the same species. There is a bit of a debate in botany as to what constitutes a species. With animals, the biological species concept is fairly well established. If two animals can mate and reproduce viable offspring, they are the same species. If they can't do that, they're not the same. Plants are a bit more promiscuous in terms of reproduction, which could account for so much variety and eventually speciation in plants.

Some plants have developmental delays to prevent self-fertilization. For example, the pollen would mature and get blown away before the stigma is ready to accept pollen. Other plants will spatially segregate the pollen from the stigma. If the stigma protrude far outside of the flower, but the anthers are deeper inside the petals, then there is less chance that the pollen from the same flower will land on the stigma.

But, still, sometimes pollen from the same individual will land on the stigma. How do plants recognize pollen that is self or non-self? It's in the genes, as they say. There are a group of genes responsible for recognizing pollen as self or not self. When plants cannot self-fertilize, they are called self-incompatible, and about 50% of eudicot species fall into this category.

The genes then provide a biochemical mechanism to prevent fertilization by pollen from the same individual. Collectively, these genes are called the S locus. Even though it isn't just one gene, it's still called a locus. S is for self-incompatibility. So, when these S genes are all the same, the pollen tube doesn't grow or the stigma prevents it from growing.

There are a number of different mechanisms, but if there are mechanisms to prevent self-fertilization, then there must be mechanisms to promote outcrossing or fertilization with non-self individuals. There are, and this is what pollination is all about.

The majority of flowering plants are pollinated by insects. Insects get nectar or pollen, or both, and in exchange, the pollen they get on their body gets transferred to another flower when they go to visit it for more rewards.

Flowers and insects can be so specialized as to have a monogamous relationship. For example, many orchids are pollinated by only one species of insect. The insect will only visit that flower and no other flowers. The flower, in turn, could only be visited by one species of insect. In this relation, the flower should have to produce less nectar because the pollinator will only be visiting that flower. However, the flower runs the risk of that particular insect becoming scarce and not pollinating a significant amount of the plant's population. Likewise, the insect will be out of luck if something bad happens to that particular species of flower,

Flowers can go a different route and be pollinated by a variety of different pollinators. The same is true for the insects of course. A plant pollinated by many pollinators, such as a dandelion, can increase its likelihood of successful pollination, but the sticky stigma could get clogged up with a lot of foreign pollen. Also, the plant will have to ensure a prodigious supply of nectar for any insect that comes to call, even if that insect next travels to an entirely different species.

Of course, plants and insects can choose an intermediate lifestyle, which is what most do. In other words, plants are pollinated by a group of insect types, such a beetles or flies. And, likewise, insects tend to prefer plants that are suited to their needs, skills, preferences and body types.

Let's start with the beetles because they are the oldest and most diverse group of insects. The biologist J. B. S. Haldane wrote a book called *What is Life* in 1947. In it he writes,

> The Creator would appear as endowed with a passion for stars, on the one hand, and for beetles on the other, for the simple reason that there are nearly 300,000 species of beetle known, and perhaps more, as compared with somewhat less than 9,000 species of birds and a little over 10,000 species of mammals. Beetles are actually more numerous than the species of any other insect order.

From the fossil record, we know that beetles were present long before flowering plants arrived on the scene, and we also know that some of the gymnosperms, particularly the cycads, are beetle pollinated. We can see this ancient pollination mechanism happening today with magnolias. They are beetle-pollinated, much like they were when they first appeared.

What are the characteristics of a beetle-pollinated flower? Well, let's think about a beetle. That a beetle can even fly is sort of engineering marvel. Think of a ladybug, which is actually a beetle, not a bug—the kind that is red with the spots—it should be called a ladybird beetle. That hard outer shell on its back is actually a pair of modified wings called *elytra*. They are there to protect the inner wings, which are the flying wings. The *elytra* will move to the side while the insect is flying. Most beetles, ladybird beetles included, are not great flyers because they are so heavy. But some beetles are very tiny, and others can be quite large, and because they don't fly too well, a flower should provide them with a steady landing site. Also, beetles don't have very long tongues, so the nectar of a beetle pollinated flower can't be buried too deep inside the flower. Lastly, beetles don't have great vision, but they do have a good sense of smell. So we would expect a beetle-pollinated flower to be cream-colored or white since this would contrast best against a canopy of green. And, we would also expect these beetle-pollinated flowers to have an attractive smell, a floral scent if you will.

Pollination by beetles is called *cantharophily*. What is the etymology or maybe it's entymology because we're talking about insects? If you ever confuse those words, remember entomology has an N for insects, and etymology doesn't—it's for words. OK, so the word comes from *canthero* for beetles and *philly* for love. Typical large flowers pollinated by beetles are magnolia and pond lilies, but also smaller flowers that are clustered together, so the beetle doesn't have to travel too far, like goldenrods.

Although beetles may be the oldest pollinators, bees are certainly some of the most important pollinators, especially since a good number of foods we eat are bee-pollinated. In light of the decline of bees from many causes, beekeepers will actually travel around with their hives to stop by an orchard or a farm and let the bees pollinate the flowers. Almonds and avocados, blackberries and blueberries, peaches and pears, cantaloupes, and watermelon are all examples of foods that rely on bee pollination.

Bees, like beetles, are also fairly heavy insects. Additionally, because bees feed their young, they need a greater amount of nectar than some other pollinators. Perhaps you've heard of Diabetes mellitus, which is a disease that disrupts sugar levels in the blood. *Mellitus* is Greek for sweet, so bee pollination is called *melittophily*.

Bees also collect pollen as a protein source. They can store pollen in pollen baskets on their rear set of legs. Some flowers actually won't release their pollen until a bee buzzes the pollen off of the anther. This buzz pollination is also called *sonication*. According to Dr. Anne Leonard at the University of Nevada, Reno, about 8% of the world's flowers are pollinated by bees using sonication. This method is very specialized so that the flowers don't just give pollen away to any old insect coming for a visit. Often times, flowers that are buzz pollinated don't have nectar. The pollen is the only reward. Cranberry and blueberry flowers are both buzz pollinated.

Bees see very well, especially in the ultraviolet, which is the wavelength shorter than violet, so it's just out of the spectrum for our eyes. But, flowers that are pollinated by bees exploit the bee's ultraviolet abilities by becoming extra visible. Many bee-pollinated flowers are yellow or blue, such as lupine. But there are also some that are red or pink show up as particularly visible because of UV modifying pigments in the petals, such as roses. Bees also have longer tongues, so bee-pollinated flowers can have spurs or nectar pockets. So, flowers pollinated by bees should still be large enough with a decent place for the bee to be while gathering nectar and pollen. But, unlike beetle-pollinated flowers, those pollinated by bees will be more colorful, but not as heavily scented. A quintessential bee-pollinated flower would be a pea flower.

This is a good time to look at a pea flower because it's very distinctive, and once you recognize the peas, you recognize one of the most diverse families of plants. The pea family is called the *Fabaceae*. I like to say peas are fabulous to help students remember the name. A pea flower has five petals, so it's a eudicot. These petals are modified so that the flower is very recognizable. One of the petals is larger and stands upright—above the rest of the flower, and it is called the banner. Two petals are modified that stick out on either side, and these are called the wings. The last two petals look like one petal and they form a curved structure in the middle of the flower called the keel. So, there's a little dance that goes with this banner, wings, and keel equals fabulous. This helps you remember the flower structure of the *Fabaceae*—the pea family.

Flies are another common insect pollinator, but they are smaller and lighter, in general than bees or beetles. They also do not feed their young, so flowers pollinated by flies need not produce as much nectar or pollen as those pollinated by bees. Although flies can see a bit better than beetles, the flowers they pollinate still tend to be white, cream, or yellow, and typically smaller than those pollinated by bees. Cocoa trees, for example, have small, inconspicuous white flowers pollinated by tiny flies called midges.

Flies are important pollinators to plants that may be flowering at times of the year when there are not as many bees or beetles. There tend to be more flies throughout the year, so if a flower is particularly early or late in its blooming, it may be fly pollinated.

As such, we expect fly-pollinated flowers to be small, without much scent and of a cream-colored hue. This is a good time to learn a whole family that is mainly pollinated by flies. This family is fairly easy to distinguish because it is named after the inflorescence type that most member of the family display. The inflorescence type is called an umbel. It's basically the shape of a wild carrot or poison hemlock. And because most of the members of this family have this particular type of inflorescence, the family is called the *Umbelliferae*. However, you may recall that in order to be an accepted family name, the family has to be named after a genus within that family followed by the suffix *aceae*. So, *Apium* is the celery genus, which is in this family, so the new name is *Apiaceae*, but many botanists will still just say it's in the umbel family. Members of the *Apiaceae* are often typically fly-pollinated.

The *lepidopterans* are insects that are moths and butterflies, and even though these two insects are in the same order, and they look superficially alike, they pollinate different types of flowers. Butterflies are typically day active, they hold their wings up when at rest, and their antennae are thinner and have a small club at the end. Butterflies have good color vision, too, so they tend to prefer flowers that are yellow, red, or orange. Because butterflies don't have much sense of smell, flowers they pollinate don't have much scent.

Moths are generally active at night, hold their wings out or folded on their backs at rest and have thicker antennae that lack a club at the end. Since moths are active at night, they tend to pollinate white or cream colored flowers. However, moths have a good sense of smell, and a strong scent can help them find flowers at night, so moth-pollinated flowers will have a strong scent, usually at night.

Both moths and butterflies have long proboscises, which are the tubular structures through which they suck up nectar. As such, flowers that are pollinated by moths and butterflies have long nectar spurs. These are elongated parts of the flower that have nectar at the bottom. Moths have even longer proboscises than butterflies. In fact, there is a good story about Darwin viewing an orchid from Madagascar with nectar spurs up to 30 cm long. When he saw these very long spurs, he predicted that there must be a moth with a proboscis long enough to reach the nectar at the base of the nectar spur. He made this remark in 1862, and in 1903, his prediction was validated when the hawk moth was discovered. In later observations, this moth, was indeed, found to pollinate that particular orchid.

Since butterflies are smaller than moths, they have lower energy requirements, so there is generally less nectar produced in butterfly-pollinated flowers. Despite their larger body size, moths prefer to hover while gathering nectar, which means that moth-pollinated flowers must produce even more nectar. Also, because of the hovering, moth-pollinated flowers tend to be bilateral in shape and have a corolla and nectar spur that are easily accessed by a hovering moth. A gardenia, with its heavenly fragrance and white blossoms, is a typical moth pollinated flower. A butterfly pollinated flower would be smaller and colorful, like a lavender flower.

Insects aren't the only animals to pollinate flowers. Since most flowers evolved with insect pollinators, other animals were later to the game and may have co-opted, especially at high altitudes, or on remote islands, or other places where insects are scarce. Like butterflies, birds also have good vision but lack a well-developed sense of smell. So, if you had to guess, what would a bird-pollinated flower look like?

If you said, red, you're right. In addition, these bird-pollinated flowers are going to be larger than butterfly-pollinated flowers, and what would you say about the amount of nectar? Well, all birds, even tiny hummingbirds, are larger and have more energy requirements than butterflies, so these flowers will produce a lot of nectar. The shape will depend on the type of bird performing the pollination.

Hummingbirds are only in the Americas, and they do pollinate many red flowers in this region, such as the century plants found in the Sonoran desert. These hummingbird pollinated flowers will be bushy, almost brush shaped and also have a nectar tube in many cases, a salvia is a good example. In other regions of the world, other birds will pollinate, such as sunbirds in South Africa and honeyeaters in Australia. In New Zealand, the beautiful brilliant red pōhutukawa tree is pollinated by birds, including the endemic tui.

Other animals besides birds also pollinate. Bats, like moths, are major nocturnal pollinators. As the saying goes, blind as a bat, vision is not so important to bat pollinators, but like many mammals, bats have a well-developed sense of smell. So, like moths, bats will pollinate white or cream colored flowers that have a strong scent. A beautiful and sought after bat-pollinated plant is the night-blooming cereus. This cactus has a lovely, strong scent that is especially released at night. By opening its flowers only at night, the cactus can avoid visitation from other pollinators or nectar seekers.

In addition to bats, there are non-flying mammal pollinators, too. About 50 other species of mammal, usually rodent type animals, pollinate flowers. In Madagascar, lemurs pollinate traveller's palms. Typically, flowers pollinated by non-flying mammals will have flowers close to the ground with a strong odor. These flowers also have to be sturdy to hold up to mammals.

Lest it seems like the plants are merely manipulating the animals to do their bidding, keep in mind that many times, the animal receives either nectar or pollen as a reward for their services. Sometimes, however, there is no reward. This is called deceit pollination. Those lying, cheating plants are fooling the insects to stop by and do some work for nothing at all.

We've already met a couple of these cheaters when we talked about the corpse flowers. As their name suggests, these flowers give off an odor of rotting flesh. This is to lure flies that normally feed on rotting flesh and lay their eggs in such flesh so that the maggots have food when they hatch. When flies land on these corpse flowers, they are sometimes tricked so much that they do in fact lay their eggs in what they think is indeed rotting flesh. The flowers continue the ruse by being a dark red color and even having hairs that resemble those found on an animal. Of course, the maggots won't make it, but the fly will go on to visit another flower, perhaps to feed, and the pollination deed has been done.

Another fascinating type of deceit is when a flower takes on the shape of a female insect to lure males to come mate with it. While the male believes he is mating, he is actually getting pollen all over his body, only to drop that pollen off when he mistakenly mates with another flower. The physical likeness of these flowers to the females is impressive enough, but the flowers also release scents that mimic those of the female insect as a perfume to further trick the male into thinking he is mating with an actual female instead of a flower. One species that does this incredibly well is the fly orchid. In fact, many species of orchids have some sort of tight relationship with an insect.

This is one reason why the orchids are so diverse. In fact, the orchid family, *Orchidaceae*, is the most speciose family of all the plants, with between 25,000 and 30,000 species. The evolution of the insect and a plant together is coevolution. One flower started looking a bit like the female fly, then through time, those flowers looking most like the female were selected for by males visiting them and spreading their pollen. Over time, the orchid isn't pollinated by any other insect, and so it becomes its own species. Other than horticulture, the only economically useful orchid is the vanilla orchid, which is pollinated by a stingless bee. The coevolution between the orchid and the bee was so tight that when vanilla was first introduced to areas other than its native Mexico, it would not produce fruit because there were none of the native bees to pollinate it.

We use pollination generally for many crops. A huge variety of fruits depends on pollinators to reproduce. Bees, in particular, are valuable pollinators, but because of colony collapse disorder, there are fewer honey bees to pollinate. This is why some beekeepers transport their hives to farms and orchards and let their bees forage and pollinate, performing a valuable ecosystem service.

The exact cause of the decline of the bees, or of colony collapse disorder is unknown, but if we had to pollinate crops by hand, it would be extremely time consuming and expensive.

Let's see some pollination in action. So in nature, this would be done by insects, but we can mimic that action by using a paintbrush. What I'm going to do here is take some pollen from this flower—so here we have the anthers—we can see this yellow pollen is getting all over my paintbrush. And then I'm just going to pretend that I'm a bee and I'm going to fly to another flower. So then I'm going to fly over here and I'm going to do the same thing. I want the nectar, but I've also got pollen all over my body. So then as I try to get the nectar, the pollen rubs off on the sticky stigma. And now I have cross-pollinated these flowers.

OK, so once the pollen has landed, and it's viable pollen, and the pollen tube starts growing, what happens next. Eventually, the pollen tube will grow all the way into the ovule, the unfertilized egg. This can happen quickly, around 15 minutes in some plants, like corn. Or it can take up to a year in others, such as pine.

There is a special fertilization process that only occurs in flowering plants. This fertilization process is called double fertilization. Essentially, the name says it all—there are two fertilizations taking place. The first is the old-fashioned one of the sperm meeting the egg, leading to a zygote. The second fertilization is a second sperm fertilizing what are called the polar nuclei of the egg cell. Moreover, there are two polar nuclei for the second sperm, so when this second fertilization takes place, it creates a tissue that has three copies of the genetic code, so it is a triploid tissue, and it's called the endosperm. This triploid endosperm provides nutrients to the growing embryo. Once the ovules are fertilized, the flower can release its petals, while the ovary and carpels focus on becoming a fruit. Think of a coconut, which is a fruit, and not a nut at all. The hard outer brown part is the seed coat and the inner liquid part, the coconut milk inside, is actually the triploid endosperm.

In the next lecture, we'll learn how to make sense of all that—and the many other kinds of fruits.

LECTURE 14

THE MANY FORMS OF FRUIT: TOMATOES TO PEANUTS

Fruits are how plants move—at least how they move their progeny. You probably know that plants have a myriad of fruit types, but what you may not realize is that they are all based on how the carpels were organized in the flower. Of course, botanists have names for all of these fruit types.

FRUITS

- Some fruits that we think are fruits are actually not fruits. For example, a strawberry is botanically not a berry. In fact, it's not really a fruit because it's not the swollen ovary of the plant. But there are fruits on a strawberry.

- What might appear to be seeds on the outside of the strawberry are actually the fruits. They are a small type of fruit called an achene, which is a simple fruit, meaning that it is derived from one carpel, within a flower that has a superior ovary. All of the other flower parts will insert into the receptacle below the ovary in a flower with a superior ovary.

STRAWBERRY
Fragaria sp.

- What one might call the actual berry of the plant is not botanically a berry; it is a receptacle, which in most plants is simply where the petals, sepals, and other flower parts insert into the stem. In the strawberry, this receptacle becomes red and swollen when ripe, exposing the tiny fruits on the outside. This type of fruit, where the actual fruit part isn't an ovary, is called an accessory fruit. The swollen receptacle is an accessory to the actual fruit, the achenes on the surface.

- What many people might call a vegetable—the tomato—is a fruit, and it's botanically a berry. Botanically, a berry is a fruit with several carpels, each of which has many seeds. Blueberries and cranberries fit this definition, as do grapes, although many grapes in stores are now seedless.

FRUIT
(A BERRY OF TOMATO)

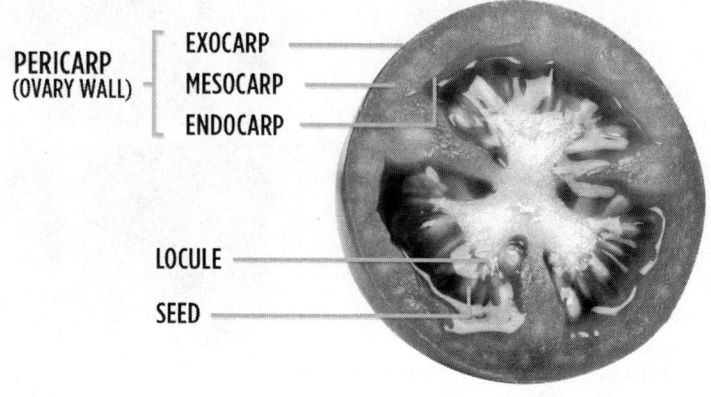

PERICARP (OVARY WALL)
- EXOCARP
- MESOCARP
- ENDOCARP

LOCULE

SEED

◇ Inside a tomato, there's the gooey part with tiny seeds, and there's the wall part. There's a fruit wall on the outside, and this is called the pericarp generally. In a tomato, the outermost part of the pericarp is the exocarp, the middle is the mesocarp, and the inside is the endocarp. The endocarp is pretty difficult to distinguish in a tomato, but the exocarp is clearly the skin, and the mesocarp makes up most of the fruit wall.

◇ The runny parts are called locules, but usually each locule started off as a carpel, so there are several different carpels within the ovary. There are also several seeds within each carpel.

◇ A true berry has a fleshy inner fruit wall. A tomato is a true berry because the whole fruit wall is fleshy in a tomato.

◇ There are 2 other variations of berries. If a berry has a leathery rind, it's called a hesperidium. An example of this is an orange. There are many carpels: Each of the sections of the orange, the orange slices, is a carpel. There are typically many seeds inside, and there is a leathery fruit wall instead of a fleshy wall.

- The watermelon is a second type of berry. Although it's difficult to see the carpels when looking at the cross-section of a watermelon, they are there. It's easier to see them in a related plant, the cucumber. Both of these plants are in the Cucurbitaceae family. Other members of this family are cantaloupe, squashes, and pumpkins. The fruits of this family are called pepos. A pepo is a berry with a hard outer fruit wall.

- Blackberries and raspberries aren't botanically berries. They are what botanists call aggregate fruits, which have several carpels aggregated together in one flower, so that the carpels are separate but still together in one fruit.

- A blackberry is an aggregate of many small balls, each with a seed inside. Each of those small round balls with a seed inside came from a carpel. Those small balls are a fruit type that botanists call a drupe, which is any fruit that has one to several carpels but only a single seed. Each one of those blackberry or raspberry fruits has one seed inside. Blackberries and raspberries are called an aggregate of drupes.

WINE RASPBERRY

Rubus phoenicolasius

- Drupes also include all of the stone fruits, such as peaches, plums, apricots, and cherries, which are in the Rosaceae family. Drupes also include avocados and mangoes. Botanists have not been able to develop a seedless stone fruit, but they're getting closest with plums.

- A coconut is another drupe. When the coconut is on the tree, the brown round ball is encased in a big husk, which helps protect the coconut that has the seed on its journey through the world's oceans. The husk is generally removed before being shipped to temperate countries.

- Despite their common name, coconuts aren't nuts. For that matter, neither are almonds, walnuts, or pecans. Like the coconut, all of these are drupes—specifically the hard, stony seeds of a drupe. For botanists, true nuts are fruits, whereas a lot of what the grocery store calls nuts are really seeds.

- A peanut is a legume, which is a dry—as opposed to fleshy—fruit that splits along both sides at maturity. A legume is the botanical name given to all the plants of the Fabaceae, or pea, family: The fruit of a bean plant is the pod, and the seeds are inside the pod.

- To make matters more confusing, there is also a peanut butter fruit from the peanut butter tree, *Bunchosia argentea*, that produces cherry tomato–sized fruits, which are berries that are somewhere between a date and a peanut.

- Botanically, a nut is a fruit with a hard shell surrounding the seed. Hazelnuts, acorns, and chestnuts are true nuts. A true nut cannot be encased in any other kind of fruit. The almonds we eat are seeds inside a hard-fruit shell. Walnuts are seeds encased inside an even bigger fruit.

PEANUT BUTTER TREE
Bunchosia argentea

◇ An oddity among the would-be nuts is the cashew, which is in the same family as mango, which is a drupe. The infamous poison ivy also belongs to this family, Anacardiaceae. What we call the cashew is actually produced on the bottom end of a larger fleshy receptacle (the base of the flower) called the cashew apple. The cashew nut, which is a nut because it's in a hard shell, attaches to the bottom of the cashew apple. For this reason, some botanists call it a drupaceous nut.

◇ A cashew apple is not an apple because it is the swollen receptacle and doesn't have seeds. A real apple is botanically called a pome. The only fruits that are pomes are apples and pears.

◇ The pome develops from a compound ovary that is inferior, which means that the sepals and petals attach above the ovary instead of below it. The part of the apple we eat is actually the receptacle. The ovary begins where we stop eating, at the exocarp, or the papery membrane that surrounds the core.

CASHEW
Anacardium occidentale

- The pome is interesting in terms of ripeness. Apples are a long-lasting fruit because they stay ripe for a long time. A pear, on the other hand, has a narrower window of ripeness, and unlike apples, many people enjoy pears when they're soft.

- While opinions of the perfect ripeness of fruit may vary, figuring out cues to ripeness is a big business. A fruit that will stay unripe during transport and then become ripe just when it goes for sale at the grocery store is probably the best kind of fruit from a grower's perspective.

FRUITING BODIES

- Although not a fruit at all, a mushroom is a kind of fruiting body. The mushroom we see above the ground is the reproductive organ of the fungus that lives mostly underground. Because it's difficult to transport spores underground, when conditions are right, the strands of fungus, called hyphae, all come together to form the structure we recognize as a mushroom.

- This is why the mushroom is called a fruiting body. It doesn't produce fruits with seeds, but it is a structure that forms to produce spores, which are a form of reproduction. The spores in mushrooms are much like those of mosses and ferns. They are single-celled, very light, and very easy to disperse via wind.

- Fungi are more closely related to animals than plants; however, unlike animals, they have cell walls, like plants, but unlike plants, they don't photosynthesize. They are not plants, but they are not animals, and because of that, they have traditionally been grouped with the plants—not taxonomically, but informally.

- Most mushrooms are formed by one particular fungal group, the Basidiomycota. But there are some delicious notable exceptions. The morel mushroom happens to belong to another group, the Ascomycota. Because the morel can't be cultivated, the gathering of morels is a big business.

◇ The same is not true for the truffle, which is also an Ascomycota. Truffle farms have been appearing all over, but it takes a while for them to be productive. One must first inoculate a tree with the fungus and then wait. In France, however, 80% of truffles come from truffle orchards.

◇ Truffles are mycorrhizal fungus, which means that they form a symbiotic mutualism with the roots of trees. The fungus increases the plant's ability to absorb minerals and water because it greatly increases available surface area of the roots. The plants provide the mushrooms with sugars.

◇ Like ferns and mosses, mushrooms reproduce by spores, which are microscopic and haploid. The fungal life cycle is fairly complicated, but suffice it to say that these spores are released from the gills under the cap of the mushroom. Once the mushroom begins to grow, it can double in size in 24 hours.

READINGS

Blackburne-Maze, *Fruit*.

Chase and Llewellyn, *Seeing Seeds*.

Stuppy, *Fruit*.

QUESTIONS

1   When you think of your favorite fruit, can you locate the exocarp, mesocarp, and endocarp? What is its botanical name?

2   How are mushrooms like fruits?

LECTURE 14 TRANSCRIPT
THE MANY FORMS OF FRUIT: TOMATOES TO PEANUTS

Fruits are how plants move, at least how they move their progeny. Let's start with an unusually big example. In my very first year as a Ph.D. student, my advisor, Dr. Phil Rundel, gave me a great opportunity to travel to Thailand as a teaching assistant on a field biology quarter.

During one of his lectures, Phil was telling everyone about the many exotic tropical fruits we would find in Thailand. He told us that one of these fruits was so pungent, there were signs on the hotels asking guests not to even bring the fruit inside. Maybe you already know what I'm talking about, but at the time I couldn't imagine what is this fruit? What would it smell like if it wasn't welcome inside a hotel? I also couldn't imagine anyone wanting to eat something that would smell so strong.

When we got to Thailand, I never saw the infamous durian fruit growing on the tree, but I did finally get to taste it. Our Thai cooks had some fresh and cut it up for us to try. Thankfully, we were eating outside. The smell is a bit like garlic and rancid butter, mixed with a cloyingly sweet smell, like the fruit is overripe. I was assured that this particular fruit was at the peak of its deliciousness. That may be the case, but I can tell you that I had two bites—first and last, and I'm happy to say I've had the experience, and I'm also happy that there are many other fruits to eat.

Durian grows in a rather interesting way called cauliflory. This is when the fruit grows right out of the trunk under the main canopy of leaves and not out on the branches. Since the average durian can weigh about five pounds, it's a good thing that it doesn't grow on the edge of the branches, because it might break those branches.

Many tropical fruits grow this way both for pollination and later dispersal. It's easier for the animals to find the fruits if they're not hidden in a canopy of leaves. The durian has large, white flowers that are open at night with a strong smell. Can you guess the pollinator? If you said bat, you're right. The really large flower is a give away that it probably isn't a moth. Later, the fruits have to be dispersed by a large animal, like a rhinoceros or an elephant, and, again, these animals don't want to search the leafy canopy for their treat.

But before the fruit can be dispersed, it has to develop. Plants have a myriad of fruit types, but what you may not realize is that they are all based on how the carpels are organized in the flower.

Botanists, of course, have names for all of these fruit types. Some fruits that we think are fruits are actually not fruits. For example, think of a red, juicy, ripe strawberry. It's botanically not a berry at all, and in fact, it's not really a fruit because it's not the swollen ovary of the plant. There are fruits on a strawberry, however. What one might appear to be the seeds on the outside of the strawberry are actually the fruits. They are a small type of fruit called an *achene*, which is a simple fruit, meaning it is derived from one carpel, within a flower that has a superior ovary. All the other flower parts will insert to the receptacle below the ovary in a flower with a superior ovary.

So what one might call the actual berry of the plant is not botanically a berry at all. It is a receptacle, which in most plants is simply where the petals, sepals, and other flower parts insert into the stem. In the strawberry, this receptacle becomes red, and swollen when ripe, exposing the tiny fruits on the outside. This type of fruit, where the actual fruit isn't an ovary, is called an accessory fruit. The swollen receptacle is an accessory to the actual fruit, the achenes on the surface. Of course, people have selected for larger and larger strawberries, so that the current item we buy in the grocery store is barely recognizable as its wild ancestor.

So, if a strawberry isn't a berry, what is a berry? A tomato. That's right. What a lot of people might call a vegetable is a fruit and it's botanically a berry. So, what makes it a berry? Botanically speaking—and is there really any better way to speak—a berry is a fruit with several carpels, each of which has many seeds. Blueberries and cranberries fit this definition. So do grapes, although a lot of grapes in stores are now seedless. To understand a true, botanical berry, picture the inside of a tomato, a cross section.

There's the gooey part with the tiny seeds, and there's the wall part. There's fruit wall on the outside, and this is called the pericarp generally. In a tomato, the outermost part of the pericarp is the exocarp, the middle is the mesocarp, and the very inside is the endocarp. The endocarp is pretty difficult to distinguish in a tomato, but the exocarp is clearly the skin and the mesocarp makes up most of the fruit wall.

The runny parts are called locules, but usually, each locule started off as a carpel. So, there are several different carpels within the ovary. There are also several seeds within each carpel. There you have it, several carpels, each with many seeds equals berries.

We're not quite done with that definition because a true berry will have a fleshy inner fruit wall. So, is a tomato a true berry? If you said yes, you're right. The whole fruit wall is fleshy in a tomato. But, there are two other variations of berries. If a berry has a leathery rind, it's called a *hesperidium*. An example of this would be an orange. There are many carpels—each of the sections of the orange—the orange slices—that's a carpel. There are typically many seeds inside, and there is a leathery fruit wall instead of a fleshy wall. Now, if you're thinking that your oranges don't have seeds, you're right because there are many seedless varieties now, but in nature, a fruit should produce seeds, or it won't be selected for because it won't reproduce without seeds.

So, how do we get seedless oranges? Citrus fruits generally have a high degree of what botanists call *parthenocarpy*. This word comes from the Greek words *Parthenos* for virgin and *karpos* for fruit. So, seedless oranges are literally virgin fruits. The miracle of an orange—who knew.

These parthenocarpic fruits develop without seed formation. This can happen because pollination fails due to the genes of the pollen and the stigma being incompatible. Such is the case with oranges. The seedless varieties that farmers grow cannot self-fertilize. And, in order to make sure the fruit is the same and the oranges continue to be seedless, the orchards are clones—they are all genetically identical. So, farmers usually use a graft of the top of the orange tree they want on the rootstalk of another orange tree. Citrus grafts really well, which is one reason we have so many seedless varieties.

Parthenocarpic fruits can also form because of chromosome imbalances. When a plant is triploid, meaning it has three copies of the genome, the chromosomes will not match up correctly, and so seeds will not be produced. This is what happens with seedless watermelons. Oddly enough, these fruits have to be produced anew each time. Growers have to have a tetraploid plant cross with a regular diploid plant. The sperm and egg of the tetraploid plant will end up having two copies of the genome, and the diploid plant will produce sperm and egg that have only one copy of the genome. When these egg and sperm meet, the resulting offspring will be triploid, and it will produce fruit but not seed—the seed won't come to fruition. Get it, fruition? These fruits have to be grown from scratch each time, but even though it's time-consuming, the seedless variety is coveted by consumers.

The watermelon is actually a second type of berry. Although it's hard to see the carpels when looking at the cross-section of the watermelon, they were there. It's easier to see them in a related plant, the cucumber. Both of these plants are in the *Cucurbitaceae*. It's pretty easy to spot the fruit of a member of the *Cucurbitaceae* because it always will be what we think of as some type of melon. Other members of this family are cantaloupe, squashes, and pumpkins. That's right zucchini is a fruit. Botanists don't call the fruits of this family squashes or melons, however, we call them pepos. A pepo is a berry with a hard outer fruit wall.

So, we've met the berries, including the true berries, the hesperidium, and the pepo. Perhaps you noticed that I didn't mention blackberries or raspberries. Think about what's different, These fruits aren't botanically berries. They are what botanists call aggregate fruits. These fruits have several carpels aggregated together in one flower so that the carpels are separate but still together on one fruit. Think of a blackberry here. It's actually an aggregate of many small balls, each with a seed inside. Those are the seeds that get caught in your teeth when you eat them. And each one of those small round balls with a seed inside came from a carpel. Those small balls are actually a type of fruit that botanists call a drupe.

A drupe is any fruit that has one to several carpels but only has a single seed. Each one of those blackberry fruits has one seed inside. The same is true for raspberries, but blackberries and raspberries are called an aggregate of drupes.

Drupes also include all of the stone fruits—peaches, plums, apricots, cherries—all of these are in the *Rosaceae*—the rose family. Drupes also include avocados and mangoes. Try as they might, botanists have not been able to develop a seedless stone fruit, but they're getting closest with plums.

The famous horticulturist Luther Burbank created varieties of plums that were almost pitless. He found a variety from France that had a very small pit and grafted this onto another rootstock. He got close, but part of the problem of creating a pitless plum is that it normally takes four or five years for a tree to produce fruit. So, experiments with different crosses take a long time to bear fruit. There is a professor at Syracuse, Sam Van Aken, grafted 40 different varieties of stone fruit—plums, apricots, peaches, and almonds, but even grafts will take two or three years to bear fruit.

Botanists working at the ARS Appalachian Fruit Research Station in Kearneysville, West Virginia, have developed a fast growing plum by inserting a FLOWERING LOCUS T gene from a poplar tree. With this inserted gene, the plums can bloom and bear fruit continuously in the greenhouse. This allows a generation every year instead of the normal 4–7 years in the field. This will certainly speed up the search for a pitless plum. Using breeding programs, genetic engineering, and the plant growth hormone gibberellic acid—or GA—which can induce fruiting, the botanists are hopeful that they will be able to develop a new variety of plum that lacks a pit. After development in the plum, they could move on to other stone fruits. Can you imagine pitless cherries? Wow.

One more drupe to talk about is a coconut. When the coconut is on the tree, the brown round ball that we associate with coconut is actually encased in a big husk. This husk helps protect the coconut that has the seed in it on its journey through the world's oceans. The husk is generally removed before being shipped to temperate countries.

There is another type of coconut that we don't see here in North America because it's native to the Seychelles Islands in the Indian Ocean, northeast of Madagascar. A type of palm tree there produces the double coconut also called the coco de mer. This specific palm produces the world's largest seed measured up to 38 pounds inside nature's largest fruit, weighing up to 92 pounds, which is exactly what my 9-year-old son weighs.

Now, this fruit record is for a naturally occurring fruit. Pumpkin breeders have managed to grow a pumpkin weighing over 2,000 pounds. That's a lot of pumpkin pie.

Despite their common name, coconuts aren't nuts. For that matter, neither are almonds, walnuts, or pecans. Like the coconut, all of these are drupes, specifically the hard, stony seeds of a drupe. For botanists, true nuts are fruits, whereas a lot of what the grocery store calls nuts are really seeds.

What about peanuts? A peanut is a legume, which is a dry, as opposed to fleshy, fruit, that splits along both sides at maturity. A legume is the botanical name given to all the plants of the Fabaceae, the pea family–green beans, snow peas, sugar snap peas. The fruit of a bean plant is the pod, and the seeds are inside the pod.

To make matters even more confusing, there is also a peanut butter fruit from the peanut butter tree, *Bunchosia argentea*, which produces cherry tomato sized fruits, which are berries, that are somewhere between a date and a peanut.

So, what food is a nut? A hazelnut is a nut. Botanically, a nut is a fruit with a hard shell surrounding the seed. Acorns are a nut and they have an involucre bract. Chestnuts are also true nuts. A true nut cannot be encased in any other kind of fruit. The almonds we eat are seeds inside a hard fruit shell. Walnuts are seeds encased inside an even bigger fruit.

An oddity amongst the would-be nuts is the cashew. The cashew is in the same family as mango, which is a drupe. The infamous poison ivy also belongs to this family, the *Anacardiaceae*. What we call the cashew is actually produced on the bottom end of a larger fleshy receptacle called the cashew apple. Recall that the receptacle was the base of the flower. So, the cashew nut, and it is a nut because it's in a hard shell, attaches to the bottom of the cashew apple. For this reason, some botanists call it a drupaceous nut, and drupaceous would be a great band name.

Of course, by this time, you're probably aware that a cashew apple is not an apple because it is the swollen receptacle, and doesn't have seeds. A real apple is botanically called a *pome*, which is handy since that's the French word for apple. I love the French word for potato, too, *pomme de terre*. Of course, potatoes are not fruits at all, and because they are from vegetative growth, i.e. without sex and non-seed producing, potatoes are vegetables, as would be lettuce and spinach. The only fruits that are pomes are apples and pears.

The pome develops from a compound ovary that is inferior, which means that the sepals and petals attach above the ovary instead of below it. The part of the apple we eat is actually the enlarged base of the flower, the receptacle The ovary begins where we stop eating, at the exocarp, or the papery membrane that surrounds the core. Even though we don't eat the core, we probably could. There is a substance in apple seeds called *amygdalin* that breaks down in the intestine to produce cyanide. For this reason, most people avoid apple seeds, but if you don't chew the seeds, the cyanide has a harder time forming. You'd also have to eat over 100 apple seeds to get close to the amount of cyanide that would be toxic.

The Bible talks about a forbidden fruit in the Garden of Eden, leading to much speculation. Apples are indigenous to Central Asia, so it's possible that they were in the Holy Land. But if we think of the modern day Middle East, it's a Mediterranean climate at best and a desert climate at worst. Michelangelo chose to paint a fig in the ceiling of the Sistine Chapel. I think it was most likely a pomegranate or a persimmon. But I digress.

The pome itself is also interesting in terms of ripeness. Apples are a long lasting fruit because they stay ripe for a long time. A pear, on the other hand, has a narrower window of ripeness, and unlike apples, many people enjoy pears when they're soft. While opinions of the perfect ripeness of fruit may vary, figuring out cues to ripeness is big business. A fruit that will stay unripe during transport and then become ripe just when it goes up for sale at the grocery is probably the best sort of fruit from a grower's perspective.

One way that fruit will ripen quickly is from exposure to the plant hormone ethylene, which is a hydrocarbon gas—H2C=CH2. Ethylene turns on the genes that are then transcribed and translated to make enzymes that will cause ripening in the fruit. An enzyme is a protein that will catalyze a biological reaction, and they always end with the suffix -ase. Hydrolases are enzymes that degrade the bonds holding together bitter chemicals inside the fruits, and amylases are enzymes that break down starch into sugar. Both of these reactions will result in a sweeter tasting fruit.

In order for the fruit to become softer, pectin has to be broken down. Pectin is a bit like our friend lignin, which is responsible for the toughness of some leaves and wood. Pectin is a complex molecule that is located between the plant cells and holds them together. Pectin is also the basis of fruit jellies and jams. So, in the fruit, ethylene will cause enzymes called pectinases to speed up the breakdown of pectin. If the pectin breaks down too much, then the fruits get mushy.

Supermarkets will use ethylene to their advantage. They will keep produce in ethylene-free environments to delay ripening until the fruit is put out, or they will actually gas unripe fruit with ethylene to hasten its ripening. To use ethylene to your advantage, you can store an unripe fruit, say an avocado, with a ripe fruit, such as a banana, in a paper bag. The ethylene production by the ripe fruit will speed up the ripening of the unripe fruit. When the fruit reaches its peak ripeness, you can store it in the refrigerator, which will slow down the ripening process immensely.

Ethylene will also cause the breakdown of chlorophyll and the onset of other pigments to advertise ripeness. Still, the fruit isn't truly ripe until it's ready to be picked, and the process that facilitates our picking of the fruit is fruit abscission. The pectinases break down the pectin that holds the cell walls together where the fruit is attached to the plant, and as this area degrades, the fruit becomes easier to pick.

Although not fruits at all, this is a good time to talk about another kind of fruiting body, and that is a mushroom. The mushroom we see above the ground is the reproductive organ of the fungus that lives mostly underground. As it's difficult to transport spores underground, when conditions are right, the strands of fungus, called hyphae, will all come together to form the structure we recognize as a mushroom. I can just hear John Lennon singing "Come Together" as the hyphae assemble into their reproductive structure.

This is why the mushroom is called a fruiting body. It doesn't produce fruits with seeds, but it is a structure that forms to produce spores, which are a form of reproduction. It should really be called a sporing body, but that's just not as familiar to people, I guess. The spores in mushrooms are much like those of mosses and ferns. They are single-celled, very light, and very easy to wind disperse.

The fungi are sort of weird in that they are more closely related to animals rather than plants, and yet, unlike animals, they have cell walls, like plants, but unlike plants, they don't photosynthesize. Really weird. But they're definitely not plants. But, they are most definitely not animals, and because of that, they have traditionally been grouped with the plants, not really taxonomically, but sort of informally.

Most mushrooms are formed by one particular fungal group, the *Basidiomycota*. But, there are some delicious notable exceptions. The morel mushroom happens to belong to another group, the Ascomycota. Since the morel can't be cultivated, the gathering of morels is big business.

The same is not true for the truffle, which is also an *Ascomycota*. Truffle farms have been appearing all over, but it takes awhile for them to be productive. First, one must inoculate a tree with the fungus and then wait. In France, however, 80% of truffles come from truffle orchards. Truffles are actually a mycorrhizal fungus, which means they form a symbiotic mutualism with the roots of trees. The fungus increases the plant's ability to absorb minerals and water because it greatly increases the available surface area of the roots. The plants provide the mushrooms with sugars.

Some truffles even produce chemicals that mimic mammalian steroids. Truffles produce a compound that is very similar to one found in the testes of pigs. This steroid appears in the male pig's saliva and causes the female pig to become receptive to sex, and this is why female pigs have been used to hunt for truffles, which form underground. Truffles are dispersed by animals, which is the probable reason they have this odor. Recently, more dogs are starting to be used for truffle hunting because they can be trained to sniff the truffle but not eat it, and they are easier to transport. Have you ever tried to get a pig in the back of your car?

Like ferns and mosses, mushrooms reproduce by spores, which are microscopic and haploid. The fungal life cycle is fairly complicated, but suffice it to say that these spores are released from the gills under the cap of the mushroom. Once the mushroom begins to grow, it can double in size in 24 hours. So, if it seems like mushrooms appear in your yard overnight, they do.

A common lawn mushroom is the fairy ring, which is a series of mushrooms that grow in a ring. Because the mushrooms are taking up many of the nutrients, the grass in the middle of the ring can be stunted or a lighter color. These rings continue to grow outward and hyphae in the soil will actually secrete chemicals to enrich the soil for further fungal growth. There is a fairy ring in France that is almost a half mile in diameter. This ring is estimated to be 700 years old.

In addition to fairy rings, there are earth stars. It's sort of like a fantasy land of fungus. When dry, these fungal fruiting bodies look like an onion with an outer layer composed of rays. When they become wet, the rays unfurl in a star shape.

It gets even stranger. There is a genus of mushrooms called *Phallus*, for reasons that I hopefully don't have to explain. Thankfully, their common name refers to their odor, as they are actually called stinkhorns. Like the corpse flowers, these mushrooms have an odor of rotting flesh to attract flies. When the flies visit the smelly fungi, they get spores all over themselves, which they then disperse. Stinkhorns are cooked and eaten in Asian cultures.

And then there are the colors. Although we typically think of mushrooms as brown or cream colored, mushrooms come in as many colors as wildflowers. The *Amanita muscaria*, or fly agaric mushroom, is bright red with polka dots. I can just imagine the caterpillar sitting atop of it, telling Alice that one side makes you smaller and the other side makes you grow larger. This mushroom is poisonous, though. The *Lactarius indigo* is, as its species name suggests, bright blue. Even stranger, it produces a blue milky latex when the mushroom is punctured. And, this blue, milky latex mushroom is edible.

The amethyst deceiver is a brilliant violet. It's called deceiver because it loses this color as it gets older, making it hard to identify. There are numerous species of orange mushroom, one of which, commonly called the orange peel fungus does look for all the world like discarded orange peels. And, even though it doesn't photosynthesize, there is a brilliant green mushroom from New Zealand. All shapes and sizes and colors. The mushrooms are fascinating, tasty, and weird, kind of like a durian if you're the right sort of person.

Just keep in mind, what the fruiting bodies of all mushrooms lack are seeds. Next time, we'll turn to how plants disperse their seeds, including plants whose fruits and seeds fly through the air.

LECTURE 15

PLANT SEEDS GET AROUND

The evolution of the seed was a major advantage for land plants. But unlike gymnosperms, which had a naked seed, the flowering plants produce a fruit around that seed, which aids in germination or dispersal—or both. The seed includes an entire embryonic plant, complete with an immature leaf, the cotyledon, which also stores starch so that the embryonic leaves can grow. Some flowering plants, the eudicots, have 2 of these immature leaves. The embryonic plant also has an immature root, called the radicle. A seed has germinated when the radicle first emerges from the seed. But before germination, the seed is dispersed. Like pollination, there are many ways that fruits, with the seed or seeds inside of them, can be dispersed.

EXPLODING SEEDS

◇ Fruits come in many forms. If fruits are fleshy, then they are botanically called fleshy fruits. However, not all fruits are fleshy. Many fruits are dry at maturity. When they are completely ripe, dry fruits either split open or don't. Dry fruits that split apart at maturity are called dehiscent, and only dry fruits will split, as fleshy fruits will simply decay. Dry fruits that never split are called indehiscent.

◇ Launching the fruits a far distance ensures that the seeds won't try to germinate right beneath the parent tree and then compete with it.

◇ *Hura crepitans* is a tree that is native to the American tropics and produces a fruit that is botanically classified as a capsule, which means that at maturity, it's a dry dehiscent fruit that develops from 2 or more carpels, which are divisions of the ovary. A capsule also has many seeds and splits along or between the carpels. There are also capsules that have pores at the top, where the seeds come out, and these are called poricidal capsules.

SANDBOX TREE
Hura crepitans

◇ The fruits look innocuous enough, but as they dry, they begin to split along the carpels, and the cells become smaller and smaller because the water has left. The seeds will split from the wall of the fruit, and if harvested at this time, the fruit will sound like a rattle with the loose seeds inside. Eventually, the cells will break, and the seeds will shoot out of the fruit. The seeds are discoid shaped, which reduces their wind drag and helps them go even farther. These ballistic seeds could do some damage if you, or other plants, got in their way.

- There is also a plant called the touch-me-not that, when disturbed by a breeze or a touch from an animal, will also explode its seeds. However, these seeds are much smaller and pose no threat of injury. In fact, a number of plants employ this drying-out mechanism, called xerochasy, for ejecting their seeds.

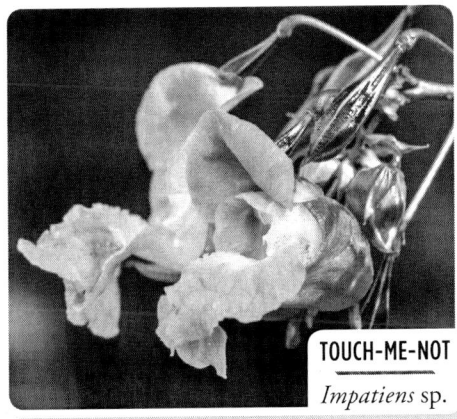

TOUCH-ME-NOT
Impatiens sp.

- Plants can also explode by using hygrochasy, which is opening by water. This mechanism still involves the loss of water. It occurs when enlarged cell walls shrink with water loss and then expand with its uptake. Then, these increases in the swelling tissue create pressure because this swelling tissue is attached to lignified, non-swelling resistance tissue. When the pressure becomes too great, the fruit explodes.

- Hygrochasy doesn't just occur in dry fruits. There can also be fleshy fruits that explode, and even different fruit types in the same family can explode.

- This is just one of the many ways that plants disperse the seeds. There are negatives and positives for each type of seed dispersal. If the seed doesn't go too far, as in the case of exploding seeds, then chances are that where it lands is a pretty good place to germinate, because it's not too far from the favorable habitat where its parent germinated. This is why we sometimes see clumps of a certain type of plant in nature.

- On the other hand, too many individuals in one spot can lead to competition in that spot. Plus, maybe the parent plant is in a spot that might be good for an adult but not so great for germination. In this case, the seeds might be better off if they are dispersed a greater distance. For these greater distances, plants can't rely on fruit explosions alone, and there are a myriad of ways that plants disperse their seeds.

DISPERSAL BY WIND AND WATER

◇ In addition to explosions, the other abiotic, or nonliving, dispersal methods are wind and water.

◇ Many fruits have very small seeds that are dispersed by wind. In addition, a plant that has an animal pollinator might also have a different dispersal method. For example, the orchid family, Orchidaceae, is animal pollinated but wind dispersed.

◇ A dandelion is a wind-dispersed fruit. One reason a dandelion can produce so many seeds is that it can produce seeds without pollination, a process called apomixis. This form of reproduction is asexual, but the cloned seeds are derived from ovules that would have given way to sexual reproduction had they not skipped pollination and gone straight to seed development.

◇ Dandelions are part of the sunflower family, Asteraceae. The name of the family comes from the genus *Aster*, which means star, because these flowers look like stars. The main fruit type of the sunflowers is called a cypsela, and they are perfect for riding on the wind with their own miniature parachutes. A cypsela is basically like an achene—a fruit that is derived from one carpel, within a flower that has a superior ovary—except that it is derived from an inferior ovary.

◇ A parachute is just one way to use the wind. A plant could also produce cottony seeds that are easily blown around. Cotton is a great example of this dispersal method, and so is the Kapok tree, which produces so much cottony fiber that it was used as mattress filling.

ALPINE ASTER

Aster Alpinus

- Another dispersal method is the winged fruit. A maple tree produces a winged fruit called a double samara. The botanical description of a samara is an achene with a wing on it.

- Another means of wind dispersal is found in the tumbleweeds. This dispersal mechanism is convergent, meaning that it's found in a number of different plant families. About a dozen families have plants that employ this method of moving the seeds around as the bush-like plant gets blown across the landscape. Baby's breath forms a tumbleweed as a dispersal mechanism.

- After wind, the other abiotic disperser is water. Plants don't have to grow in water to be dispersed by water. Plants that grow near water often use water as their dispersal agent. These plants will often produce fruits that are buoyant because of their small size or because of their structure. The coconut has an outer seed coat that is impermeable to water, so the coconut fruit can travel across oceans.

COCONUT TREE
Cocos nucifera

- Sea beans is a common name that refers to seeds and fruits that are produced in one place and ride the ocean currents to end up somewhere entirely new. They are also called drifter seeds. There are more than 200 species of plants that are dispersed in this way. Most times, these fruits are produced by trees in the tropics. Many people enjoy collecting these fruits and using them for art; they have also been used by oceanologists to map out currents.

DISPERSAL BY ANIMAL

- Perhaps the most complex mode of dispersal involves animals. There are 2 major strategies for plants: Be a delicious, irresistible fruit that animals can't wait to eat, or fashion yourself into some kind of hook or barb to attach to animal hair and hitch a ride.

- Even though insects do the lion's share of the work in pollination, vertebrates, especially birds and mammals, do the majority of the dispersal effort carried out by animals.

- The seeds produced by many fleshy fruits will pass through the digestive tract of the bird or other vertebrate without being harmed. Some plants with fleshy fruits actually have seeds that must pass through an animal's digestive tract to weaken the seed coat.

- This is the process that will start germination. As the seed coat weakens, it becomes permeable to water, and this will cause the cells inside to swell and grow, and the radicle will emerge, and germination has occurred.

- An example of a seed that must pass through an animal is a blackberry or a raspberry. The fruits are each individual spheres, and inside each one is a tiny seed, which needs to be roughed up in a bird's gizzard before they can germinate. In fact, gardeners who grow these plants from seed have to scratch these seeds in some way, usually using sandpaper, to weaken the seed coat and ensure germination.

- Plants that hitch a ride often have barbs or hooks to ensure that they hold for a little while. In fact, it was burdock, a biennial aster plant, that was the inspiration for Velcro.

- Birds and mammals seem likely fruit dispersers, but other vertebrates, such as tortoises and iguanas—and even some fish—eat fruits and disperse seeds, too.

- Even though vertebrates do most of the fruit dispersal by animals, a good number of plants also employ ants for this purpose. Plants that have ant-dispersed fruits have a special growth on the outside of their seeds called an elaiosome. This small, but visible, fleshy appendage contains proteins, sugars, fats, and vitamins.

- The ants carry the seed and elaiosome back to their nests and use the elaiosome to feed other workers, or their developing larvae. While the elaiosome gets eaten, the seed is generally unharmed, and the seed will germinate in the waste pile of the ants' nest, where the seed is protected from predators and already has a source of nutrients surrounding it. Common examples of plants dispersed by ants in North America include many violets and hyacinths.

READINGS

Fenner and Thompson, *The Ecology of Seeds*.

Hanson, *The Triumph of Seeds*.

Kesseler and Stuppy, *Seeds*.

QUESTIONS

1. How does a seed know when to germinate?

2. Find some seeds in your neighborhood. Can you think of how they are dispersed? If it's wintertime, look on the Internet for pictures of seeds of local plants and determine their dispersal method.

LECTURE 15 TRANSCRIPT
PLANT SEEDS GET AROUND

Botany is not typically a violent field of academic study, and yet, there has been a plant that has been described as a botanical hand grenade. When this grenade goes off, it can damage animals and structures up to 30 meters, which is about 98 feet.

Steven Vogel, who virtually started the field of biomechanics, describes the tree in his paper entitled, "Living in a Physical World: The Bioballistics of Small Projectiles." He writes, "The *Hura crepitans* launches with an audible pop at prodigious speeds, as high as 70 meters per second." He goes on to note that, "Curiously, this fastest speed known in the plant kingdom is indistinguishable from the maximum speed in the animals, the dive of a falcon."

To understand why this tree would possibly launch its seeds like a cannon ball, we need to understand some botanical terms about seed dispersal. Fruits come in many forms. If fruits are fleshy, like a peach, then they are called that botanically, fleshy. The peach is a drupe, which is a particular kind of fleshy fruit with one seed, the pit, in one carpel in one ovary.

However, not all fruits are fleshy. Many fruits are dry at maturity. When they become completely ripe, these dry fruits either split open or they don't. Dry fruits that split apart at maturity are called dehiscent, and only dry fruits will split, as fleshy fruits will simply decay. Those dry fruits that never split are called indehiscent. To keep this straight—which is which—if you can imagine hearing da hissing sound as the dry fruit opens, then that's a dehiscent fruit that does split apart.

Launching the fruits a far distance ensures that the seeds won't try to germinate right beneath the parent tree and then compete with it. You may be wondering, what about apple trees? They've been demonstrating gravity for a long time by dropping straight down. Yes, but those apples are so tasty that some animal is going to come by and eat them up and then get on with its life, and eventually do its business somewhere else. And later, those apple seeds are ready to germinate, away from the parent tree, with a heaping pile of compost.

Our *Hura crepitans* tree, however, doesn't have tasty fruits to eat. It produces a fruit that is botanically classified as a capsule, which means that at maturity, it's a dry dehiscent fruit that develops from two or more carpels, and you may recall from the fruit lecture that carpels are divisions of the ovary. A capsule also has many seeds and splits along or between the carpels.

There are also capsules that have pores on the top, where the seeds come out, and these are called poricidal capsules, a poppy is an example. In fact, our explosive tree, *Hura crepitans*, is commonly called the sandbox tree because the fruit was used to make small bowls to hold sand as a blotter for pens before the existence of blotting paper. Nowadays, they are purported to be filled with molten lead and used as a paperweight, but I couldn't find one on Amazon. So, the fruits look innocuous enough, in fact, they've been described as looking like small pumpkins, but we know the pumpkin is not a dehiscent fruit in that it isn't dry at maturity. The pumpkin is a fleshy fruit, a type of berry, called a pepo.

These fruits on the sandbox tree, though, will dry out, and as they dry out, they begin to split along the carpels. As the fruit dries, the cells become smaller and smaller because the water has left. The seeds will split from the wall of the fruit, and if harvested at this time, the fruit will make a sound like a rattle with the loose seeds inside. Eventually, the cells will break and the seeds will shoot out of the fruit. The seeds are discoid shaped, which reduces their wind drag and helps them go even farther.

The explosion is strongest where the weather is dry because that will enhance the breaking of the cells. Thus, the fruits will dehisc and disperse seeds in the dry season. And even though this tree is native to the American tropics, the record flight was actually recorded in Ghana, in West Africa, which also experiences a dry season, like many of the American tropics.

Clearly, these ballistic seeds could do some damage if you, or some other plant, got in their way, but the sandbox tree has other ways to be dangerous, too. The fruit, before it dehisces, is poisonous to eat. This tree is in the *Euphorbiaceae* family, and most plants in this family produce a milky latex when the punctured. This latex is also poisonous, causing an extreme rash if it gets on the skin, or death if it gets under the skin, which is why indigenous peoples used it for poison arrows. The moral of this story is no matter how cool this tree might sound, it's not a good tree for landscaping.

The sandbox tree isn't the only plant to have explosive seeds, though. Violets are another plant that has dehiscent capsules, though their projectile capacity isn't nearly as great. There is also a plant called the touch-me-not that when disturbed by a breeze or a touch from an animal, will also explode its seeds. These seeds, though, are much smaller and pose no threat of injury. In fact, a number of plants employ this drying-out mechanism for ejecting their seeds. This mechanism is given the botanical name *xerochasy*. The prefix xero- means dry in Greek.

But, there's another way plants can explode. They can use *hygrochasy*, and this is opening by water. This mechanism still involves the loss of water. It occurs when enlarged cell walls shrink with water loss and then expand with its uptake. Then, these increases in the swelling tissue create pressure because this swelling tissue is attached to lignified, non-swelling resistance tissue. When the pressure becomes too great, the fruit explodes. Interestingly, many plants in arid lands have this dispersal mechanism. Although this seems counterintuitive, if there is adequate rainfall for *hygrochasy* to occur, then seeds will be dispersed when there is enough water for seeds to germinate.

Hygrochasy doesn't just occur in dry fruits. There can also be fleshy fruits that explode and even different fruit types in the same family. For example, we just said that the pumpkins, which are in the cucumber family, the *Cucurbitaceae*, are fleshy fruits, but there are two species in this family, which also have exploding fruits. There is the exploding cucumber and the squirting cucumber.

The squirting cucumber is like something out of a movie. The scientific name of the squirting cucumber is *Ecballium elaterium*. This is from the Greek *ekballein*, meaning to throw out, as the fruit throws out the seeds when it squirts. The cells in the wall of the fruit are under a lot of turgor pressure, they are jam packed full of water. These cells then expand rapidly when the fruit breaks its connection with the stem. Botanists know that the plant hormone ethylene, which is released as a gas, is responsible for the ripening of fruit, and it is also responsible for this particular fruit exploding off the stem. Ethylene causes the cells of the abscission zone at the end of the stem to detach so that when the hygrostatic pressure is strong enough, the cells are not held together, and the fruit leaves with a strong ejection.

These are just the way they sound, the exploding cucumber is touched, and then explodes, while it still looks like an innocent wild cucumber. The exploding cucumber has an interesting looking seed, like a jigsaw puzzle piece. In this fruit, the cells with high pressure are on the inside of the fruit, so that when touched, it explodes from the inside ejecting its seeds.

This is just one of the many ways that plants disperse the seeds. There are plusses and minuses to each type of seed dispersal. If the seed doesn't go too far, as in the case of our exploding seeds, then there is a chance that where it lands is a pretty good place to germinate, since it's not too far from the favorable habitat where its parent germinated. This is why we sometimes see clumps of a certain type of plant in nature. The conditions in that area are good for growth of that species, and the seed didn't have to travel very far.

On the other hand, too many individuals in one spot can lead to competition in that spot. Plus, maybe—and of course, the plants don't know this—but maybe the parent plant is a spot that might be good for an adult, but not so great for germination. In this case, the seeds might be better off if they are dispersed a greater distance. For these greater distances, plants can't rely on fruit explosions alone, and there is a myriad of ways that plants will disperse their seeds.

Before we look at those ways, let's investigate the structure of the seed. As we mentioned when we discussed gymnosperms, the evolution of the seed was a major advantage for land plants. But unlike gymnosperms, which had a naked seed, the flowering plants will produce a fruit around that seed, which will either aid in germination, or dispersal, or both. The seed includes an entire embryonic plant, complete with an immature leaf, the cotyledon. The cotyledon also stores starch so that the embryonic leaves can grow. Some flowering plants, the eudicots, will have two of these immature leaves. The embryonic plant also has an immature root, called a radicle. A seed has germinated when the radicle first emerges from the seed.

But, before germination, the seed is dispersed. Like pollination, there is a myriad of different ways fruits, with the seed or seeds inside of them, can be dispersed. For a close up look at the physical beauty and diversity of seeds, Rob Kesseler and Wolfgang Stuppy have a lovely coffee table book entitled *Seeds: Time Capsules of Life*. The book not only describes the formation, evolution, and dispersal of seeds, it also depicts them with artistic grandeur.

We'll begin with the abiotic, or nonliving, dispersal methods, wind and water, and we already talked about explosions, Many fruits have very small seeds that are dispersed by the wind. Also, just because a plant has animal pollinator, that same plant may have a different dispersal method. For example, in the orchid family, the *Orchidaceae*, is animal-pollinated but then wind dispersed. Think of the tiny seeds inside the vanilla pod, a fruit that is botanically a dehiscent capsule, just like our sandbox tree. But, just because the fruit dehisces doesn't mean that it explodes, the capsule dries up and opens, and then the wind carries the seeds away.

A wind-dispersed fruit that everyone knows is the dandelion. One reason a dandelion can produce so many seeds is that it can produce seeds without pollination, and this is called *apomixis*. This form of reproduction is asexual, but the clones seeds are derived from ovules that would have given way to sexual reproduction had they not skipped pollination and gone straight to seed development.

This is also a good time when we talk about the dandelion to introduce its family, the *Asteraceae*, also known as the sunflower family. The name of the family comes from the genus aster, which means star. And, these flowers look like stars, This family is very recognizable because it has a very distinctive flower type, called a head.

A daisy is another example. Let's look at the white petal-looking structures that surround the yellow center. Each one of these petals look-alikes is actually a whole flower, and if you look at it closely at that yellow center, you'll see that it is actually a collection of small flowers. The petal-like flowers are called rays, and the tightly-packed flowers in the middle are called tubes.

Sometimes, members of this family will only have tube flowers, like thistles. Other times, they will have only rays, like chicory. The *Asteraceae* rivals the orchid family for total numbers of species. It's estimated that there are around 23,000 different species in the sunflower family. Aside from sunflower seeds, lettuce and artichokes are the main agricultural products of this family. Of course, these flowers are grown as ornamentals as well.

There is a great thing and a terrible thing about the asters. On the one hand, they are very recognizable. Only a few other families have the composite of flowers arranged in a head the way the sunflowers do. So, to a botanist, a member of this family is instantly recognizable. However, keying out these plants to species level is difficult. There are many, many genera—over 1,600 genera—so even that sometimes can be a challenge.

For this reason, some botanists refer to these plants as DYCs for Damn Yellow Composites. Many of the most difficult ones to identify are, in fact yellow, and composite refers to the fact that there is a composite of flowers on the head. In fact, the old name for this family used to be *Compositae*.

The main fruit type of the sunflowers is called a cypsela, and they are perfect for riding on the wind with their own mini-parachutes, A cypsela is basically like an achene, except that it is derived from an inferior ovary, whereas an achene is derived from a superior ovary. The classic cypsela is what we call a sunflower seed. When it's in the shell, it should be called a sunflower fruit. The seed is the tasty nugget inside the hard shell.

The sunflowers produce a lot of weedy species in addition to dandelion because of their very efficient dispersal mechanism. Also, the sunflowers tend to be drought tolerant, which makes them pretty hardy growers in disturbed areas. Incidentally, there is no botanical definition of weed. That word is a social construct. A weed is a plant that's anywhere that you don't want it to be. Botanists do think about invasive plants, those that are not native to an area and spread rapidly.

Of course, sunflowers aren't the only plants that are dispersed by the wind. And a parachute is just one way to use the wind. A plant could also produce cottony seeds that easily blown around. Clearly, cotton is a great example of this dispersal method, and so is the kapok tree, which produces so much cottony fiber that it was used as mattress filling. Where I live, in Colorado, there is a native tree called the cottonwood, and in the early summer, it's really easy to see why it got that name. It produces copious cottony seeds that get everywhere. The milkweed is another great example of wind-dispersed cottony plant.

Next, to blowing cypselas off a dandelion and throwing a milkweed pod into the air to watch it release its seeds, a favorite dispersal method for kids to play is the winged fruit.

Who hasn't seen the winged fruit of a maple tree and wanted to release it from some height to watch it whirlybird down to the ground. The botanical name for the maple fruit is a double samara. I think samara would make a great band name or a company name. It's not taken that I know of, but there is a visually interesting documentary by the same name,

There are other winged fruits too. Elms also have winged fruits. Since the botanical description of a samara is an achene with a wing on it, elms also have samaras, even though they are round and not elongated like a maple, and they aren't a double samara, just a single one. Maybe the largest winged fruit is from the cuipo tree in South America. These are also giant trees, often left uncut after forest clearing because they grow so large.

One last means of wind dispersal is found in the tumbleweeds. This dispersal mechanism is convergent, meaning it's found in a number of different plant families—about a dozen plant families—that employ this method of moving the seeds around as the bush like plant gets blown across the landscape. The well-known baby's breath from flower arrangements actually forms a tumbleweed as a dispersal mechanism.

Ironically, the most well-known tumbleweed in western landscapes and films is Russian thistle, which is from Russia, but it's not a thistle, not even in the sunflower family, where thistles are located. This annual plant does have spiny growth on it, making it prickly, like a thistle. When it dies, the base of the plant breaks off and forms a tumbleweed, dispersing its seeds as the wind rolls it along. Russian immigrants probably brought it here in flax seed they brought over during the 1870s. Though it's considered an invasive weed, it probably saved numerous cattle during the dust bowl when no other forage was available.

After the wind, the other abiotic disperser is water. Plants don't have to grow in water to be dispersed by water. Plants that grow near water often use water as their dispersal agent. These plants will often produce fruits that are buoyant because of their small size or because of their structure. The coconut has an outer seed coat that is impermeable to water, and so the coconut fruit can travel across the oceans.

Sea beans are a common name that refers to seeds and fruits, like the coconut, that are produced one place and ride the ocean currents to end up someplace entirely new. They are also called drifter seeds. I'm not crazy about either name, as many of these fruits are not botanically beans, such as the coconut, which is a drupe, that means it's one seed, what we think of as the whole brown coconut inside a larger husk, which we don't see. Drifter seeds isn't so great either because mostly it's the fruits that get dispersed. There are actually over 200 species of plants that are dispersed in this way—that is, they'll form what I'll call drifter fruits. Most times, these drifter fruits are produced by trees in the tropics. These trees live beside rivers and are big enough to produce large seeds that are carried downriver, and then out onto the currents.

Apparently, many people enjoy collecting these drifter fruits and using them for art. There is a whole organization devoted to this interest with a newsletter called *The Drifting Seed*, and a website, too, Drifter fruits have also been used by oceanologists to map out the currents.

Perhaps the most complex mode of dispersal involves animals, and there are two major strategies for plants here. The first, be a delicious, irresistible, fruit that animals can't wait to eat. And, the second strategy is to fashion yourself into some sort device able to attach to animal hair and hitch a ride.

Now, even though insects do the lion share of the work in pollination, vertebrates, especially birds and mammals, are doing the majority of the dispersal effort carried out by animals. The seeds produced by many fleshy fruits will pass through the digestive tract of the bird or other vertebrate without being harmed. Some plants with fleshy fruits actually have seeds that must pass through an animal's digestive tract to weaken the seed coat.

That is the process that will start germination—the weakening of the seed coat. As the seed coat becomes weaker and weaker, it will become permeable to water, and this will cause the cells inside to swell and grow, and then the radicle will emerge. And that's when germination has occurred.

An example of a seed that must pass through an animal is a blackberry or a raspberry. The fruits are each individual spheres, and inside each one is a tiny seed. These are the ones that get caught in your teeth when you eat a blackberry. These seeds need to be roughed up in a bird's gizzard before they can germinate. In fact, gardeners who grow these plants from seeds, have to scratch these seeds in some way, usually using sandpaper, to weaken the seed coat and ensure germination. Scratching—whether in nature, or artificially—is called scarification, and many seeds have to be scarified in some way before germinating.

Additionally, some seeds have to go through a period of chilling before they will germinate. This is called vernalization. In temperate regions, where vernalization is common, seeds are sent off in the spring, summer, and sometimes late summer to find their way in the world. In most temperate regions, mid to late summer is not a great time to germinate because there won't be enough moisture. When precipitation comes in abundance in the fall, the temperatures are too cold to get a good start. So, seeds have a large dose of a hormone called abscisic acid in their tissues. This hormone prevents germination. With a period of prolonged cold, enzymes slowly degrade the abscisic acid, and the seeds are ready to germinate in the spring.

Plants that hitch a ride often have barbs or hooks to ensure that they hold for a little while. In fact, it was burdock, a biennial aster plant that was the inspiration for velcro. Apparently, a Swiss engineer, Georges de Mestral, went walking in the woods and received a bunch of the fruits all over his pants. We would call them burs, and, he thought that maybe there was something to learn from the burrs—something useful. De Mestral successfully reproduced the two fabric strips that would stick together. He called it velcro—a blend of velvet and crochet.

Another hook plant that is very unusual is the Devil's Claw plant in the genus *Proboscidea*, which grows in the desert Southwest. It looks almost like a steer's head with two very long curving spikes. The fruit of this plant is a capsule. As the fruit dries, the capsule splits and two of these long cylindrical splits curve back to become the horns. These horned fruits are pretty large, so they're definitely not dispersed by jackrabbits.

Therein is the problem. There's rarely a native animal that is big enough to disperse the Devil's Claw. But 13,000 years ago, all sort of large animals roamed the plains of North America. Horses and camels both evolved here but then became extinct. There were other sorts of animals too, a giant sloth, and an elephant-sized animal called a gomphothere.

The same hypothesis is presented for large fleshy fruits, too. Consider the osage orange. It's the size of a softball. What animal disperses the osage orange? None, though squirrels will eat it, they don't do much dispersing. And the same can be true for the Kentucky Coffee Bean tree. Dr. Dan Janzen was the first to hypothesize about why some fruits are so huge, and yet have no dispersers. Dr. Janzen spent most of his time in Costa Rica, where he noticed a group of mid-sized fruits was ignored by the native mammals but eaten with relish by introduced livestock. He then came up with the theory of ecological anachronism. This means the large, undispersed fruits, are simply a holdover from a time in the past.

Birds and mammals seem likely fruit dispersers. Mammals are like us, and we like fruit. We've seen birds eating fruits, and we've seen seeds in their droppings. Other vertebrates, like tortoises and iguanas, eat fruits and disperse seeds, too. And so do fish. Fish? Yes, there was a seminal paper in 2008 that described just how important a fish—a pacu—was for dispersal of many trees. These fish live in the Brazilian Pantanal—the largest freshwater wetland in the world. These fish will swim far inland during the rainy season to eat fruits dropped by trees along the wetlands. Researchers found the seeds of 141 trees in the guts of the animals they examined.

Even though vertebrates do most of the fruit dispersal by animals, a good number of plants also employ ants for this purpose, so we would be remiss if we didn't mention such plants here. Plants that have ant-dispersed fruits have a special growth on the outside of their seeds called an elaiosome. This small, but visible, fleshy appendage contains proteins, sugars, fats, and vitamins—it's a complete meal. The ants carry the seed and elaiosome back to their nests and use the elaiosome to feed other workers or their developing larvae. While the elaiosome gets eaten, the seed is generally unharmed, and the seed will germinate in the waste pile of the ants' nest. Here, the seed is protected from predators, and already has a source of nutrients surrounding it.

Some plant communities have up to 30% of their seeds dispersed in this way, particularly forest understorey plants. Common examples of plants dispersed by ants in North America include spring beauty—*Claytonia virginica*—many violets and hyacinths, and Trillium.

It's interesting to point out that in evolutionary terms, there isn't a hierarchy of seed dispersion methods. That is to say, it's not as if the wind is a primitive way to spread seeds, and fruit production is a more evolved dispersal mechanism. It's true that the ferns, which are definitely more primitive than the flowering plants, do use wind dispersal for their spores. But gymnosperms use wind and animals for dispersal. Basal angiosperms, those thought to be the oldest of the flowering plants, mostly use animal dispersal, not wind at all, while grasses, which are much more recent, do use wind dispersal. The compound flowers of *Asteraceae* are thought to be one of the more recent families to evolve, but these are very recent and very complex plants that are almost exclusively wind dispersed.

So, in terms of seed dispersal, there isn't any older way or newer way, there is the way that works, even it means exploding to get your seeds out there.

LECTURE 16

WATER PLANTS CAME FROM LAND

Plants started in shallow water, like a freshwater algae, and then eventually moved to land, where flowering plants dominated—but then a few came back to the water. The approximately 70 species of seagrasses are distributed in 4 different families of flowering plants. Most likely, these groups diverged from a single monocot flowering plant around 100 million years ago.

SEAGRASSES

- Despite being in different families with growth forms that vary from grasslike to very leafy, all seagrasses grow in similar habitats. They grow in sandy or muddy bottoms of shallow coastal waters. They all deal with the same stresses, including low light levels, sediments that are anaerobic and nutrient poor, and the difficulty of getting carbon dioxide into the leaves and stems.

- Part of the reason that seagrasses face low light levels even though they're in shallow water is because areas just off the coast usually receive a fair amount of sediment, which diffuses light as soon as it enters the water. Some of the sediment will land on the seagrass and limit light levels even further.

- Light levels are also low because there are small algae growing on the seagrasses. Algae that grow on seagrasses will also grow on almost anything else, living or dead, within the ocean. This type of algae is called periphyton, and it absorbs sunlight, further reducing the amount of light that makes it to the seagrass.

PERIPHYTON GROWING ON SEAGRASS

- Seagrasses deal with these low light levels by putting their chloroplasts, which is where photosynthesis occurs, in the epidermis, the outermost layer, whereas most plants have their chloroplasts internal to the epidermis. Seagrasses have chloroplasts in their stems, too, not just their leaves.

- As a further adaptation to low light levels, seagrasses are rhizomatous, like many terrestrial grasses. The advantage of the rhizome, an underground stem, is that it can store sugars for times when light levels are so low that photosynthesis can't take place. In fact, seagrass roots undergo fermentation, just like yeast, after they've been in the dark for a few hours. And, like brewer's yeast, they release alcohol, ethanol, as a waste product into the water around them.

- Another issue that seagrasses have to face is an anaerobic substrate. It's not actually soil because it's underwater. The substrate around the roots is anaerobic—it has no oxygen. This is a problem because roots don't photosynthesize. The cells in the roots have mitochondria, just like human cells do, and these mitochondria need oxygen, just like humans do. If roots become waterlogged and can't get oxygen, they die, and the plant dies.

- Seagrasses have an amazing adaptation to deal with this. The leaves and stems of seagrasses are photosynthesizing during the day, and photosynthesis produces oxygen as a waste product. Seagrasses release this oxygen back into their own tissues, which are cavernous and full of small spaces. The oxygen then diffuses down to the roots because the roots don't have any oxygen, so there's a lower concentration of oxygen there. This transport ceases at night, though, because photosynthesis doesn't work in the dark, which is why the roots undergo fermentation after a while.

- Not only is the substrate devoid of water, but it also lacks nutrients. Marine phytoplankton tend to absorb nutrients in the seawater. As plankton die and fall to the seafloor, their decomposition provides some nutrients in the substrate, but these are often consumed by bacteria. To maximize nutrient uptake, seagrasses absorb minerals from ocean water and the substrate. Seagrasses also have bacteria that live in their roots that can break down nitrogen into a usable form.

- There is yet another hurdle that seagrasses have to jump to live underwater. Seawater doesn't have much dissolved carbon dioxide. Most of the carbon dioxide that is in seawater is in the form of the bicarbonate ion, so seagrasses use active transport to get the bicarbonate ions into the tissues, where an enzyme called carbonic anhydrase converts it into carbon dioxide so that it can be run through the Calvin cycle to make sugars as part of photosynthesis.

- Seagrasses, like all plants living in a marine environment, also have to deal with copious amounts of salt. Seagrasses cope with high salt levels by increasing the solute concentration in their tissues. By concentrating salts in their tissues, it can be even saltier than the ocean water, which ensures that it doesn't lose water to the surrounding environment.

- Seagrasses are mostly dioecious, with separate male and female plants, so pollen has to travel through the water column to get to a flower on a female plant.

- One genus of seagrass, *Amphibolis*, has some of the largest pollen among the flowering plants. This long pollen can travel to the female plant in 1 of 3 ways.

    - Pollen can float to the surface of the water when tides are their lowest and drop pollen from there onto (hopefully) female plants below.

    - Pollen can conglomerate on the top of the substrate, whether it is sand or mud, and form a collective tower, which can reach a meter in length. Because many seagrasses grow in colonies, a meter is enough distance to ensure meeting a female flower and completing pollination—again, hopefully.

    - A third method, which is the same mechanism used by many marine animals, involves broadcasting sperm into the ocean water and, again, hoping for the best. This is called hydrophilous pollination.

- Once pollination has occurred, fruits develop and then need to be dispersed. In the seagrasses, dispersal entails floating fruits, mainly achenes or capsules, being carried in water currents. Oddly, none of the seeds float, so once the fruit settles, or is eaten, the seed will attempt germination where it lands. Germination can happen within a few hours or a few years.

MANGROVES

- Another flowering plant that lives in the ocean—but isn't submerged and is a tree—is the mangrove. There are about 55 different species of trees and shrubs that grow as mangroves do, in the ocean, and don't occur elsewhere.

- As with seagrasses, the condition of being a mangrove appears to be convergent; that is, the character to live this way is found in 16 different families. Because these families are unrelated, the traits enabling a tree to be a mangrove arose in evolution multiple times as an adaptation for living in seawater.

- Mangroves have a more restricted range than seagrasses. They are limited to tropical and subtropical coasts, whereas seagrasses can extend into the temperate ocean as far as the southern coasts of Greenland and Norway.

- Mangroves also live in shallow water because their crowns are exposed to air. Even though they can do normal gas exchange like terrestrial plants, they still have a suite of unique features to live in the ocean.

- Like seagrasses, mangroves have to deal with anaerobic substrates and roots. Some species have the advantage of living in tidal areas that are periodically free of water, and in these habitats, mangroves produce growths from their roots that go straight up. Most roots grow down, but these special structures in mangroves are negatively geotropic, so they grow upward. The growths are pneumatophores, and they carry oxygen to the rest of the root system.

- The pneumatophores are covered with small openings called lenticels. Because there is a lower concentration of oxygen inside the root than outside, the oxygen diffuses into the pneumatophore. This only works when the tide is out and the pneumatophores aren't covered with water.

- Mangroves have other root adaptations besides pneumatophores. They have prop roots and drop roots. Prop roots grow from the trunk and look like they are propping up the tree, and they are helpful for support, but these roots also transport oxygen to the submerged roots, like the pneumatophores do. Drop roots drop down from the branches of the tree and serve the same function as prop roots—to anchor the tree and increase oxygen to the submerged roots.

- Like seagrasses, mangroves also have much aerenchyma tissue, which means that the internal tissue contains many spaces for holding air, both in the roots and in the stem. In addition, many mangroves have rather spread out and shallow root systems so as to avoid very anoxic, deeper substrates.

RED MANGROVE
Rhizophora mangle

- Also like seagrasses, mangroves have ways of coping with the salinity of ocean water. Mangroves are facultative halophytes, which means that they can grow in either freshwater or ocean water. However, because of their unique abilities to live in saltwater, there is less competition there, and saltwater is where they are mostly found.

- Facultative halophytes, including mangroves, can either be salt excluders, which don't take salt into their roots, or salt secretors, which do take in the salt but then excrete it, usually on the leaves. To exclude salt from entering the roots, mangroves concentrate other ions in their root tissues that will enable the water to enter the root, but not the salt. The excretion of salt may be one reason why mangroves don't keep their leaves very long.

- Perhaps the most interesting adaptation of mangroves is that they are viviparous, which means that the fruits of mangroves have a seedling that is already developed. In this way, they don't disperse fruits or seeds—they disperse seedlings.

- Mangroves do this because of the harsh conditions of germination. A mangrove seed would have to land in an appropriate substrate and anchor in by the time the tide came back in or it would float away. Seedlings can grow as much as 70 centimeters before they drop off the parent plant. By that time, they are ready to hit the substrate and root. In some species, these seedlings are contained in a long cylindrical casing that can float, and they can root and be upright in the substrate before the next tide.

- The seagrasses and the mangroves make up the only 2 groups of truly marine vascular plants. Interestingly, the ferns and gymnosperms don't have members that live in the ocean. They are confined to the land or freshwater. Many angiosperms live in freshwater systems and are usually called aquatic plants or water plants.

ALGAE

- All of the other plantlike organisms in the ocean are not true plants. They are algae, which are in the protist kingdom. Algae are not considered as part of the plant kingdom, primarily because the zygote, also known as the fertilized egg, does not develop in a special chamber. That said, algae are often grouped with plants because they both photosynthesize.

- Algae come in different colors—green, brown, and red—and this is one way that they are grouped taxonomically. Red algae are an important food source in Asia and have a high protein content. Giant kelp is a brown algae, and next to coral reefs, kelp forests make up the second most diverse and productive ocean community.

- As interesting as these marine organisms are, in terms of vascular plants, there are many more species that live in freshwater. Though some families are dominated by aquatics, there are many more aquatics scattered throughout other groups of ferns and flowering plants.

READINGS

Hogarth, *The Biology of Mangroves and Seagrasses*.

Mabey, *Cabaret of Plants*.

QUESTIONS

1   Why do seagrass and mangroves have different global distributions?

2   Why do you think there aren't more flowering plants, or any conifers at all, living in the ocean?

LECTURE 16 TRANSCRIPT
WATER PLANTS CAME FROM LAND

When I was starting out as a beginning faculty member, I received a fellowship to spend two weeks at the National Tropical Botanic Garden in Coconut Grove, FL. And, it turned out to be a great experience. On this trip, we went snorkeling on seagrass beds, and this was when I finally got to understand seagrasses. I assumed it was some sort of algae, like a kelp, only smaller. I was shocked to find out that this was a flowering plant that grew underwater. I was shocked both at the uniqueness of a plant doing this, and that I had gone to graduate school on a coast, and I didn't know about seagrasses. Such was my terrestrial bias.

What's amazing is that plants started in shallow water, like a freshwater algae, and then eventually moved to land, where flowering plants dominated, but then a few came back to the water. So, in a sense, seagrasses are like whales. The closest living relative to whales are hippos. So, sometime around 50 million years ago, an ancestor of the whale, a terrestrial mammal, went back into the ocean.

Some seventy species of seagrasses are distributed in four different families of flowering plants. Most likely, these groups diverged from a single monocot flowering plant around 100 million years ago.

Despite being in different families with growth forms that vary from grass-like to very leafy, all seagrasses grow in similar habitats. They grow in sandy or muddy bottoms of shallow coastal waters. They all deal with the same stresses, low light levels, sediments that are anaerobic and nutrient-poor, the difficulty of getting CO_2 into the leaves and the stems.

Why would seagrasses be facing low light levels if they're in shallow water? Part of the reason is areas just off the coast usually receive a fair amount of sediments. Think about being by the coast near a river after a rainfall. There is sometimes a plume of sediment visible in the water. This sediment diffuses light as soon as it enters the water. Some of the sediment will land on the seagrass itself and limit light levels even further.

Light levels are also low because there are small algae growing on the seagrasses. This is like the algae that can accumulate in a fish tank. Algae that grow on seagrasses like this will also grow on coral and most anything else living or dead within the ocean. This type of algae is called periphyton, *peri* from the Greek word that means around or about. *Phyto* is Greek for plants, and even though algae aren't plants, they do photosynthesis. These periphyton algae will absorb sunlight too, further reducing the amount of light that makes it to the seagrass.

But, seagrasses have a way of dealing with these low light levels. They put their chloroplasts in the epidermis, the outermost layer. Most plants have their chloroplasts, which is where photosynthesis takes place, internal to the epidermis in a layer called the mesophyll. Seagrasses have all but done away with their mesophyll—there's just a tiny bit of it. All the action is in the epidermis. And, seagrasses have chloroplasts in their stems, too, not just their leaves.

As a further adaptation to low light levels, seagrasses are rhizomatous, like many terrestrial grasses. The advantage of the rhizome, an underground stem, is that it can store sugars for times when light levels are so low that photosynthesis can't take place. In fact, seagrass roots will actually undergo fermentation, just like yeast, after they've been in the dark for a few hours. And, like brewer's yeast, they will release alcohol, ethanol, as a waste product into the water around them. Though I've not heard of a seagrass ale just yet, I'd be willing to try it.

Another issue that seagrasses have to face is an anaerobic substrate. I call it a substrate because it's not actually soil because it's underwater. The substrate around the roots is anaerobic—it has no oxygen. And this is a problem because roots don't photosynthesize. The cells in the roots have mitochondria just like we do, and these mitochondria need oxygen, just like we do. If roots become water-logged and can't get oxygen, they will die, and the plant will die. You may have seen this in action when an area is flooded and the plants with their roots submerged in the water can die if the water stands for too long.

Seagrasses have an amazing adaptation to deal with this, too. The leaves and stems of seagrasses are photosynthesizing during the day, and photosynthesis produces oxygen as a waste product. The seagrasses release this oxygen back into their own tissues, which are cavernous and full of all these small spaces called lacunae. The oxygen then diffuses down to the roots because the roots don't have any oxygen, so there's a lower concentration of oxygen down there. This transport ceases at night, though, since the photosynthesis doesn't work in the dark, which is why the roots will undergo fermentation after a while.

Not only is the substrate devoid of oxygen, it also lacks nutrients. Marine phytoplankton tend to absorb nutrients in the seawater. As plankton die and fall to the sea floor, their decomposition provides some nutrients in the substrate, but these are often consumed by bacteria. To maximize the nutrient uptake, seagrasses absorb minerals from ocean water and the substrate. The seagrasses also have bacteria that live in their roots that can break down nitrogen into a usable form, just like soybeans and all the other members of the pea family, the *Fabaceae*.

There is still yet another hurdle seagrass have to jump in order to make this whole living underwater thing work for them. Seawater doesn't have much dissolved carbon dioxide. Most of the carbon dioxide that is in seawater is in the form of the bicarbonate ion, or HCO_3. So, seagrasses use active transport to get the bicarbonate ions into the tissues where an enzyme called carbonic anhydrase converts it into CO_2 so that it can be run through the Calvin Cycle to make sugars as part of photosynthesis.

Lastly, seagrasses, like all plants living in a marine environment, have to deal with copious amounts of salt. Imagine what leftover salad looks like after it's already been dressed. The dressing has a lot of solutes in it. Think about the solute concentration in the salad dressing versus that in the lettuce. Since the lettuce is mostly water, it's water concentration is pretty high. So, how is the water going to move? Does it move into, or out of, the lettuce leaf?

If you've ever had day old dressed salad, then you know the lettuce leaf is wilted all to heaven, so the water has left the lettuce. Now, think of our seagrass, which is growing in the ocean, which is even saltier than a delicious garlic Caesar. We would have to think the water would move out of the seagrass, but that can't be because then the seagrass wouldn't survive.

The seagrass copes with high salt levels by increasing the solute concentration in its tissues. By concentrating salts in its tissues, it can be even saltier than the ocean water, which ensures that it doesn't lose water to the surrounding environment.

All this existence is well and good, but we know plants also have to reproduce to persist in the world. How seagrasses reproduce is pretty interesting. How would pollen travel in water? Seagrasses are mostly dioecious, with separate male and female plants, so pollen has to travel through the water column to get to a flower on a female plant. The flowers are only about one or two centimeters long. These are not the smallest flowers in the world, although the smallest flowers in the world do belong to another species of aquatic plant, the tiny little duckweed, a flowering plant that lives on the top of ponds with flowers smaller than one millimeter in diameter.

But, how does that seagrass pollen move from the male plant to the female plant? It can't swim. Pollen grains don't swim. Remarkably, one genus of seagrass, *Amphibolis*, has some of the largest pollen amongst the flowering plants. The pollen strands have been recorded up to 3 mm long. Most land plants have pollen about one tenth of a millimeter long. This long pollen can travel to the female plant one of three ways.

First, the pollen can float to the surface of the water when tides are at their lowest and drop pollen from there onto, hopefully, the female plants below. Secondly, pollen can conglomerate on the top of the substrate be it sand or mud and form a collective tower, which can reach a meter in length. Imagine a meter tall, underwater, pollen tower—amazing. As many seagrasses grow in colonies, a meter is enough distance to ensure meeting a female flower and completing pollination, again hopefully. A third type of pollen dispersal—and this is the same mechanism used by many marine animals—involves broadcasting sperm into the ocean water and, again, hoping for the best. This is also called hydrophilous pollination.

Once pollination has occurred, fruits develop and then need to be dispersed. In the seagrasses, dispersal entails floating fruits, mainly achenes or capsules, being carried on the water currents. Oddly, none of these seeds float, so once the fruit settles, or is eaten, the seed will attempt germination where it land. Germination can happen within a few hours—or a few years. It's hard to have rules in botany.

If cows eat grass, what eats seagrass? Sea cows. But, it's true, actually. Manatees, in Southeast Asia, dugongs, are large sea mammals that graze on the seagrass beds. Although grasses and seagrasses are not at all related, they do have some ecological similarities, mainly that they support other creatures, like the manatees, and that both communities are highly productive. In botanical language, productive means that collectively the plant community is doing a lot of photosynthesis and producing a lot of biomass over a given time. In my Master's defense, when I said biomass, one of my committee members, Bill Lunch, quipped, "Is that a bunch of biologists going to Catholic Church?" It is just what it sounds like, the mass produced by living organisms.

The production of all this biomass means photosynthesis. It's been estimated that one square meter of seagrass can produce 10 liters of oxygen a day. Seagrasses can also filter nutrients out of the ocean, which is good because sometimes, there is a nutrient overload where rivers meet the ocean. These rivers accumulate excess fertilizers that run off farms and even though sewage waste gets treated for bacteria, it will often have high levels of nutrients.

While a lot nutrients sounds like a good thing, too many of them can cause a process called eutrophication. This occurs when a lot of nutrients in the water stimulate algal growth. The algae grow and grow and get so thick they can block the sun to other organisms. Then, when the algae have consumed all the nutrients, they start to die. Next, bacteria start to proliferate as they decompose the dead algae. But, as they do this, they respire, a lot. That means that they will use up all the oxygen in the water, and the water becomes anoxic. When this happens, fish and other animals living in the water can die because there isn't enough oxygen. Eutrophication like this can happen in freshwater and marine systems.

Collecting excess nutrients isn't the only way seagrasses improve the water quality. They also help trap sediments, which helps to improve the clarity of the water. In addition to that, their roots can help keep sediments on the ocean floor during storm surges.

And, seagrasses provide a safe haven for small animals during storms, too. Seagrasses, like kelp forests and coral reefs, provide habitat and protection for animals. When storm surges occur, small animals can avoid drifting out to sea if they are within the protection of the seagrass bed. Also, like coral reefs and kelp forests, seagrass beds are nurseries to protect animals, especially babies. If your choice is to hide from a big fish on the open sandy seafloor or within the seagrass, you're going to get a lot more protection in the seagrass.

So these seagrass beds are important marine plant communities, but there is another flowering plant that lives in the ocean, and this plant isn't submerged, and it's a tree. It's a mangrove. It's actually about 55 different species of trees and shrubs that grow as mangroves do, in the ocean, and don't occur anywhere else. As with seagrasses, the condition of being a mangrove appears to be convergent, that is, the character to live this way is found in 16 different families. Because these families are unrelated, the traits enabling a tree to be a mangrove arose in evolution multiple times as a good adaptation for living in seawater.

Mangroves have a more restricted range than seagrasses. They are limited to tropical and sub-tropical coasts; whereas, seagrasses can extend into the temperate ocean as far as the Southern coasts of Greenland and Norway. Mangroves also live in shallow water because their crowns are exposed to air. Even though they can do normal gas exchange like terrestrial plants, they still have a suite of unique features to live in the ocean.

Like seagrasses, mangroves have to deal with anaerobic substrates and roots. Some species have the advantage of living in tidal areas that are periodically free of water, and in these habitats, mangroves produce growths from their roots that go straight up. Most roots grow down obviously, but these special structures in mangroves are negatively geotropic, so they grow upwards. These growths are *pneumatophore*, from the prefix pneumo- for lung and the suffix -phore for carrier because these structures will carry oxygen to the rest of the root system.

The pneumatophores are covered with small openings called lenticels. Because there is a lower concentration of oxygen inside the root than outside, the oxygen diffuses into the pneumatophore. Of course, this only works when the tide is out, and the pneumatophores aren't covered with water.

And mangroves have other root adaptations besides pneumatophores. They have prop roots and drop roots. Prop roots grow away from the trunk and look like they're propping up the tree, and they are helpful for support, but these roots also transport oxygen to the submerged roots, just like the pneumatophores do. Drop roots drop down from the branches of the tree and serve the same function as the prop roots—to anchor the tree and increase oxygen to the submerged roots.

Like seagrasses, mangroves also have much *aerenchyma* tissue, which means that the internal tissue contains many spaces for holding air, both in the roots and in the stem. In addition, many mangroves have rather spread out in sort shallow systems so as to avoid very anoxic, deeper substrates.

Also like seagrasses, mangroves must have ways of coping with the salinity of ocean water. Mangroves are facultative halophytes, which means that they can grow in either freshwater or ocean water; however, because of their unique abilities to live in saltwater, there is less competition there, and saltwater is where they are mostly found.

Facultative halophytes, including mangroves, can either be salt excluders or salt secretors. Salt excluders don't take salt into their roots where the salt secretors will take in the salt but then excrete it, usually on the leaves. To exclude salt from entering the roots, mangroves concentrate other ions in their root tissues that will enable the water to get into the root, but not the salt. The excretion of salt may be one reason why mangroves don't keep their leaves very long. One report noted that in Venezuelan mangroves, leaf half-lives were 60 days in *Laguncularia racemosa*, 100 days in *Rhizophora mangle*, and 160 days in *Avicennia germinans*. The half-life is when a leaf reaches the middle of its age.

But, I think the most interesting adaptation of mangroves is that they are viviparous. This literally means live birth, and it's true. The fruits of mangroves have a seedling that is already developed. In this way, they don't disperse fruits or seeds, they disperse seedlings. Imagine an apple tree with each apple growing a small seedling, and you get the picture. Why would mangroves do this?

The answer lies in the harsh conditions of germination. A mangrove seed would have to land in an appropriate substrate and anchor in by the time the tide came back in or it would float away. Seedlings can grow as much as 70 centimeters before they drop off the parent plant. By that time, they are ready to hit the substrate and root. In some species, these seedlings are contained in a long cylindrical casing such that they can float, and amazingly, they can root and be upright in the substrate before the next tide.

The seagrasses and the mangroves make up the only two groups of the vascular plants that are marine. Interestingly, the ferns and gymnosperms don't have members that live in the ocean. They are confined to the land or freshwater. Many angiosperms live in freshwater systems and can be called aquatic plants or water plants.

This might be a good time to remind ourselves why the other plant-like organisms in the ocean are not true plants. They are all algae, which are in the protist kingdom, the Kingdom Protista. Algae are not considered as part of the plant kingdom, primarily because the zygote, also known as the fertilized egg, does not develop in a special chamber. That said, algae are often grouped with plants because they both photosynthesize.

Algae come in a number of different colors—green algae, brown algae and red algae—and that is one way they are grouped taxonomically. Some people think red algae is bad because the type of algae causes a red tide. But, red tide is only caused by a few species of dinoflagellates, which are red algae. However, more broadly red algae are an important food source in Asia and have a high protein content. Giant kelp is a brown algae, and it's important to talk about here because next to coral reefs, kelp forests make up the second most diverse and productive ocean community.

Kelp look a lot like plants because they sort of look like they have leaves, stems, and roots. The leaf-looking structures are called blades. What looks like a stem is called a stipe. There are no roots, but there is tissue at the base of the plant designed to hold on to the substrate, and this is called the holdfast. Giant kelp actually make floats out of their stipe tissue that help the leaves stay buoyant with the blades floating a hundred and seventy feet above the holdfast.

As interesting as these marine organisms are, in terms of vascular plants, there are many more species that live in freshwater. Though some families are dominated by aquatics, there are many more aquatics scattered throughout other groups of ferns and flowering plants.

If we take an evolutionary approach, we can begin with the aquatic ferns. Maybe one of the most common is *Azolla*. This is actually a whole genus of aquatic ferns that contains seven species. They get the common name duckweed, but you have to careful with common names because almost any smallish water plant gets called a duckweed. Watch out, especially whenever the common name for a plant involves a very common animal, such as a duck. There are five different genera of water plants that are also called duckweed.

There are two really interesting observations about *Azolla*. The first is that *Azolla* has been grown in rice paddies for centuries. *Azolla* is called a companion plant in these rice paddies because it improves the yield of rice. It does this because it can provide extra nitrogen to the rice from the symbiotic bacteria that it houses within its tissues. The bacteria convert atmospheric nitrogen to a form that's usable by plants, which is good for the *Azolla* but also good for the surrounding rice. Since the *Azolla* float on the surface of the water that surrounds the rice, they also shield the substrate from light, so that competing plants don't have a chance.

The second interesting thing about *Azolla* is that it has a whole phenomenon named after it—the *Azolla* effect. What, pray tell, is the *Azolla* effect you ask? Basically, it's a theory that these aquatic ferns—and they're tiny, just millimeters across for the leaves—had such a massive population explosion, called a bloom, that they, single-handedly gulped so much carbon dioxide out of the air, that they made the once-tropical poles turn frigid. This would have happened maybe about 49 million years ago. The evidence for this theory was the discovery of *Azolla* fossils in sediment cores and that the bloom of *Azolla* coincided with a decline in carbon dioxide. During this time, carbon dioxide fell from 3,500 ppm in the early Eocene to 650 ppm after the bloom. For reference sake, we're sitting at about 400 ppm now. This decrease didn't happen overnight, but the correlation with the presence of *Azolla* is pretty interesting.

For this reason, some researchers are investigating whether *Azolla* could solve the current climate change problem. If *Azolla* are so good at taking up CO_2, we just need more *Azolla* everywhere to suck up the CO_2. Of course, there are some issues with this idea. *Azolla* can't actually stand temperatures below freezing. Plus, it's invasive plant in some areas, and we certainly don't want *Azolla* to outcompete other aquatic plants. Still, the concept is interesting as a way to use plants to increase the CO_2 draw down out of the atmosphere.

A flowering aquatic plant that many people recognize is the water lily. These plants all belong to the same family, the *Nymphaceae*. They are thought to be some of the most primitive flowering plants, which makes sense if we consider that the early ancestor of the angiosperms, the *Archaefructus*, may have been an aquatic plant.

The water lilies are not related to land lilies, and I don't think their flowers look alike much at all. If I had to name them, I would call them the water roses or water peonies, but suffice it to say, that unlike lilies, they have many petals and pistils, and stamen.

The star attraction of many a botanic garden is the giant water lily, usually the Amazon lily in the genus Victoria. These plants form giant lily pads that can easily reach a diameter of two meters. The leaf veins underneath these giant pads are comprised of lots and lots aerenchyma tissue with many air spaces to ensure floating. Additionally, these leaves will drain water that gathers on the top of the pads through small pores present in the leaf surface. The point of having such large leaves shades out competitors and maximizes the photosynthetic area. It's reported that if the weight is evenly distributed, the pads of this plant can easily hold a hundred pounds.

The water lily is not to be confused with the sacred lotus, though you would be forgiven if you did place them in the same family. It's been done. My 1980 copy of *The Complete Guide to Water Plants* puts the sacred lotus, familiar from the Asian pictures with the Buddha, in the same family as the water lilies. Earlier botanists thought that the two water plants must be related, as they both form flowers and have leaves and float on the water. But, with the advent of molecular analysis, the lotuses, and there are only two species, have been placed in their own family, and they are more recently evolved than are the water lilies.

Even looking at the leaves, there is a difference in that the lotus plant has truly peltate, or circular leaves, whereas the leaves of the water lilies have a characteristic notch in the margin, so as not to be a complete circle. The leaves of the lotus will typically emerge from the surface of the water while those of the water lily will lay flat on the surface. The leaves of the lotus have a notable ability to have muddy water roll right off of them. It's a characteristic called superhydrophobicity, and this is sometimes referred to as the lotus effect.

Lotuses are also eaten. It's not the fruit, but the underground stem, or the rhizome, that is consumed in Asian cuisine. The dried fruit is sometimes used in flower arrangements. Lotuses can regulate the temperature of their flowers. They are able to do this by having a different enzyme receive the electron during the electron transport chain of the light reaction of photosynthesis. When the enzyme receives the electron, it gives off heat, warming the flower. Botanists suspect this sauna effect is attractive to pollinating insects. A heat generating flower? No wonder it's sacred.

So, plants can grow in water. Some can grow in the ocean like seagrasses and mangroves. More plants live in freshwater because there is no challenge of extreme salinity. Although there are not a great many species of water plants, they have developed a fascinating adaptation to living a wet life.

In the next lecture, we'll consider plants in only a slightly drier habitat—the tropical rainforest.

LECTURE 17

WHY THE TROPICS HAVE SO MANY PLANT SPECIES

Tropical regions are the most diverse ecosystems on land. Rapoport's rule, named after Eduardo Rapoport, who provided some of the first evidence for this trend, states that diversity increases toward the tropics, which is with decreasing latitude. The tropics are defined as any geographic region between the Tropic of Cancer, at 23.5 degrees north, and the Tropic of Capricorn, at 23.5 degrees south.

BIODIVERSITY IN TROPICAL REGIONS

- There are several reasons for the rich biodiversity that is found in tropical regions. For example, biodiversity can also refer to genetic diversity within a population. Like any population, there is generally variety among the individuals. And genetic diversity within a population is also important. But, at its simplest definition, biodiversity refers to the number of species within a given area, and ecologists refer to this as species richness.

- Although the tropics include seasonally dry forests and grasslands, and even mountains, the amazing diversity occurs in tropical rain forests, which are the most diverse terrestrial ecosystems on the planet. Typically, tropical rain forests are very wet and pretty warm. Ironically, it's this narrow temperature range that may have led to such high biodiversity.

- In the temperate zone, a plant must tolerate freezing cold temperature, either by being dormant, or an annual, or simply toughing it out. Tropical plants don't have to do this, but this might mean they have a very narrow temperature range in which they can operate. This would lead to species only being able to occupy a small range, so as soon as one travels up a mountain within a tropical rainforest, the temperature will become slightly cooler, and new species will be there.

- Warm and wet could also affect biodiversity through many parasites and diseases that thrive under warm and wet conditions. If a parasite can only infect one particular species of host, then it would be called a specialist. If neighboring seedlings had slight mutations that prevented infection from the parasite, they might thrive and eventually become their own species—increasing the amount of biodiversity.

AMAZON RAINFOREST

◇ Another hypothesis for the diversity has to do with insect herbivores. Many plants in the tropics have leaves with various chemical compounds that are unpalatable to certain insects, or even poisonous. An insect couldn't possibly adapt to all the chemical compounds that exist in leaves in the tropics, so most insects specialize their feeding on one species, or perhaps one genus, of plants. But the plants need to stay ahead of the insects, and one way to do this is through mutations that lead to different chemical structures in the leaves.

◇ Another feature of speciation, which isn't restricted to the tropics, is the spatial separation of species. If there was one species—for example, a small terrestrial herb that is dispersed by ants—and this species has a population that gets separated by a river, now there are 2 different populations of the herb. Over time, these populations might become different enough so as to be different species. Perhaps there are very different soil types on either side, or perhaps the ant that used to perform dispersal didn't make it to the other side, so now a different type of insect does the dispersal.

◇ This type of speciation caused by a population being physically separated is called allopatric speciation, and it may have occurred a lot in the tropics because of past climate changes. The example of the river may have happened a few times, but river separation isn't thought to have spurred much diversification in the Amazon, which has many rivers.

◇ Although the tropics were never fully glaciated, we can suppose that some glaciers came down and retreated from mountains, which would cause separation of populations on either side of a glacier or mountain range.

◇ Changes in sea levels and lake levels could cause physical separation, too. If the sea level goes down, land bridges can form between islands and mainlands and between islands of archipelagos. As ocean levels rise, populations become isolated and speciation can occur. Rising and falling lake levels can act the same way.

◇ Being near the equator means that there is a lot of sun energy hitting the tropics, and this is another hypothesis for why these latitudes are so diverse. More energy means that more plants can use this energy.

- Another idea about why the tropics are so diverse has to do with available habitat. The tropics are certainly expansive in space, but the boreal forest region just below the North Pole takes up much more area than the tropical forests. Rain forests have several layers of vegetation, and this is probably possible in the rain forests because of the intense sunlight, which shines for about 12 hours a day. Because of this energy input, sunlight can penetrate through the layers, and a small amount can reach the forest floor.

- Getting sunlight is a major theme in the tropics. With so many plants around, it's a cutthroat game to get to the top or deal with the shade, so many tropical plants are shade-adapted plants. This is why so many houseplants are tropical plants—because they thrive in low light levels.

- Tropical rain forests are shady, not too hot, and pretty moist—even when it's not raining. The ground is covered with plants of all shapes and sizes. If there is a very thick middle layer, there will be fewer plants, and it won't be quite so dense on the forest floor, but there are still many plants thriving in those low light levels.

TROPICAL PLANTS

- The understory of the canopy is where many animals—particularly arboreal, or tree-dwelling, animals—live. The canopy in general in rain forests has many more species than temperate forests do; by some estimates, about 40% of all plants are in the canopy of rainforests.

- The plants that live in the canopy on the trees are called epiphytes. Temperate latitude forests have epiphytes, too, many of them mosses, but the tropics have a greater diversity and a larger quantity of epiphytes.

- One temperate epiphyte that is pretty common is mistletoe, which is parasitic. It grows rootlike structures into its host tree and uses the host's sugars. But most epiphytes are not parasitic; they just use the tree for real estate.

◇ There are 3 major types of epiphytes common in the tropics, and they are all pretty much absent from temperate canopies. The major epiphytes are bromeliads, orchids, and lianas.

◇ Bromeliads are often called tank plants, because their overlapping leaf bases form a small tank where water can collect. These type of plants, including the pineapple, are all in the Bromeliaceae, or bromeliad, family. The tanks of bromeliads form their own microecosystems of algae, insects, and amphibians. Bromeliads aren't limited exclusively to the tropics.

BROMELIAD
Bromeliaceae

◇ Today, orchids, in the Orchidaceae family, are arguably the most diverse plant family on Earth. Orchids are famous in the botanical world for having unique relationships with pollinators; that is, there are many orchids that are pollinated by a single species of insect. Not all orchids are epiphytic, and not all orchids are tropical, but many orchids in the tropics are epiphytic. Many people think of orchids as fragile flowers, but epiphytic orchids are able to survive on little water and no soil in the canopy.

◇ The liana is a woody vine. There are many different types of lianas found in different families, so it's a convergent form. A familiar liana is the rattan. There are several genera of rattan, and they are all in the palm family, Arecaceae. Palms, including rattans, don't have true wood. The rattan, as a palm, doesn't produce wood like an oak, but the stems are very woody nonetheless, and they can be woven into furniture.

◇ Rattan is a fairly sustainable forest product where it grows in Southeast Asia. Because it's a liana, it grows faster and can be harvested easier than trees. But if it's overharvested, then there won't be any left.

- The palm family is very large and mostly restricted to the tropics. It also provides the coconut palm; the date palm, which grows in deserts; and the oil palm. Unlike the sustainable rattan liana, there is some controversy over the sustainability of the oil palm, because large parts of tropical rainforests are being destroyed to make way for palm oil plantations. Palm oil and palm kernel oil are used for all kinds of things, from cosmetics to food to lubrication for machinery.

- The most exported tropical crop, and the second most traded commodity in the world after oil, is coffee. This is a shrub that is native to the Ethiopian highlands, which aren't tropical rainforests, but this area falls geographically within the tropics. Coffee can be somewhat sustainable because the best kinds of coffee are shade grown. This is good because forests in the tropics have many layers, especially the canopy and the understory. If the canopy is left intact, and coffee is grown in the understory, then there is still some habitat left.

- An even better tropical crop for sustainability is chocolate, *Theobroma cacao*, which is in the same family as common hibiscus and cotton, Malvaceae. Chocolate flowers are produced on the trunk of the tree, and despite all the flavor and smells of chocolate, the flowers have very little scent. Chocolate trees are understory trees that are native to South America and are adapted to shade, and as of yet, there isn't a full-sun variety, so by definition, chocolate is shade grown.

COCOA TREE
Theobroma cacao

WHY THE TROPICS HAVE SO MANY PLANT SPECIES

READINGS

Attenborough and Graham, *The Private Life of Plants*.

Bernhardt, *Wily Violets and Underground Orchids*.

Royte, *The Tapir's Morning Bath*.

QUESTIONS

1 Why are there more plant species in the tropics than anywhere else on Earth?

2 Why are the orchids, in particular, so diverse?

LECTURE 17 TRANSCRIPT
WHY THE TROPICS HAVE SO MANY PLANT SPECIES

Terry Irwin is an entomologist with the Smithsonian Institution. In 1982, Dr. Irwin wrote a two-page paper in a journal devoted to beetles. In this paper, he estimated that the total number of species that might be present in the world was based on the number of insects that he found on 19 individual trees in an unrich forest in Panama. These 19 individual trees produced almost a thousand species of beetles. Not a thousand beetles—a thousand species of beetles. This is remarkable because it shows just how diverse the tropics are. These trees occurred in an area of just about an acre.

So, a thousand different beetles in one acre in Panama. By comparison, there are about 240 beetles in my whole state of Colorado. Although plant diversity isn't quite as astounding, the increase of diversity in the tropics is still dramatic. Let's look at a comparison of trees. Florida has about 300 native trees, which is more native trees than any other state, In fact, Florida contains about half of the tree species in the lower 48 states. The Amazon rainforest has about 11,000 species of trees. So, about 36 times the species in Florida.

This phenomenon is known as Rapoport's rule, named after Eduardo Rapoport, who provided some of the first evidence for this trend. It states that diversity will increase toward the tropics, which is with decreasing latitude. The tropics are defined as any geographic region between the Tropic of Cancer at 23.5 degrees North and the Tropic of Capricorn at 23.5 degrees South. So, there are actually very cold places in the tropics, and these are found on the tops of tropical mountains, like those in the peaks of Ecuador.

Tropical regions are the most diverse ecosystems on land. There are several reasons for this rich diversity. For example, biodiversity can also refer to genetic diversity within a population. Like any population, there is generally variety amongst the individuals. And genetic diversity within a population is also important. But, at its simplest definition, biodiversity refers to the number of species within a given area, and ecologists refer to this as species richness. Let's talk a bit about what we really mean by diversity.

So, why are there so many species in the tropics? Let's think about the climate of a tropical rainforest. Although the tropics include seasonally dry forests and grasslands, and even mountains, the amazing diversity we're thinking about occurs in tropical rainforests, which are the most diverse terrestrial ecosystem on the planet. Typically, tropical rainforests are very wet and pretty warm. Ironically, it's this narrow temperature range that may have led to such high diversity.

Why would warm and wet lead to more diversity? If we think about what a plant must tolerate in the temperate zone, it must deal with freezing cold temperature, either by being dormant, or an annual, or by simply toughing it out, tropical plants don't have to do this, but this might mean they have a very narrow temperature range in which they can operate. This would lead to species only being able to occupy a small range, so as soon as one travels up a mountain in a tropical rainforest, the temperature will become slightly cooler, and there will be a whole new species there.

Warm and wet could also affect biodiversity through many parasites and diseases that thrive under warm and wet conditions. If a parasite can only infect one particular species of host, then it would be called a specialist. This parasite would infect only one species of host, and if neighboring seedlings had slight mutations that prevented infection from the parasite, they might thrive and eventually become their own species—increasing the amount of diversity.

Keeping in the vein of natural selection, another hypothesis for the diversity has to do with insect herbivores. Many plants in the tropics have leaves with various chemical compounds that are unpalatable to certain insects, or even poisonous. An insect couldn't possibly adapt to all the chemical compounds that exist in leaves in the tropics. So, most insects specialize their feeding on one species, or perhaps one genus, of plants. But the plants, of course, need to stay ahead of the insects, and one way to do this is through mutations that lead to different chemical structures in the leaves.

Another feature of speciation, which isn't restricted to the tropics, is the spatial separation of species. If there was one species, let's say it's a small terrestrial herb that is dispersed by ants, and this species has a population that gets separated by a river, now there are two different populations of our herb. Over time, these populations might become different enough so as to be different species. Perhaps there are very different soil types on either side or perhaps the ant that used to perform dispersal didn't make it to the other side, and so now a different type of insect does the dispersal.

This type of speciation caused by a population being physically separated is called allopatric speciation. Allo- is a prefix meaning apart. This type of speciation may have occurred a lot in the tropics because of past climate changes. Our example of the river above may have happened a few times, but river separation isn't thought to have spurred much diversification in the Amazon, which has a lot of rivers. Let's think about some other ways that allopatric speciation could occur.

Although the tropics were never fully glaciated, we can suppose that some glaciers came down and retreated from mountains, which would cause separation of populations on either side of a glacier or a mountain range. Changes in sea levels and lake levels could cause physical separation, too. If the sea level goes down, land bridges can form between islands and mainlands and between islands of archipelagos. As ocean levels rise, populations become isolated and speciation can occur. Rising and falling lake levels can act the same way. There is good evidence that the cichlid fishes in Lake Victoria are so diverse because of this phenomenon.

Another hypothesis for high diversity in the tropics, specifically, is increased energy. Being near the equator means there is a lot of sun energy hitting the tropics, and this is another hypothesis for why these latitudes are so diverse. More energy means more plants can use this energy.

The last idea I'll present about why the tropics are so diverse has to do with available habitat. Certainly, the tropics are expansive in space, but if you look at a globe, the boreal forest region just below the North Pole takes up much more area than the tropical forests. What I mean by available habitat is in the vertical strata.

Rainforests have several layers of vegetation, and this is probably possible in the rainforests because of the intense sunlight. Temperate forests certainly have tall trees, consider the world's tallest trees, the Redwoods in California. Yet, the temperate forests don't have the year-round intense sunlight for about 12 hours a day as the rainforests do. Because of this energy input, sunlight can penetrate through the layers and a small amount can reach the forest floor.

Getting sunlight is a major theme in the tropics. With so many plants around, it's a cutthroat game to get to the top or deal with the shade, so to speak. So a lot of tropical plants are actually shade-adapted. This is why so many of our houseplants are tropical plants—because they thrive in low light levels.

If you've ever been to the rainforest, it's dark. It's shady and not too hot. It's pretty moist, too, even when it's not raining. The ground is covered with plants of all shapes and sizes. If there is a very thick mid-layer, there will be fewer plants, and it won't be quite so dense on the forest floor, but there are still many plants thriving in those low light levels.

Moving up to the understory of the canopy, this is where many animals, particularly arboreal, or tree-dwelling animals will live. The canopy in general in rainforests has many more species than our temperate forests do. By some estimates, about 40% of all plants are in the canopy of rainforests. Wow. The plants that live in the canopy of the trees are called epiphytes—epi- for on top of, like the epidermis of your outer skin on top of your dermis, your deeper skin layer. Temperate latitude forests have epiphytes too, many of them mosses, but the tropics have a greater diversity and a larger quantity of epiphytes. One temperate epiphyte that is pretty common is mistletoe, which is parasitic. It grows root-like structures into the host tree and uses the host's sugars. But, most epiphytes are not parasitic at all, they are just using the tree for real estate.

There are three major types of epiphytes common in the tropics, and they are all pretty much absent from temperate canopies. The major epiphytes are bromeliads, orchids, and lianas.

I have here a bromelia, and the bromelias are of the pineapple family, but they are also called tank plants. Let me show you why this is called a tank plant. This is not because they might be grown in a tank, it's because their overlapping leaf bases will form a small tank where water can collect. There is actually very many tanks in here where the leaf meets the base, or where the leaf meets the stem, each forms a tiny little tank, and that's where insects and small invertebrates can live. Normally, these plants wouldn't have a giant flowering stalk here, so there might even be one larger tank in the middle, but that's why they're called tank plants. If you've ever seen the green top of a pineapple, then you've seen a tank plant. These type of plants, including the pineapple, are all in the same family, the *Bromeliaceae*, or the bromeliad family.

The tanks of bromeliads in the tropics have been studied extensively. These tanks form their own micro-ecosystems of algae, insects, and amphibians. Bromeliads aren't limited exclusively to the tropics. The southern states have Spanish Moss, *Tillandsia usneoides*, which is not a moss but a bromeliad. Instead of having one large tank, each leaf base is formed to be able to store water, and the leaves themselves can absorb the water. This is a pretty incredible adaptation because it means that the tillandsia can live in the canopies of trees, without roots, surviving on mere humidity.

To have a small reservoir of water in a plant is called phytotelmata. Many of these tanks have species of animals that aren't found in other tanks, so there is a lot of specificity. These tanks are also an important source of nitrogen for the plant, as waste products from the animals collect in the water. It might seem that the bromeliads would be carnivorous, catching these animals in their tanks, but this is generally not the case.

Only three species of bromeliads are carnivorous, and all of these tanks have bright yellow leaves. This might be so that the leaves appear like flowers to would-be pollinators. Two of these species are in the genus *Brocchinia* and have scales on their leaves that are coated with wax. This wax serves two important purposes. The first is that the wax is slick and makes it hard for insects to escape. The second purpose is that the wax is highly reflective of ultraviolet radiation, and ultraviolet radiation attracts insects.

The third bromeliad that is carnivorous is the *Catopsis berteroniana*, but this is more in the category of possibly a carnivorous plant. Its leaves are covered with hairs, or trichomes, on the surface that give the plant a white appearance. These trichomes serve the same purpose as the wax; that is, the trichome covering makes it difficult for the insects to escape the tank. These trichomes also reflect ultraviolet radiation, like the wax, which attracts insects. It's also been demonstrated that the trichomes help the leaves shed water, and this is helpful because if water stays on a leaf too long, it can be a good place for pathogens to take hold.

The two genera where these tank plants are found each consist of around 20 species. And, all the rest of these species are non-carnivorous regular tank bromeliads. Given that tank plants seem like such natural carnivores, it's interesting that there isn't more carnivory in this family. Where carnivorous plants exist in other genera, almost the entire genus is made up of carnivorous plants. Stewart McPherson, a carnivorous plant naturalist, suggests that the lack of carnivory in the bromeliads suggests that this trait may be a recent adaptation.

And, speaking of recent adaptations, a group of plants that was thought to have evolved recently is the orchid family or the *Orchidaceae*. However, in a paper in the journal *Nature* from 2007, Santiago Ramirez from Harvard University, described a fossilized orchid from around 15–20 million years ago. Then, using molecular techniques, they suggested that the origin of orchids likely goes back to the Cretaceous. This finding means that orchids were around during the time of the dinosaurs and probably radiated around the so-called K-T boundary when the dinosaurs died. This suggests, that after the extinction of the dinosaurs and many other organisms at the K-T boundary, orchids began to radiate, or evolve into many other different types of orchids. And today, orchids are arguably the most diverse plant family on Earth.

When orchids came onto the scene is also interesting in terms of what came first—the orchid or its pollinator. The orchids are famous in the botanical world for having unique relationships with pollinators. That is, there are many orchids that are pollinated by a single species of insect. What scientists have found, though, is that the insects almost always came first in an evolutionary sense, so the orchids need the insects more than the insects need the orchids. This is important because when thinking about orchid conservation, we need to think about what the insects need to survive.

Not all orchids are epiphytic and not all orchids are tropical, but many orchids in the tropics are epiphytic. And, because the Orchid family is the most diverse family of plants, that means there are a lot of epiphytic orchids. These epiphytic orchids have some unique adaptations to help them exist in the canopy. This is pretty interesting because many people think of orchids as a very fragile flower. However, epiphytic orchids are able to survive on very little water and no soil in the canopies—since they're living in the trees—so they are hardier than one might think. Most orchids probably die from overwatering and fertilizing than from underwatering and lack of fertilization.

One major adaptation that epiphytic orchids have is the development of pseudobulbs. As these structures are not botanically bulbs, which are comprised of leaves, these structures are for storage of water and carbohydrates for dry times. Although we tend to think of tropical forests as wet year-round, many parts of the tropics are seasonally dry. Instead of winter, these areas have a dry season.

Insects pollinate and disperse orchids, and there are many, many different shapes and sizes of orchids that have evolved for specific pollinators. There are orchids that produce pheromones that attract insects so much that the insects will rub the pheromones on their body to attract mates—like perfumes. There are other orchids that smell of rotting flesh to attract flies, and many orchids have no smell at all, but still manage to attract pollinators with nectar.

The other major epiphyte in the canopy of the tropics is the liana, which is a woody vine. The grape vine genus, *Vitis*, can form lianas in the temperate forests, but this lifeform is way more common in the tropics. These are the plants that Tarzan would have used to travel through the canopy. There are many different types of lianas found in different families, so it's a convergent form.

Lianas, like vines, start in the ground, then they climb their way up trees to the canopy where they can get sunlight. Sometimes, the seeds will actually germinate in the canopy, then the roots will grow down into the soil, and the shoot will start climbing. They keep their roots in the ground, and so most lianas are not parasitic at all.

A vine, and not a liana, that grows in the tropics and subtropics is the luffa. This is a genus of plants in the cucumber family, and as the name suggests, this is the where we get loofah from—those scratchy exfoliating shower scrubbers. It's the fruit that dries out, and then we use it. The seeds were inside those empty spaces.

There is one very unique and common liana in the tropics that begins life as a harmless little liana in the canopy, then it grows down to the ground and roots, and then it climbs up its host tree, usually the same tree where it landed and then ever so slowly, this liana begins to grow around its host, and then what was just a woody vine, has now encircled its host. Then this unwelcome neighbor begins to strangle its host to death.

Eventually, the tree that was the host will die and leave a tree that has no inside, and this is the strangler fig. The strangler fig is classified as a hemiparasite since it's not parasitic for the first part of its life. Only after it starts the climb back up does it start to kill the tree. Because it never actually steals sugars from its host, it's mainly a parasite of real estate or space in the canopy.

A liana that may be a bit more familiar to us here in the temperate region is the rattan. There are several genera of rattan, and they are all in the palm family. We typically don't think of palms as lianas, we generally think of them as trees, but the palm family, *Arecaceae*, is large and diverse. Palms, including rattans, don't have true wood because they don't have a vascular cambium that produces that secondary xylem that becomes the wood.

That secondary xylem was the tissue that conducted water from the roots to the shoots. That tissue gets old as new tissue is laid down near the vascular cambium, which is close to the bark. So, the oldest part of the wood is in the middle of the tree. As the xylem in the middle dies out, it becomes wood. And if a nail is driven into the trunk of a tree with true wood, the secondary xylem, the tree will grow around it eventually completely engulf the nail.

But, this secondary growth is absent in palms. If a nail is driven into the trunk of a palm, that's not what happens. Instead, a nail driven into the trunk of a palm tree will remain where it's driven, and not become embedded within the trunk. This is because palms grow differently. When the seed of a palm germinates, the seedling first grows to the full width of the trunk to be grown. After the seedling has reached an adequate girth, it begins to grow vertically. But because palms don't have a vascular cambium, there is no continuous production of new radial xylem that would increase the trunk girth.

So, the rattan, as a palm, doesn't produce wood like an oak, but the stems are very woody nonetheless, and it can be woven into furniture. The stems are so woody because the cells are impregnated with a lot of lignin, and this substance makes them pretty tough. Rattan is actually a fairly sustainable forest product where it grows in Southeast Asia. Because it's a liana, it grows faster and can be harvested easier than trees. I'm not sure how people figured out they could use rattan to make things—many of the species have spikes and hooks on the outside to facilitate its hanging on in the forest canopy.

Rattan can be used to make wicker products. The term wicker means products that can be woven from any sort of plant material. Interestingly, the early Egyptians made wicker furniture from reeds, which are grass-like plants.

The palm family is very large and mostly restricted to the tropics. It also provides the coconut palm, the date palm, which grows in deserts, and the oil palm. Unlike the sustainable rattan liana, there is some controversy over the sustainability of the oil palm. Indigenous to Africa, the oil palm produces fruits that are drupes, as do many palm trees. A drupe is—as you now may know from the foods like peaches and cherries—a fleshy fruit with a hard seed that is sometimes called the pit or the kernel. It's this kernel that produces palm kernel oil, but it's actually the mesocarp of the fruit that produces palm oil. The mesocarp here would be similar to the flesh of a peach, where the peach skin is the exocarp, and the peach pit is the kernel. These palm oil fruits are about the size of small apricots, and the fruits ripen continuously throughout the year.

Palm oil and palm kernel oil are used for all sorts of things from cosmetics to food to lubrication for machinery. Like coconut oil, palm oil is highly saturated. A saturated fat means that there are no double bonds between the carbon atoms in the molecule, which means the fat is saturated with hydrogens. This saturation means that there are no double bonds to break, so the melting point is typically higher, which means that the palm oil is solid at room temperature. These qualities make it desirable as an additive to anything that requires an oil.

The reason it's controversial is because large parts of tropical rainforests are being destroyed to make way for palm oil plantations. Of course, this is the same problem with any sort of tropical crop, pineapples, mangoes, bananas, and most anything else that comes from the tropics. There are a few exceptions. As we mentioned, rattan is a sustainable product because trees don't have to be cut down, but if it's over harvested, then there won't be any left.

The most exported tropical crop, and the second most traded commodity in the world after oil, is coffee. This is a shrub native to the Ethiopian highlands, which aren't tropical rainforests, but this area falls geographically within the tropics. Coffee can be somewhat sustainable because the best kinds of coffee are shade grown. This is good because forests in the tropics have many layers, especially the canopy and the understory. If the canopy is left intact, and the coffee is grown in the understory, then there is still some habitat left, especially for birds and epiphytes and primates that use the canopy. In some parts of Central America, the only forests left are the shade grown coffee forests.

As you can imagine given our love of coffee, there is quite a bit of research into why these crops might taste better when grown in the shade. One hypothesis is that there are fewer fruits in shade-grown crops, but these fruits tend to be of a more uniform size. This uniformity is better when roasting because it leads to a more even roast, which leads to a tastier cup. There are also studies that suggest that sun-grown coffee is bitter, and I am hypothesizing here, but I would guess that might be because the plant can produce more sugars so it can make more secondary compounds, which is one of the compounds that tannins are that would give coffee that bitter flavor.

An even better tropical crop for sustainability is chocolate. *Theobroma cacao* is the scientific name for chocolate. Theobroma is Greek for god food—food of Gods. Of course, to a chocolate lover, this name is not at all surprising. Chocolate is in the same family as the common garden variety hibiscus and cotton, the family is *Malvaceae*.

Chocolate flowers are produced right on the trunk of the tree and despite all the flavor and smells of chocolate, the flowers have very little scent. They are pollinated by midges, which are tiny little flies and they're dispersed by mammals. Unlike us, the mammals are eating the pulp of the fruit. Botanically, the fruit is a berry, like a tomato. There is a skin, called an exocarp, then the pulp, and then the seeds. What we roast and turn into confection are the seeds. Sometimes these are called cocoa beans, but they are not botanically beans since this plant is not in the bean family. Animals eat the fruit and disperse the seeds.

Chocolate trees are understory trees native to South America that are adapted to shade, and as of yet, there isn't a full-sun variety, so by definition, chocolate is shade grown. Chocolate has a deep taproot for a tropical tree, although the tree itself doesn't grow very tall. In nature, these trees can reach about 50 feet, but they are usually kept much shorter in cultivation, to aid in the harvesting of the fruits.

Brazil nuts from Brazil are slightly radioactive because Brazil—just like parts of the American West—happens to have more uranium and radium in the ground than other places. So, Brazil nuts not grown in Brazil won't be radioactive, and the largest exporter of Brazil nuts is actually Bolivia. Brazil nuts don't grow well in cultivation, so they come primarily from pristine forests. Because of this, Brazil nuts are harvested within forests, but tropical forests are not cut down to grow Brazil nuts.

Brazil nuts are closely related to another tropical forest tree called the gutta-percha. This was an early connection in my study of botany to my dad, who was an endodontist, that's a dentist who specializes in root canals. Gutta-percha material is still used to fill the hole in the tooth that is created when the diseased tooth material is removed. Although plastics and artificial polymers are catching up, gutta-percha is still the standard, just as natural rubber is still favored over synthetic rubber for many uses.

Clearly, the tropics are diverse. This idea of biodiversity is linked to an idea called resilience, which means that an ecosystem can bounce back from a disturbance, or return to what it was like before that disturbance.

One final point, biodiversity seems to be tied to human health. In a recent study, the more small rodents that were present in the tropical region of Panama, the lower the incidence of hantavirus. So, the tropics give us diversity, coffee, chocolate, and good health.

LECTURE 18

THE COMPLEXITY OF GRASSES AND GRASSLANDS

Grasses are not only the basis of human diets today, but they have also played a major role in the development of human society. In addition, botanically, grasses are very interesting, and they are very easy to recognize. Even though the orchids, sunflowers, and peas might have more species, the grass family is the most important plant family in terms of crops for energy, forage for animals, building material, and even delicious seasonings.

GRASSES

⬧ Grasses are interesting in that they are defined ecologically, functionally, and taxonomically. This means that a grass isn't in any other family, but the grass family and grasses are pretty unique in terms of their botany.

⬧ Grasses are all nonwoody, except for the bamboos. They are monocots, which means that they have parallel veins in their leaves, which are long and slender. These plants are visibly different from broadleaf plants, which have netted or palmate venation in the leaves. Grasses have a very unique, small, but highly complex, flower morphology. Most grasses are wind pollinated.

⬧ Grass botany is difficult because of the complexity and scale of how to identify grasses. Identifying a plant in this family is pretty easy, but identifying it down to species is much more difficult.

GIANT TIMBER BAMBOO
Bambusa oldhamii

⬧ There are 2 other families that can sometimes be mistaken for grasses. One family is the sedges, known botanically as the Cyperaceae family, after the most characteristic genus, *Cyperus*. A fairly notable member of this genus is *Cyperus papyrus*, which is the source of papyrus. The other family is the Juncaceae family, which are the rushes. A common rush is bulrush, which is the common name for many aquatic rushes in the *Scirpus* genus.

PAPYRUS SEDGE

Cyperus papyrus

- Sedges, like papyrus, often have triangular stems. Rushes have round stems and often have stems that are solid. Grass stems are typically hollow.

- Grasses typically have leaves coming off of the main stem at several junctions. The node is where a leaf comes off the stem. Often at this junction there will be a small membrane called the ligule, which is just one of the structures that are unique to grasses.

- It's hypothesized that the ligule's function may be either physical or physiological. The hypothesis that refers to the ligule's physical function is called the passive hypothesis. This idea suggests that the ligule prevents dirt, water, and pathogens from getting to the vulnerable parts of the culm, which is a special name for the stem of a grass. The physiological hypothesis about the ligule, called the active hypothesis, suggests that the ligule may play a role in the storage of starch produced in the blades.

- There is a special vocabulary just for the grasses; the blade, culm, and ligule are vegetative vocabulary. The grass flowers have their own vocabulary, too. In grasses, the flowers are called florets because they are small. A few florets make up a spikelet, but the exact number of florets in a spikelet depends on the species. Each flower is subtended by 1 or 2 modified leaves called palea, and above the flower is another modified leaf called a lemma.

- Several spikelets make up a spike, which has a different arrangement on the stem based on the species type. Typically, spikes are solitary or arranged on a panicle. These floral spikelets have special structures above and below them. Each spikelet is subtended by a certain number of modified leaves called glumes.

- There are suppositions that these structures might provide certain adaptations to drought or heat, but they're certainly not to attract animals. The grasses are almost exclusively pollinated by wind and dispersed by wind. There is no need for the flower to invest in showy flowers that would cost a lot of carbon, or in beautiful scents, to attract animals. The flowers of the grasses have a utilitarian beauty.

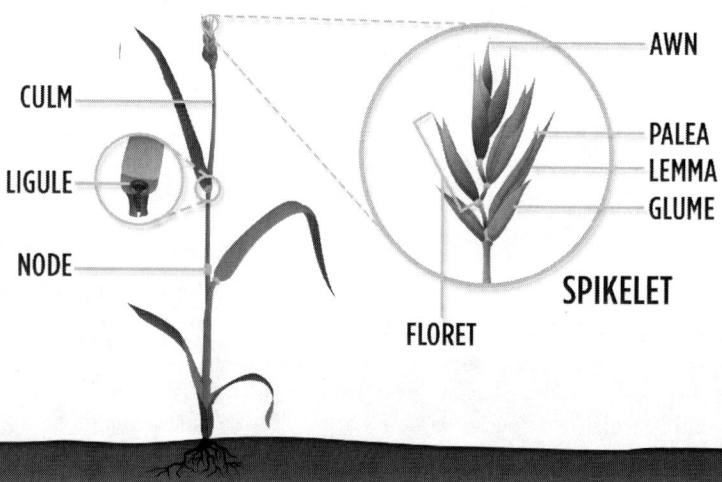

ADAPTATIONS OF GRASSES

- The grassland ecosystem is on every continent except Antarctica; though, interestingly, there are only 2 vascular plants on Antarctica, and one of them is a grass. A small bunch grass called *Deschampsia antarctica* grows on the western peninsula of Antarctica, but this growth isn't prolific enough to be called a grassland.

ANTARCTIC HAIR GRASS
Deschampsia antarctica

- Grasslands are estimated to make up about ⅓ of the terrestrial land area of the planet. Grasslands are broad in that they can be cold or hot, wet or dry. Typically, though, they are drier than forests. Rainfall may be seasonal, falling at one time during the year. Some areas that are dominated by grasses may not be what we commonly think of as grasslands. Habitats such as tundra, which is too cold for trees, and wetlands, which are too wet for trees, come to mind.

- Grazers are typically animals that eat grass. A browser is an animal that browses on leaves of shrubs or trees. The grasslands have evolved with grazers. There are several adaptations that grasses have employed to exist in landscapes full of hungry grazers.

- One of the most interesting adaptations of grasses is that they grow from the base. Trees and most other plants, by contrast, have their growing tissue at the top of the tree and the tips of the branches.

- In all plants, the growing tissue is called the meristem. If the plant has determinate growth, it will grow until the plant reaches the size that it has genetically defined and then stop. In plants that have indeterminate growth, the plant will continue to grow until the plant dies.

- The meristem is a region of cells that are constantly dividing. The meristem of the grasses is found at the base of the blades, so it's called a basal meristem—which in grasses is a great adaptation to grazers. The grazer comes along to eat the top of the grass, but that doesn't kill it; in fact, most times, it will actually promote growth of the plant.

- This is the same principle behind mowing our grasses. When we mow, we promote the grasses to reproduce vegetatively, which means that the grass will produce more rhizomes, which are underground stems. The promotion of more of these underground stems will eventually send up more shoots, and the grass will look fuller.

- Another unique and adaptive feature of grasses is the presence of very small bits of silica, called phytoliths, in the epidermal cells, which are the outermost cells of the blades. There are 2 hypotheses about the presence of phytoliths in grasses: that grasses developed these phytoliths to inhibit grazing and that grazers developed teeth that could withstand silica in the blades.

- Grass leaf blades are not only adapted to grazers, but they are also adapted to the climate in which they live. Grasses from humid environments typically have larger, wider leaf blades, while those from drier regions have blades that are more linear. Grasses in the most arid environments have blades that can roll up during extreme sunlight and heat or leaves that are almost needlelike.

- Although grasses are now found in a variety of habitats, they most likely evolved in hot places.

- Many grasses have another unique adaptation to a hot environment: a special kind of photosynthesis called C4 photosynthesis, which was described in Lecture 8.

TYPES OF GRASSES

- Even though grasses probably evolved in warm regions, there are grasses that are accustomed to growing in cool climates. We tend to think about grasses as being warm-season or cool-season grasses. Warm-season grasses will become green and flower when it's warm. They will continue their life cycle through the warmest part of the summer. A cool-season grass is one that will become green and flower early in the spring, and by midsummer, these grasses will start to turn brown and finish their growing season.

- Examples of warm-season grasses are Bermuda grass and zoysia grass. Common cool-season grasses are Kentucky bluegrass and perennial ryegrass.

- This isn't the only way that grasses are classified. They are also grouped as bunch grasses versus turf grasses. A bunch grass will confine its growth to a small clump, or bunch, while a turf grass will spread out.

- Almost all grasses will send stolons or rhizomes out to promote vegetative growth, but the rhizomes in turf grasses will typically be longer, so that the individual culms of the grass can spread out. In this way, a whole field of grasses might be genetically the same, what botanists would call one genet, while each individual culm of grass would be a ramet.

- Grasses typically have much deeper roots than we might expect from just looking at the above-ground growth. Grass roots are not only very deep, but they are also very dense; that is, they are fibrous and have many root hairs that go in between the major roots. In grasses, these root hairs arise from trichoblasts, which are specialized cells in the roots.

- Grasses also have many more root hairs than other kinds of plants, which is one reason why grassland soils are so rich. Not only is there sparse rainfall, so nutrients don't dissolve away in the rainwater, but the grasses, with their deep roots and grass hairs, prevent erosion and soil runoff when it does rain.

◇ Because many grasslands have been used for crops, grasslands are declining. Not all of the grasses grown for crops are used for food; many grasses are grown for use as biofuels. Anything that was alive contains carbon, and when it dies, it can be burned for fuel. This is true of wood, paper, and even manure.

SUGARCANE
Saccharum officinarum

◇ Corn is a common biofuel that is used to make ethanol, which is most commonly used as an additive to gasoline. In addition to corn, sugarcane is another popular grass that is used to make biofuels, but now native prairie grasses are also being used.

PURPLE CONEFLOWER
Echinacea purpurea

◇ Despite looking monotonous to the untrained eye, grasslands are actually very diverse. Most grasslands have a number of other species, most notably members of the sunflower family, Asteraceae.

READINGS

Bone, *Steppes*.

Cooke, *Grasses and Bamboos*.

Darke, *The Encyclopedia of Grasses for Livable Landscapes*.

Gibson, *Grasses and Grassland Ecology*.

QUESTIONS

1   Some botanists are working on changing rice, a C3 plant, to a C4 plant so that it could be more efficient. Why aren't all grasses C4? Why would plants that grow in hot climates still maintain the C3 photosynthesis pathway?

2   How are some herbicides able only to kill eudicot, or broadleaved, species? Would it be useful to have an herbicide that only killed grasses? Why or why not?

LECTURE 18 TRANSCRIPT
THE COMPLEXITY OF GRASSES AND GRASSLANDS

In 1992, a University of California Berkeley Ph.D. student named Mark Blumler published his dissertation about the biogeography of grasses. He was looking at seeds, and he looked at 56 of the heaviest grass seeds from species around the world. Why was he looking at the weight of grass seeds?

In grasses, the seed takes up most of the fruit. Typically, the fruits of grasses are called caryopses. These are small fruits made up of the seed of the grass and a seed coat. For this reason, in the grasses, there is little difference between the fruit and the seed. In popular terms, the fruit of the grass is called a grain. Sometimes, the word cereal is used to denote a crop that is used like a cereal but isn't actually a grass. Crops like buckwheat and quinoa fall into this category, but virtually every other grain we eat is the fruit of a grass species.

So, in Mark Blumler's dissertation, he reports that over half of the 56 heaviest grass species occur in the Mediterranean region or the Fertile Crescent. This was the birthplace of wheat and barley, which are used as the primary source of calories in many human diets. This information was later made famous by Jared Diamond's 1997 book, *Guns, Germs, and Steel—The Fates of Human Societies*.

In this book, Jared Diamond outlines how different cultures domesticated different cereal grains, which then set them on different paths of development. He writes about the idea that it was wheat, which was easily cultivated and had a naturally large grain, that contributed to the birth of agriculture and thus civilization 11,000 years ago.

Looking at Mark Blumler's dissertation again, sub-Saharan Africa only had four of the 56 heaviest species of grass, plus moving these grasses around the continent would be difficult because it's much easier, according to Jared Diamond, to move grasses across the same latitude with similar climates and day lengths. Trade and movement of agriculture are more difficult north to south because of the change in climate and day length. So, grasses aren't just the basis of human diets today, they have also played a major role in the development of the entire human society.

In addition, botanically, grasses are super interesting. They are very easy to recognize because, they, well, they look like a grass. Grasses are interesting in that they are defined ecologically, functionally, and taxonomically. This means, a grass isn't in any other family, but the grass family and grasses are pretty unique in terms of their botany. They are all non-woody, except for the bamboos, which we'll get to in a minute. They're all monocots, which means they have parallel veins in their leaves, which are long and slender. In lay terms, these plants are visibly different from broadleaf plants, which have netted or palmate venation in the leaves.

Grasses have a very unique, small, but highly complex, flower morphology. Most grasses are wind pollinated. Grass botany is so complex that if you study grasses, you get a special name, you're an agrostologist. What makes grass botany so difficult is the complexity and scale of how to identify grasses. See, like the sunflower family, identifying a plant in this family is pretty easy, but identifying it down to species is much more difficult.

That said, there are actually two other families that can sometimes be mistaken for grasses. These are the sedges, known botanically as the Cyperaceae, after the most characteristic genus, *Cyperus*, which comes from the ancient Greek name for sedge, *kypeiros*. A fairly notable member of this genus is *Cyperus papyrus*, which is the source of the papyrus.

To make papyrus, the plant is harvested and the skin of the stem is cut off. The inside pulp is rolled flat and then soaked in water for about six days. This watery pulp is pounded flat and then laid out to dry—and eventually ready to write on.

The other family that might be confused with grasses is the Juncaceae. These are the rushes. Interestingly, the word reed doesn't have a botanical definition. It's a lay term for any plant that grows in water that might be a bit stiff and could be used to make reeds. The interesting thing about that, however, is that reeds for instruments, like a clarinet, are actually made from a grass called

Arundo donax. The rushes, however, do make up their own family. A common rush is bulrush, which is the common name for many aquatic rushes in the *Scirpus* genus.

There is a great rhyme that beginning botany students everywhere learn to remember the differences between these plant families. It goes like this. Sedges have edges; rushes are round, and grasses have nodes from their tips to the ground. This means that the sedges, the family Cyperaceae, like the papyrus, often have triangular stems. Not all the time, but it is a distinguishing feature. Rushes have round stems and will often have stems that are solid; whereas grass stems are typically hollow.

Grasses typically will have leaves coming off of the main stem at several junctions, which is why we say they have nodes from the tips to the ground. The node is where the leaf comes off the stem. Often at this junction, there will be a small membrane called the ligule. Grasses have structures with special names, which is one reason why it's difficult to identify grasses to species. The ligule is just one of the structures unique to grasses.

It's hypothesized that the ligule's function may be either physical or physiological. The hypothesis that refers to the ligule's physical function is called the passive hypothesis. This idea suggests that the ligule prevents dirt, water, and pathogens from getting to the vulnerable parts of the culm, which is a special name for the stem of a grass. The physiological hypothesis about the ligule is called the active hypothesis, and this suggests that the ligule may play a role in the storage of starch produced in the blades.

As I've mentioned, there is a special vocabulary just for the grasses. The blade, culm, and ligule are vegetative vocabulary, and the grass flowers have their own vocabulary, too. In grasses, the flowers are called florets because they are small. A few florets make up a spikelet, but the exact number of florets in a spikelet depends on the species.

Each flower is subtended by one or two modified leaves called palea and above the flower is another modified leaf called a lemma. Sometimes that [glume?] has a long needle-like structure called an awn. In grasses, the flowers are called florets and several florets make a spikelet, and then several spikelets make a spike. These spikes have different arrangements on the stem based on the species type. Typically, the spikes are solitary or arranged on a panicle.

These floral spikelets have special structures above and below them, too. Each spikelet is subtended by a certain number of modified leaves called glumes. There are suppositions that these structures might provide certain adaptations to drought or heat, but we do know for certain that they're not there to attract animals. The grasses are almost exclusively pollinated by wind, and dispersed by wind. There is no need for the flower to invest in showy flowers that would cost a lot of carbon, or in beautiful scents to attract animals. The flowers of the grasses have a utilitarian beauty.

How would these much-reduced wind pollinated and wind dispersed grasses have evolved? The grassland ecosystem is on every continent except Antarctica, though interestingly, there are only two vascular plants on Antarctica, and one of them is a grass. A small bunch grass called *Deschampsia antarctica*. It grows on the western peninsula of Antarctica. But this growth isn't prolific enough to be called a grassland.

As we said, grasslands are found on every continent and are estimated to make up somewhere around one-third of the terrestrial land area of the planet. Grasslands are broad in that they can be cold or hot, wet or dry. Typically, though, they are drier than forests. That rainfall may be seasonal, falling at one time during the year. Of course, some areas that are dominated by grasses may not be what we commonly think of grasslands. Habitats like tundra, which is too cold for trees, and wetlands, which are too wet for trees, come to mind.

Grasslands are known by many names around the world. In South America, they are called pampas. In Central Asia, they are the steppes. In Africa, they are called savannas. In North America, we call them prairies but also rangelands, which is what they are called in Australia. To me, the name range implies animals. Think of the song, *Home on the Range*. The lyrics begin, and I won't sing it to save our ears, "Oh give me a home, where the buffalo roam and the deer and the antelope play." Rangeland implies grazing.

Grazers are typically animals that eat grass. You can think Grazer eats Grass. A browser is animals that browses on leaves of shrubs or trees. So, let's test your understanding here. Would a giraffe be a grazer or a browser? A giraffe could be a grazer, but it'd be a real pain in the neck. Get it? Yes, a giraffe is indeed a browser. But, the grasslands have evolved with grazers. Let's look at some of the adaptations that grasses have employed to exist in landscapes full of hungry grazers.

One of the most interesting adaptations of grasses is that they grow from the base. Trees and most other plants, by contrast, have their growing tissue at the top of the tree and the tips of the branches. Remember the apical buds?

In plants, the growing tissue is called the meristem. If the plant has determinate growth, it will grow until the plant reaches the size it's genetically defined growth, and then stop. In plants that have indeterminate growth, the plant will continue to grow until it dies.

So, this growing tissue, the meristem, is a region of cells that are constantly dividing. The meristem of the grasses is found at the base of the blades, so it's called a basal meristem. The basal meristem in grasses is a great adaptation to grazers. The grazer comes along to eat the top of the grass, but that doesn't kill it, in fact, most of the time, it will actually promote growth of the grass. This is the same principle behind mowing our grasses. When we mow, we promote the grasses to produce vegetatively, which means the grass will produce more rhizomes, which are the underground stems. The promotion of more of these underground stems will eventually send up more shoots, and the grass will look fuller.

This is how it works for turf grass, but not all grasses come from turfs—like the bluegrass that is commonly used for lawns. Other grasses are bunch grasses that will still send out more rhizomes, but they will stay in a small bunch rather than spreading out.

Grazing still promotes growth in bunch grasses, too, because when the top parts of the grass are removed, the bottom parts will receive more sunlight and this will speed up the growth rate.

Another unique and adaptive feature of grasses is the presence of very small bits of silica, called phytoliths in the epidermal cells, which are the outermost cells of the blades. Phytolith literally translates as *phyto*, plant, and *lith*, rock, so these are plant rocks in the leaf cells. There are two hypotheses about the presence of phytoliths in grasses. One is that grasses developed these phytoliths to inhibit grazing. The other hypothesis is that grazers developed teeth that could withstand silica in the blades.

In laboratory studies, plants tend to produce more phytoliths in response to being physically cut, which is what grazing does. But, in the field, it may be that the silica inhibits insect grazers, but not large mammal grazers, which, as we said, act as mowers, and actually promote growth in grasses.

Grass leaf blades are not only adapted to grazers, they are also adapted to the climate in which they live. Grasses from humid environments will typically have larger, wider leaf blades while those from drier regions will have leaf blades that are more linear. Grasses in the most arid environments will have blades that can actually roll up during extreme sunlight and heat or leaves that are almost needle-like. What we can see then, is that although grasses are now found in a variety of habitats, they most likely evolved in hot places.

Many grasses, though not all of them, have another unique adaptation to a hot environment. That adaptation is a special kind of photosynthesis called C4 photosynthesis.

The type of photosynthesis that occurs in most plants is typically called C3 and this is because the first product of the Calvin cycle, that biochemical process of turning plain old atmospheric carbon into sugar, is a six-carbon sugar that immediately breaks up into two three-carbon sugars, thus, C3 photosynthesis.

C4 photosynthesis, as one might guess, involves a sugar with four carbon atoms. To understand why C4 photosynthesis is so advantageous, we have to think about what happens in the beginning of the Calvin cycle, which is also C3 photosynthesis.

In all plants, the stomata are the pores in the leaf through which the carbon dioxide and the oxygen diffuse. In C3 plants, the carbon dioxide diffuses in and is picked up by the enzyme RuBisCO. That name, RuBisCO, is short for ribulose, bisphosphate, carboxylase, oxygenase. The name is so important in understanding the function, so I'm going to say it again, ribulose, bisphosphate, carboxylase, oxygenase.

Basically, this is the enzyme that will pick up a molecule of carbon dioxide and stick that one carbon from that CO_2 molecule onto a five-carbon sugar called ribulose bisphosphate, and voila. We've created our six-carbon sugar. Now, it splits apart into two three-carbon sugars, and there's our C3 photosynthesis and the Calvin cycle continues.

Let's look at that enzyme name again—ribulose bisphosphate. OK, that's basically the starting sugar with the five carbons. Now, let's look at the end of it—carboxylase/oxygenase. Hmm. The carboxylase part seems more descriptive of what it's doing, which is adding a carbon onto an existing sugar.

But, the other part of the name, oxygenase, is also important because this part tells us what the enzyme could be doing. The enzyme could be taking up an oxygen molecule and trying to put it on the existing five-carbon sugar. This, however, would be bad, and it wouldn't work because the oxygen isn't going to attach to that five-carbon sugar because there isn't a place for it, but the RuBisCO enzyme doesn't know that—it's just an enzyme, and herein lies the rub. The affinity of RuBisCO to pick up an oxygen molecule, which is bad because it won't attach to the 5-carbon sugar. But, picking up a carbon molecule is good, since it will attach to the 5-carbon sugar. And the RuBisCO enzyme will more likely pick up a bad oxygen molecule with increasing temperature.

Let me say that again. The affinity for RuBisCO to pick up an oxygen molecule will increase with increasing temperature. So, let's test your understanding of this mechanism. If I give you a regular C3 plant, will its efficiency in picking up carbon during the first phase of the Calvin cycle become more efficient or less efficient with increasing temperatures? If you said that it would be less efficient, you're right.

Because the RuBisCO enzyme is more likely to pick up oxygen at higher temperatures, the efficiency of the C3 photosynthetic system declines as temperatures get warmer and warmer. So, the advantage of the C4 photosynthesis machinery is that the RuBisCO, the ribulose bisphosphate carboxylase/oxygenase never actually gets to see the oxygen. How could this be? The C4 plants have a way of capturing carbon in the cells that are closest to the stomata, the pores where the diffusion of the CO_2 will take place, and then storing it in a particular way until it comes into contact with the RuBisCO. In fact, the CO_2 won't even come into contact with the RuBisCO at all because the carbon gets stored as malate and then moved into the bundle sheath cells, where regular old C3 photosynthesis can then take place.

Even though we just talked about the idea that grasses probably evolved in warm regions, there are actually grasses that are accustomed to growing in cool climates. We tend to think about grasses as being warm-season or cool-season grasses. A cool-season grass is one that will green up and become green and flower early in the spring and by the time mid-summer rolls around, these cool-season grasses will start to turn brown and finish their growing season.

Warm season grasses, on the other hand, will green up and flower when it's warm. They will continue their life cycle through the warmest part of the summer. Examples of warm-season grasses are bermudagrass and zoysiagrass. Common cool-season grasses are Kentucky bluegrass and perennial ryegrass.

Warm and cool season aren't the only way that grasses are classified. They are also grouped as bunch grasses versus turf grasses. A bunch grass will confine its growth to a small clump, or bunch, while a turf grass will spread out. Almost all grasses will send stolons or rhizomes out to promote vegetative growth, but the rhizomes in turf grasses will typically be longer so that the individual culms of the grass can spread out. In this way, a whole field of grasses might be genetically the same, what botanists would call one genet, while each individual stem, or culm, of grass, would be a ramet. You might remember the words ramet and genet from our discussion of aspen. Any individual member of a plant clone is the ramet and the whole clone together, the forest in our aspen discussion is a genet.

Grasses typically have much deeper roots that we might expect from just looking at the above ground growth. There was an exhibit I saw put on by Dr. Jerry Glover from the Land Institute in Kansas that showed the amazing depth of grassroots—down to 13 feet below the surface.

Grass roots are not only very deep, they are also very dense. That is, they are fibrous and have many root hairs that go inbetween the major roots. In grasses, these root hairs arise from trichoblasts, which are specialized cells in the roots. Grasses also have many more root hairs than other kinds of plants, which is one reason why grassland soils are so rich. Not only is there sparse rainfall, so nutrients don't dissolve away in the rainwater, the grasses with their deep roots and grass hairs prevent erosion and soil runoff when it does rain.

The Land Institute is trying to develop perennial species of grasses. The grasses we raise as crops are all annual. A grain is essentially the fruit of a grass, and the more specific botanical name for that kind of fruit is a *caryopsis*. Many of the world's major grain starches are fruits of the *caryopsis* kind, rice, corn, wheat, barley, sorghum, and teff. Of course, two major starch exceptions are potatoes and cassava, which are not grains. But, all of the grains we grow for crops are annuals, so we have to replant them and harvest them every year.

These crops aren't like apple trees. You plant an apple tree, and you can just wait for it to grow fruit every year. Now, there are still problems like pathogens and frost, but a farmer doesn't have to harvest the entire tree each fall and replant it in the spring like a farmer has to do with corn or wheat.

The Land Institute is trying to develop crops that would be perennial. There are already many perennial grasses that exist, so by crossing them with crops, perennial crops can be created. Their most successful creation is called kernza, which is derived from a species of wheatgrass, which is a broader group that what goes into smoothies, so it's a lot like whole wheat in its texture. The roots of kernza can grow very deep, up to ten feet, so that it uses less water, and isn't as prone to fluxes in climate from year to year. But, also because the whole plant doesn't have to be harvested and replanted, using kernza would release less CO_2 into the atmosphere.

Because many grasslands have been used for crops, grasslands are declining. Students in my courses are always surprised to learn that it is the grasslands of Colorado that are most endangered and not the alpine tundra. No one wants to live or shop or grow food above treeline, but of course, all of those activities are good in the grasslands.

And, not all of the grasses grown for crops are used for food. Many grasses are grown for use as biofuels. Anything that was alive and contains carbon, and so when it dies, it can be burned for fuel. This is true of wood, true of paper, and even true of manure. Of course, corn is a common biofuel, which is used to make ethanol. Yes, as my students always want to know, this is the same ethanol found in alcoholic beverages. Ethanol is most commonly used as an additive to gasoline.

Corn isn't the only grass used for biofuel, sugarcane is another popular grass used to make biofuels, but now native prairie grasses are also being used, like switchgrass. This grass is native to North American prairies and a major advantage of switchgrass is that it's perennial, so it doesn't have to be planted every year. This saves a lot of energy during harvest and replanting time. Still, as with all biofuels, there is carbon released, and some studies report that biofuels don't offer much better air pollution profiles than gasoline. Of course, there is the advantage of having a home-grown energy source.

Despite looking monotonous to the untrained eye, grasslands are actually very diverse. Most grasslands have a number of other species, most notably members of the sunflower family, the Asteraceae. Common members are sunflowers, prairie coneflower, also called echinacea, the same flower that provides the herbal supplement for building immunity. The cut flower alstroemeria that can be bought at grocery stores is native to the Argentine steppe.

Additional species that are not grasses are sometimes called broadleaf species. They are called broadleaves based on the venation of the leaves. Grasses have parallel leaf venation in rather linear shaped leaves. Broadleaf species are, well, broader. They are wider across than grasses and their venation follows the pattern of a feather or a palm with crisscrossing veins. Broadleaf species are what some people try to eradicate in their lawns. Dandelions, for example. The herbicide 2,4-D will kill broadleaf species but won't harm monocots, like grasses.

Botanists have yet to figure out the precise mechanism for action for this herbicide, but what we do know is that 2,4-D acts like the plant hormone auxin. Since auxin is a growth hormone, it stimulates growth in the plant. When the 2,4-D is absorbed, it's moved to the meristem, which is most eudicots is at the top of the plant, or the apical meristem. Once there, in high concentrations, the 2,4-D acts a bit like a plant cancer, causing uncontrolled cell growth, which eventually kills the entire plant.

Although we've talked a lot about grasses growing in grasslands and prairies, we also mentioned that grasses grow everywhere, even Antarctica, but we haven't discussed forests of grass. Yet, when one walks amongst the bamboo forests where the giant pandas live, one is walking amongst forests of grass. In fact, there is a species of bamboo called the timber bamboo, and it can grow up to 130 feet tall. That's much higher than a typical red oak, for comparison.

Bamboo are typically classified as their own tribe within the grass family, the Poaceae. They are actually cosmopolitan, which means that they are found around the world. In fact, there is a bamboo native to the U.S., *Arundinaria gigantea*.

Like many other grasses, bamboo have hollow stems. Despite this, they are very strong and are still used for scaffolding and bridges in parts of Asia. Bamboo also have very fast growth rates—the fastest rate for any plant according to the *Guinness Book of World Records*. Most bamboo reproduce vegetatively through underground stems called rhizomes. This is probably why they do a very interesting type of reproduction called masting. This occurs when all the individuals of a population all produce their seeds at once, then they don't produce seeds for several years in a row.

If bamboo isn't enough in the cool Asian grass category, then consider that lemongrass, the delicious ingredient in Thai dishes, is also a grass.

So, even though the orchids, sunflowers, and peas might have more species, the grass family is without a doubt the most important family in the world in terms of crops for energy, forage for animals, building material, and even delicious seasonings.

LECTURE 19

SHRUBLANDS OF ROSES AND WINE

A **shrub** is a somewhat overlooked plant growth form. Shrubs don't quite have the majesty and presence of a tree, and they also lack the economic importance of the grasses. Few people grow a shrub lawn, yet shrubs are important types of plants, and shrublands are an interesting ecosystem. Plus, there are economically important shrubs.

SHRUBS AND SHRUBLANDS

LEYLAND CYPRESS
Cupressus × leylandii

◇ The common definition of a shrub is a lignified (woody) plant that has more than one main stem (underground or above ground). It's difficult to judge a shrub by its height because some plants might be small trees that are grown and pruned to form a hedge, so they look like a shrub. The Leyland cypress is an example.

◇ Although shrubs might look like trees and might only be distinguished from herbs by the presence of a woody stem, their in-between status gives them advantages. For example, a shrub's woodiness means that it doesn't need to reestablish its entire above-ground shoot as herbaceous plants must do every spring. This continuous growth season after season means that most shrubs can grow larger than most herbs, which offers them a competitive advantage for sunlight.

◇ Also, if it's a particularly stressful season, a shrub can forego reproduction and wait for better times in the future. An herb doesn't have this advantage. If an annual herb doesn't reproduce, it doesn't get another chance. Perennial herbs, too, might try to flower during lean times and then not have enough storage to make it through the winter.

◇ Shrubs have some advantages over trees, too. Because shrubs don't have has many leaves as trees do, there is less water loss through evaporation in the shrub's canopy. This means that shrubs are generally more drought tolerant than trees. Shrubs also have an advantage in their shortness. A shorter plant can avoid greater wind stress. And shorter plants with branches closer to the ground are more likely to trap snow, which is a huge advantage in arid climates.

- Given that there are so many advantages to the shrub life-form, there are shrubs virtually everywhere. Shrubs are found in grasslands, in the understory of temperate or tropical forests, above the tree line in alpine regions, and in the desert landscape.

- There are whole ecosystems that are comprised of shrubs. These special, shrub-dominated ecosystems are referred to as Mediterranean-type ecosystems (MTEs), named after the landscapes that border the Mediterranean Sea. This region is known for its hot, dry summers and cool, wet winters. This type of climate is found around the world.

- MTEs are just to the north (in the Northern Hemisphere) or to the south (in the Southern Hemisphere) of deserts. Above the Sonoran Desert of Arizona and California is the chaparral region of California. Just below the Atacama Desert of Chile is the *matorral* (which means "shrubland") region of Chile. In South Africa, the fynbos (a word from Afrikaans that means "fine bush" and refers to the slender leaves found on the plants there) is located south of the Namib and Kalahari Deserts. In Australia, the range around Perth is just to the south and west of the Great Victoria Desert.

- Coincidentally, the 5 MTEs are also the great wine-making regions of the world: the Mediterranean region, including France, Spain, and Italy; Chile; California; South Africa; and Australia.

- Just as these landscapes are dominated by shrubs, they are also dominated by fire. This is in large part because of the shrubs themselves. The shrubs in these regions are evergreen, so most of them don't drop their leaves during the winter. That's good because that's when the rain falls.

- However, because of the long, hot, and dry summer, the leaves have to be tough. The leaves of many chaparral shrubs are called sclerophyllous. The sclerophylly is from lignin, that same substance that makes up wood. There's not as much of it in these leaves as in wood, but there's enough to make the leaves tough. This helps the leaf keep its structure without retaining a lot of water.

- A leaf with a lot of water in it would be a plant that is losing a lot of water through transpiration. Plants that live in MTEs can't afford to lose a lot of water, especially in those hot, dry summers.

- Many chaparral shrubs contain volatile organic compounds. This is one reason why so many of our kitchen herbs, such as sage, rosemary, bay leaves, and thyme, come from MTEs. All of these plants contain volatile oils that we can smell, and thus we add them to our food for flavor.

- These volatile oils also deter insect predators, and plants will readily vary the amount of oils in their leaves based on how much sunlight they receive and how much herbivory they are facing. Their water status can also determine how many oils are present. And these oils are flammable.

ROSEMARY

Rosmarinus officinalis

- These shrubs grow very close together, adding to the possibility of spreading fire. But the fire isn't meant to consume everything. These areas are adapted to fire, and plants typically have a way of coming back after the fire through various means, usually by resprouting or reseeding. The fire is meant to come in hot and fast and burn out the competition and promote this resprouting or reseeding.

- For plants that are resprouters, the soil protects the base of the shrub, which has a thick, woody, underground burl. After a fire, the shrub will resprout from this burl.

- In plants that are reseeders, the adult plants die in the fire. The seeds left in the soil from previous years respond to cues from either heat or chemicals leached from the charred organic matter in the soil. The seeds that accumulate in the soil are collectively referred to as the seed bank. Plants make deposits of seeds over the years, and then when a fire occurs, there is a withdrawal as the seeds begin to germinate.

- The traits of small, dry leaves and volatile oils are adaptations to frequent fires. These fire-adapted MTEs are fairly biodiverse. Although the 5 MTE regions make up only 2% of Earth's land, they are responsible for 20% of the named plant species.

- One reason that these regions are so diverse probably has to do with fire. When an ecosystem is very stable, the plants don't need to change much to adapt to new conditions because conditions are staying the same. Alternatively, if there is a lot of change occurring frequently, there are opportunities for new species to develop.

- Many areas where shrubs dominate experience regular disturbance, such as fires in the MTEs. Other areas are somewhat new in an evolutionary sense. One example of recent change is the intermountain West. This is the region west of the Rocky Mountains but east of the Sierra Nevadas. It's also called the Great Basin. Much of this vast region is dominated by shrubs.

EXAMPLES OF SHRUBS

- One of the major shrubs of the Great Basin is the sagebrush. There are actually a few dozen different plants that go by the name of sagebrush. They are all in the *Artemisia* genus in the Aster family, or Asteraceae, which is the sunflower family. There are about 300 more species in this genus, but those other species aren't known as sagebrushes.

- Although sagebrushes make up a large portion of the Great Basin, they are currently in decline because of invasive species such as cheatgrass, which outcompetes the slower-growing shrubs. Another factor in the sagebrush

BIG SAGEBRUSH
Artemisia tridentata

decline is overgrazing. Cows don't eat sagebrush, so ranchers destroyed sagebrush habitats to make way for grasses that cows would eat. This matters because a large number of species depend on sagebrush.

- Shrubs provide a lot of food to animals. Blackberry, raspberry, and thimbleberry are all shrubs that are economically important. Another shrub from the Great Basin that produces something people call a berry is the juniper. Although the small bluish balls on this plant are called berries, they're not technically berries. Juniper is a conifer, and what is commonly called a berry is actually a cone with fleshy cone scales.

- Unlike sagebrush, which is a common name for a number of North American shrubs growing in the *Artemisia* genus, creosote—a shrub in the North American Desert—stands alone. Also unlike sagebrush, the whole genus that creosote belongs to only contains 4 other plants, most of which are in South America. Creosote bush (*Larrea tridentata*) gets its name from the substance creosote, which is an oil distilled from coal tar, oil tar, or wood tar. Creosote is famous for being a very old shrub, or at least a very old clone.

⬥ The shrub of romance is the rose. Fossils of roses have been found all over the Northern Hemisphere and range from 3 to 20 million years ago, or more. All the cultivated roses around the world are descendants of naturally occurring species in the genus *Rosa*, which is divided into 4 subgenera.

⬥ Roses are in the family Rosaceae, which includes strawberries, apples, pears, plums, apricots, and peaches. What distinguishes a rose from these other members are its prickly stems, the way the leaves grow, and the formation of the fruit of the rose, which is botanically called a hip. Although they have a reputation for being a romantic flower, and perhaps delicate because of that connotation, roses are quite hardy.

ROSE HIPS

BEACH ROSE
Rosa rugosa

READINGS

Harkness, *The Rose*.

McKell, *The Biology and Utilization of Shrubs*.

Trimble, *The Sagebrush Ocean*.

QUESTIONS

1   What are the advantages of the shrub growth form?

2   Given the advantages of the shrub form, in what kinds of habitats would you expect shrubs to dominate?

LECTURE 19 TRANSCRIPT
SHRUBLANDS OF ROSES AND WINE

Do you know that song, *Scarborough Fair*? I'm not going to sing it, but it contains the words, parsley, sage, rosemary, and thyme. Sage, like the other three, is a herb, found in lots of Italian cooking, and it's in the mint family, Lamiaceae.

Sagebrush, by contrast, is a shrub, *Artemesia tridentata*, which is common the western U.S. This same genus is also called wormwood, and it's what flavors the distilled alcoholic drink absinthe.

Now, sagebrush has something very cool going for it. In a 2014 paper published in the journal *New Phytologist*, Richard Karban and his colleagues demonstrated that sagebrush listen to their kin. Dr. Karban's lab had previously found that sagebrush can communicate with each other through volatile organic compounds released by their leaves and picked up by the leaves of neighbors. This is basically like, one plant releasing a smell and the other plant smelling it and recognizing it as a signal.

Now, of course, sagebrush plants don't have noses, but they are emitting these chemicals and other plants are responding. They are responding to a signal of herbivory. If grasshoppers eat the leaves of one plant, it releases the volatile organic compound, the scent, if you will to tell neighboring plants, "Hey, I'm being eaten over here, so you better get ready." It turns out that neighboring plants will show less damage from herbivory if they are allowed to receive the signal. If their leaves are blocked, by being tied up in plastic bags, they don't show less herbivory damage when the plastic bag is removed.

What's even more interesting about this is that plants can recognize the scent signals of related individuals. This makes sense because if all plants recognized the same signal, the initial signal could help competitors. So, Dr. Karban and his colleagues took cuttings of one plant and grew them in the greenhouse.

Now, they had 60 individual plants all with the same genetic makeup. These clones responded to signals from other clones but didn't show as great a response when the signal came from a stranger.

And in the latest piece of work from Dr. Karban's lab, they identified the identity of some of these chemicals. It's a creative method they used to make this determination. Previous work had shown that these sagebrush plants will emit chemicals even if they are simply clipped with scissors. The clipping mimics herbivory, so the response of emitting the volatile organic compounds is the same. So, to determine what exactly these compounds were, researchers bagged leaves of the shrub and then clipped other leaves. They collected the air around the bagged leaves and put this air into a gas chromatograph and a mass spectrometer. These instruments helped to identify the chemicals being used as the signal.

The researchers found that there were primarily two different chemicals being used by the plants. They called the plants chemotypes. One chemotype produced thujone and the other chemotype produced camphor, which is well-known for discouraging moths from laying their eggs. They also found that chemotype was a highly heritable trait, which is important because plants should be able to recognize signals from their parents. So, there you are sagebrush show us how important it is to listen to our parents.

A shrub is a somewhat overlooked plant growth form. Shrubs don't quite have the majesty and presence of a tree, and they also lack the economic importance of the grasses. Few people grow a shrub lawn. Yet, shrubs are important types of plants and shrublands are an interesting ecosystem all to themselves. Plus, there are economically important shrubs. It could be argued that coffee, that second most traded commodity in the world after oil, might be a shrub.

So, what is a shrub anyway? How is it different from a tree? The common, most understood definition of a shrub is a lignified plant that has more than one main stem. The lignified part just means that it's woody. The more than one stem part of the definition could be underground, or above ground. It's difficult to judge a shrub by its height because some plants might be small trees that are grown and pruned to form a hedge, so they look like a shrub. The Leyland Cypress is a good example of this.

Although shrubs might look like trees, and only be distinguished from herbs by the presence of a woody stem, their in-between status gives them advantages. For example, a shrubs woodiness means that it doesn't need to reestablish its entire above ground shoot as herbaceous plants must do every spring. This continuous growth season after season means that most shrubs can grow larger than most herbs, which offers them a competitive advantage for sunlight.

Also, if it's a particularly stressful season, a shrub can forego reproduction and wait for better times in the future. A herb doesn't have this advantage. If an annual herb doesn't reproduce, it doesn't get another chance. Perennial herbs, too, might try to flower during lean times, and then not have enough storage to make it through the winter.

On the other hand, shrubs have some advantages over trees, too. Because shrubs don't have has many leaves as trees do, there is less water loss through evaporation in the shrub's canopy. This means that shrubs are generally more drought tolerant than trees. Shrubs also have an advantage in their shortness. A shorter plant can avoid greater wind stress. And, shorter plants with branches closer to the ground are more likely to trap snow, which is a huge advantage in arid climates.

Given that there are so many advantages to the shrub life form, there are shrubs virtually everywhere. Shrubs can pop up in grasslands. They can show up in the understory of temperate or tropical forests, they are found above treeline in alpine regions, and they can make up a majority of the desert landscape.

There are, however, whole ecosystems that are composed of shrubs. These special, shrub-dominated ecosystems are referred to as Mediterranean-type Ecosystems. Named after the landscapes that border the Mediterranean Ocean. This region is known for its hot, dry summers and cool, wet winters. This type of climate is found around the world and collectively referred to a Mediterranean-type Ecosystems, sometimes just MTEs.

Mediterranean ecosystems are just to the north in the northern hemisphere and just to the south in the southern hemisphere of deserts. So above the Sonoran Desert of Arizona and California is the chaparral region of California. Just below the Atacama desert of Chile is the matorral region of Chile. Matorral is a Spanish word that means shrubland. In South African, the fynbos is located south of the Namib and Kalahari deserts. Fynbos is a word from Afrikaans that means fine bush and refers to the slender leaves found on the plants there. In Australia, the range around Perth is just to the south and west of the Great Victoria Desert.

So, there you have it. The five MTEs of the world. Coincidentally, they are also the great wine-making regions of the world. Of course, the Mediterranean region, including France, Spain, and Italy. I've already mentioned Chile and California, but there are also sizable MTEs in South Africa and Australia. I'm fairly certain my Ph.D. advisor, Dr. Phil Rundel, recognized this as he set about to become a premier MTE ecologist. Since Phil is a genius, he would no doubt know that he would have to travel to these fine locations for work.

Just as these landscapes are dominated by shrubs, they are also dominated by fire. This is in part because of the shrubs themselves. The shrubs in these regions are evergreen, so most of them don't drop their leaves during the winter. That's good because that's when the rain falls. However, because of the long, hot and dry summer, the leaves have to be tough—literally. The leaves of many chaparral shrubs are called sclerophyllous, which is sclero for hard and phyllous for leaf, so literally a hard leaf. The sclerophylly is from lignin, that same substance that makes up wood, only there's not as much of it in these leaves as in wood, but there's enough to make the leaves tough. This helps the leaf keep its structure without retaining a lot of water.

A leaf with a lot of water in it would be a plant that is losing a lot of water through transpiration. Plants that live in these MTEs can't afford to lose a lot of water, especially in those hot, dry summers. So, the leaves don't have a lot of water, and they contain fire promoting chemicals in their tissues.

Many chaparral shrubs contain volatile organic compounds, and this is one reason why so many of our kitchen herbs come from MTEs—think sage, not sagebrush—though if you've ever smelled this plant, it has a strong odor as well, rosemary, bay leaves, and thyme. All of these plants contain volatile oils that we can smell, and thus we add them to our food for flavor. These volatile oils also deter insect predators, and plants will readily vary the amount of oils in their leaves based on how much sunlight they receive and how much herbivory they are facing. Their water status can also determine how many oils are present. And, these oils are flammable.

A dramatic example of this flammability can be seen in the Biblical story of the burning bush. Botanists think that this shrub might be *Dictamnus albus*, which grows in MTEs and woodlands. There are a number of volatile organic compounds in this shrub, among them is the chemical benzene. Now, this benzene readily degrades into chavicol, which is a flammable hydrocarbon, like methane. You can actually sort of ignite the air just outside of the flowers, which is why it's also called gas plant. It's rumored that this plant can actually combust on hot days. Although there is no evidence of this in the literature, it's a cool plant nonetheless.

So, here we have these shrubs with small, dry leaves that contain fire promoting compounds. These shrubs grow very close together, adding to the possibility of spreading fire. Like the burning bush, though, the fire isn't meant to consume everything. These areas are adapted to fire, and plants typically have a way of coming back after the fire through various means, usually by resprouting or reseeding. The fire is meant to come in hot and fast and burn out the competition and promote this resprouting or reseeding.

For plants that are resprouters, the soil protects the base of the shrub, which has a thick, woody, underground burl. After a fire, the shrub will resprout from this burl. Examples of this kind of resprouting shrubs are Chamise or *Adenostoma fasciculatum*, and Manzanita plants in the Arctostaphylos genus, and Toyon, *Heteromeles arbutifolia*, also known as California holly.

In plants that are reseeders, the adult plants die in the fire. The seeds left in the soil from previous years respond to cues from either heat or chemicals leached from the charred organic matter in the soil. The seeds that accumulate in the soil are collectively referred to as a seed bank. Plants make deposits of seeds over the years, and then when there is a fire, there is a withdrawal as the seeds begin to germinate.

What's also very interesting about these fire adaptations is that they are not seen in the Chilean matorral. This has provided ecologists with an interesting natural experiment. The Mediterranean-climate region of central Chile doesn't receive much lightening, unlike the other MTE regions. The high Andes mountains in Chile protect the matorral from summer storms and lightening. Thus, botanists can compare the native flora of Chile with the native flora from other regions to determine if these aspects of the shrubs really are adaptations to fire.

It turns out that the native Chilean flora doesn't show many of these fire adapted traits. This finding is evidence that the traits of the small, dry leaves and the volatile oils really are adaptations to frequent fires.

One thing that's also interesting about these fire-adapted MTEs is that they are fairly biodiverse. In the *Encyclopedia of Biodiversity* chapter on MTEs, Dr. Phil Rundel of UCLA, writes, "Although the combined area of these five regions is little more than 2% of the land area of the Earth, MTEs are home to 50,000 species of vascular plants, 20% of the world's total of about 250,000 named species." That puts the biodiversity here into perspective. Two percent of the land responsible for 20% of the plant species.

Why might these MTEs be so diverse? One reason probably has to do with fire. When an ecosystem is very stable, the plants don't need to change much to adapt to new conditions because conditions are staying the same. Alternatively, if there is a lot of change occurring frequently, there are opportunities for new species to develop.

Many areas where shrubs dominate experience regular disturbance, like fires in the MTEs. Other areas are somewhat new in an evolutionary sense. One example of recent change is the intermountain West. This is the region west of the Rocky Mountains but east of the Sierra Nevadas. It's also called the Great Basin. Much of this vast region is dominated by shrubs, but it also only came into being about 15,000 years ago when Lake Bonneville dried up.

Lake Bonneville was a prehistoric lake that developed at the base of the glaciers. Lakes like this are called pluvial because the cold air over the glaciers meets with warmer air over the land, and that creates rain. Pluvial comes from the Latin word pluvia, which means rain. Lake Bonneville covered much of the eastern part of the Great Basin.

Another prehistoric pluvial lake, Lake Lahontan occupied much of northwestern Great Basin. But, it's only relatively recently that the shrubs that have come to dominate this area.

We started this lecture with one of the major shrubs of the Great Basin, the sagebrush. There are actually a couple dozen different plants that go by the name of sagebrush. They are all in the *Artemisia* genus in the Aster family, the Asteraceae. Do you remember that family? Aster means star—does that help? This is the sunflower family, so sagebrushes are in the sunflower family. Remember that every petal-looking part is actually a whole flower, so this is an inflorescence. There are about 300 or so more species in this genus, but those other species aren't known as sagebrushes.

Although sagebrushes make up a large portion of the great basin, they are currently in decline because of invasive species like cheatgrass, which outcompetes the slower growing shrubs.

Another factor in the sagebrush decline is overgrazing. The overgrazing isn't because cows eat sagebrush, Cows don't care for it, so ranchers destroyed sagebrush habitats to make way for grasses that cows would eat. This matters because a large number of species depend on sagebrush, some of which, aren't found anywhere else, like the Sage Grouse, a ground-dwelling bird that feeds exclusively on sagebrush during the winter months when little else is available.

Interestingly, shrubs provide a lot of food to animals. Think of blackberry, raspberry, and thimbleberry—all shrubs. Blueberries are also smallish ground dwelling shrubs as are huckleberries, and all of these shrubs are important sources of food to many animals, including bears.

Another shrub from the Great Basin that produces something people call a berry is a juniper. Although the small bluish balls on this plant are called berries, they're not technically berries. Do you know what they might be? Let me give you a hint. The juniper doesn't flower. Have we met plants that had such structures but didn't have flowers? Yes. These are the conifers. Juniper is a conifer, and what is commonly called a berry is actually a cone with fleshy cone scales.

Those who imbibe alcohol might know that juniper cones are used to flavor gin. In fact, the word gin comes probably comes from the Dutch word for juniper, which is jenever. The Dutch popularized gin in the 17th century, but juniper was used as a medicinal long before that. Unlike other liquors, gin isn't made from a specific grain. In fact, it can be made from any sort of neutral spirit, so long as it is flavored with juniper berries, which, again, aren't berries because they're not botanically fruits—they are cones.

If we go south from the Great Basin just a bit, we'll come to the Mojave Desert. Desert plants are so interesting they'll get their own lecture, but there is one shrub in the North American desert that we should discuss here, and this shrub is creosote. Unlike sagebrush, which is a common name for a number of North American shrubs growing in the *Artemisia* genus, creosote stands alone. Also unlike the sagebrush, the whole genus that the creosote belongs to only contains four other plants, most of which are in South America.

The scientific name of the creosote bush is *Larrea tridentata*. What was once thought to be a single species is now known be three genetically different shrubs. Mojave Desert creosotes have 78 chromosomes, Sonoran Desert creosotes have 52 chromosomes, while those of the Chihuahuan Desert have just 26 chromosomes. Speciation will often occur in plants through chromosome increases. When plants have a larger than typical number of chromosomes, they are called autoploids. The largest number of chromosomes, 78, may help the Mojave Desert creosote survive less summer rainfall than creosotes get in the other two deserts.

Creosote bush gets its name from the substance creosote, which is an oil distilled from coal tar, oil tar, or wood tar. The substance creosote contains a number of phenols and other organic compounds that give it a powerful smell.

Creosote bush also contains a wide variety of phenols and organic compounds and although the smell of the leaves is strong, particularly when crushed or after a rainfall, I find it a pleasing smell, not like coal tar. One might think that wood that has been treated with creosote has been treated with compounds from the creosote bush, but this is not the case. The creosote bush is not used to make creosote the substance that is used to treat wood like railroad ties or street lamps.

Creosote is famous though for being a very old shrub, at least a very old clone. The individual stem of the creosote can live up to 90 years, but then its branches will die. The crown at the bottom will split and give rise to new branches. This continues for thousands of years. The oldest clone grows in the Mojave Desert of California. Its age was determined by a professor at the University of California, Riverside, Frank Vasek. He estimated that this clone is 11,700 years old using radiocarbon dating of the outer, and thus oldest, wood fragments. It's been given the name King Clone, because of its age. Definitely, this is one of the oldest organisms in the Mojave Desert.

One noticeable pattern about creosote bushes in the desert is their surprising regular distribution pattern. Plants inhabit the landscape in three basic patterns. They can be clumped, such as trees around a spring in the desert, they can be random when there is no discernible pattern, or they can be regular.

At first, botanists thought this was because the creosote bush was trying to maximize the amount of water to the roots, by staying far away from its neighbor. After laboratory experiments, though, botanists found that creosote exuded chemicals from its roots that actually prevented other roots from growing near it. This sort of activity is called allelopathy. The word allelopathy comes from the French word allélopathie—I know my French is atrocious-from the Greek allel meaning another and pathos, which means suffering. So, allelopathy is to make the other plant suffer.

While we're talking about suffering, we must consider the shrub of romance, the shrub full of love and pain, the rose. As the Neil Young song goes, "Love is a rose, but you better not pick it." Of course, this adage of the rose and love is ancient. The beauty of the flower, the revelry of love, the agony of love lost and the thorns that represent it.

But wait, they're not thorns. Botanically, thorns are actually modified stem tissue, which means that a true thorn is a modified branch. For example, honey locust and hawthorn have true thorns. So, what does a rose actually have that will prick your skin? They do, indeed, prick, and they are botanically called prickles. A true thorn is actually very difficult to break off, but the prickles from roses, and also blackberries, can be broken off fairly easily, especially if one is wearing heavy duty gardening gloves.

Poetry and prose of the rose go back to the ancient Greeks. Fossils of roses have been found all over the Northern Hemisphere and range from 3–20 million years ago, or more. Roses have been around awhile and they've also traveled far and wide. All the cultivated roses around the world are descendants of naturally occurring roses in the genus *Rosa*. Interestingly, the genus *Rosa* is divided into four subgenera. These are categories within the genus *Rosa*. For example, *Rosa stellata*, or the wild desert rose, is in the sub-genus *Hesperrhodos*, which is Greek for Western rose. There are only two species of Rosa in this sub-genus, and both are found in the southwestern United States.

The genus *Rosa* is the type genus for the family in which it's found, thus roses are in the Rosaceae. There are a number of economically important fruits in this family, like strawberries, apples, pears, plums, apricots, and my favorite, peaches. What distinguishes a rose from those other members are the prickly stems we just mentioned, the way the leaves grow, and the formation of the fruit of the rose, which is botanically called a hip. So, rose hips are the correct botanical definition of the rose fruit.

Roses can also widely interbreed, which is how we have so many varieties of roses. The exact number of rose species is unknown, though it's probably between 100 and 150 different species. It's not that botanists haven't tried, it's because some roses are so similar, their differences are debatable. Using genetic techniques, such as DNA fingerprinting, botanists can find different alleles, or types, for different genes. But, how many different alleles make up a new species is debatable, even amongst botanists. There are even terms for scientists that describe their tendencies to look for big differences or small ones. If a scientist thinks to put more types into one species name, they're called lumpers. If a scientist wants every little difference to be a new species, they're called splitters. The academic debates about species names can persist for decades.

Although they have a reputation for being a romantic flower, and perhaps delicate because of that connotation, roses are quite hardy. There are roses in the arctic, and in the desert. Roses are found in subalpine high elevations, too. There's another member of the rose family, the cinquefoil, which refers to the five-leaflet leaf of the shrub that I've seen growing almost to treeline.

And speaking of treeline, as you may have known or probably guessed, shrubs can grow higher than trees. Shrubs are good adapters, and because they are shorter lived than trees, their populations can change faster than trees. A 2013 paper from the journal *Biological Conservation* used satellite imagery across Northwest Yunnan, China to detect the increasing range of shrubs in the Himalayan Mountains. In the time span from 1990–2009, the researchers found that 39% of the alpine grassland had converted to shrubland. The authors suggested that this change was not only due to climate change but also a response to grazing. As climates become more unpredictable, woody plants, especially shrubs are likely to succeed. One thing is certain, though, shrubs are hardy and quick to change.

LECTURE 20

THE DESERT BONANZA OF PLANT SHAPES

Deserts are located just above tropic zones usually on the south and west of large continents. Deserts are classified by rainfall, not by temperature, so there are a number of cold deserts, such as the Great Basin Desert in the United States. Usually, deserts are areas that receive less than around 10 inches of precipitation per year. Much of the vast variety of desert plant forms are adapted to use water quickly, when it's available, and then to withstand certain periods of drought. It's this necessary adaptation that gives desert plants so many unique forms.

CRASSULACEAN ACID METABOLISM

◇ One of the most interesting plant adaptations to the desert is another pathway to photosynthesis. In the C4 pathway of photosynthesis, the outside cells have a different enzyme called PEP carboxylase that only picks up carbon dioxide. Then, that carbon dioxide is moved into the regular C3 cycle that most plants have.

◇ The new pathway that some desert plants have is called crassulacean acid metabolism (CAM). A common trait among CAM plants is that they have succulent leaves because their leaves are full of water.

◇ It seems a bit counterintuitive that a plant growing in a desert would have leaves full of water. If the plant were to have open stomata, the pores in their leaves that take in carbon dioxide, the water loss would be extreme.

◇ CAM plants deal with this by closing their stomata during the day and opening them at night. Although photosynthesis can't take place at night, CAM plants convert the carbon dioxide they pick up at night into malic acid and store it in the vacuole, the large liquid-filled sac that takes up most of the plant cell volume, until daylight.

◇ When it's sunny again, the plant goes through the light reactions just as a C3 plant would, only the carbon comes from the malic acid, not directly from the stomata. All plants can use C3 photosynthesis, and some are able to use all 3 types of photosynthesis—C3, C4, and CAM—at different times.

◇ The other advantage is that the main enzyme of C3 photosynthesis, RuBisCO, cannot come into contact with oxygen. This is good because as temperatures increase, RuBisCO's affinity for oxygen will also increase. Because the plant doesn't use oxygen, this affinity for oxygen reduces the plant's efficiency.

◇ The malic acid that the plant is storing at night makes the leaves of a CAM plant acidic, with a lower pH, at night, which is why it's called crassulacean *acid* metabolism. As this malic acid is used during the day in the light reactions, the leaves become basic, with a higher pH.

- The CAM pathway has been detected in more than 1000 flowering plants in about 17 different families—about 0.3% of all flowering plants. Plants that are able to do CAM exist in a range of CAM activity. Plants can be obligate, meaning that they are all CAM all the time. Other plants might be facultative CAM, meaning that when conditions are good, the plant would just perform normal C3 photosynthesis, but when conditions get dry or hot, the plant can switch to CAM photosynthesis.

- Plants using CAM lose about 1/10 as much water per unit of sugars made compared to those using the standard C3 photosynthesis. But there is a trade-off: the overall rate of photosynthesis is slower, so CAM plants grow more slowly than most C3 plants.

- The CAM photosynthetic pathway is also found in members of the bromeliads (Bromeliaceae) and in some orchids (Orchidaceae). The CAM pathway evolved multiple times in the history of these families.

THE CACTUS FAMILY

- The CAM photosynthetic pathway is also found in the cactus family, Cactaceae, which has about 1600 species, mostly distributed in the Western Hemisphere. All the cacti can perform CAM, though most seedlings use the C3 pathway.

- For the most part, the Cactaceae are unmistakable. Cacti are recognized by their green, photosynthetic stems, and what some people might think of as their needles are actually modified leaves called spines. The function of the spines is to deter animals from eating them. Many cacti are not too palatable because of the acids in their tissues, but those tissues do contain a lot of water.

- Despite the protection these plants have, they are important water sources to animals in deserts. And the fruits of many cacti are important food sources, not just for animals. In Mexico, nopalitos are a common vegetable. These are the pads of the *Opuntia*, or beavertail, cactus. Sheep, goats, and cows also eat the pads as fodder.

◇ It's not just the vegetative parts of cactus that are eaten. Cactus fruits are important for a variety of desert animals, and humans also eat the fruits of the *Opuntia*.

◇ Cacti, especially *Opuntia*, can grow very efficiently in arid lands. Within 2 years, a cultivated *Opuntia* can begin to bear fruit. But not all cacti grow so quickly. Saguaro cacti are very slow growing. In fact, they can take 75 years to produce one arm.

◇ Some plants in deserts are commonly called cacti but aren't. Examples of such plants include agave, aloe vera, and Joshua tree. All of these plants are monocots, so they are more closely related to lilies than they are to cacti. The Cactaceae are eudicots, and they are more closely related to carnations.

BEAVERTAIL CACTUS
Opuntia basilaris

DESERT TREES

- When people think of deserts, they don't usually think of trees, but trees do live in deserts. A number of arborescent forms grow in the desert, though they are not all cacti. A desert arborescent form is the Joshua tree, or *Yucca brevifolia*, which is related to non-tree-forming yuccas. Joshua trees are in the Asparagaceae family; they are related to asparagus. The agaves are also in this family.

JOSHUA TREE
Yucca brevifolia

- Because leaves can lose so much water, many desert plants will flower during the dry season and wait for periodic rains before they produce leaves. The elephant tree of the Baja peninsula is a good example. It produces small yellow flowers during the dry season, which are an important source of nectar for pollinators.

- This delay in leaf development also enables insects to see the very small flowers without a covering of leaves. Because flowers typically don't photosynthesize, they don't transpire nearly as much, so there's little water loss for the plant to have flowers out. This tree, which is really more of a shrub, is called the elephant tree because of its thickened tree trunk, which stores water.

- A tree that has a similar kind of water-storing mechanism is the famous baobab tree of the African plains, though they also live in India, Madagascar, Australia, and the island of Socotra. All the regions where baobab trees grow are savanna, which might be desertlike half the year and grassland-like the other half of the year when the rains come. The adaptations of the baobab tree are mostly to save water.

- Like other plants in arid regions, the baobab drop their leaves during the dry season. They then put out leaves and flowers before the rains arrive again. In studies, the dominant cue seems to be photoperiod, as it is for many plants. In short, baobab trees will synchronously flower and put out leaves when the critical day length is slightly longer than 12 hours.

- Another interesting desert tree is called the boojum tree, *Fouquieria columnaris*, which is in the Foqueriaceae, or ocotillo, family. The boojum tree is endemic to the Baja California peninsula and surrounding areas. Like the other member of this family, the ocotillo, both of these plants flush with leaves in response to rainfall. The ocotillo stems have spines, like a cactus, but the spines are made by leaves. No other plant produces spines like this.

BOOJUM TREE
Fouquieria columnaris

DESERT ADAPTATIONS

- Some plants don't tolerate dryness. Perennial plants, such as the ocotillo and the baobab, time certain events, such as flowering, based on when rainfall occurs. Even most deserts have a time of year when it rains. Annual plants time their entire lives around these rainfalls.

- Every spring in the Mojave Desert, for a few weeks, the desert blooms with many desert annuals. These small wildflowers begin to germinate hours after a rainfall and then grow quickly and produce flowers, fruits, and seeds before the really hot and dry weather of the summer begins. The seeds are then dispersed into the soil by summer and wait through the winter until the next rain.

- A different way to avoid the stress of the dry season is to become dormant. A striking example of a plant like this is the resurrection plant. This plant is a lycopod, which makes it a relative of ferns. The resurrection fern is native to the Southwest United States. This plant can look dead during the dry season, and then when the rains come, it can completely revive.

- The resurrection plant prepares to have a slowdown in metabolism by storing extra sugars, such as sucrose, and amino acids, which make up proteins. Next, it starts to produce antioxidants. Lastly, it produces proteins that help the internal organelles and cell walls toughen up for the dry season.

- Another desert adaptation is to avoid leaves altogether. Some plants don't even photosynthesize. Although these plants had the ability to make chlorophyll, over time they were selected to lose this ability and instead use sugars from a host plant. These are parasitic plants, such as dodder, which is a bright orange plant that can often grow on creosote shrubs in the desert.

READINGS

Crawford, *Plants at the Margin*.

Nobel, *Environmental Biology of Agaves and Cacti*.

Tudge, *The Tree*.

QUESTIONS

1 Why are there so many unusual forms of plants in deserts?

2 How does the CAM photosynthetic pathway differ from the C4 pathway?

LECTURE 20 TRANSCRIPT
THE DESERT BONANZA OF PLANT SHAPES

You hear the word desert, and you might imagine a lifeless place. But that's so wrong. Deserts create the largest variety of unusual plant shapes on earth. They range from tiny desert annuals to giant, towering Saguaros. Deserts have upside-down trees and ocotillos, which look like sticks in the ground. The way desert plants have adapted to a lack of water has created a myriad of shapes and forms. This makes deserts remarkably rich in plant morphologies.

For example, the resurrection plant may look quite dead, but when you put it in water, you see that why it's called the resurrection plant. So here is what the resurrection plant looks like normally during a dry season. There is no water, it hasn't rained. I can just avoid that stress by curling up into this very dry ball. But then, as soon as the rains come—and this one has been soaking for only about a day in a very little bit of water—and you can see how much greener and how the whole plant has just sort of unfurled. And now, remember, this is a fern, so this doesn't have vascular tissue, so it's going to stay very low to the ground. But this is what's going to help it collect water and live during a rainy season. And then, when the dry season comes it can fold back up again like this to avoid the stress of drought.

The Atacama Desert is the driest desert on Earth. There were adults in Arica in the 1990s who had never seen rain. The last year I was there in 2012, it actually rained—maybe a millimeter or so, and people came outside to stare at the sky.

To get to the high Andes, you drive Highway 11 in a series of perilous turns through the the hyperarid zone. This region is so dry that there are no plants—none. I'm a botanist—I was looking. For the first 2000 meters of elevation along that road, not a single plant is found. Yet, just as the road climbs higher, the increase in elevation brings enough fog water off of the Pacific Ocean to support plant life. It still almost never rains, but the moisture in the fog is enough for some plants to survive.

At 2200 meters, a spectacular arborescent cactus appears if out of nowhere, and then it dominates the slopes where little else grows. This cactus is *Browningia candelaris*. The genus name, *Browningia* is from the botanist W.E. Browning, an American, who worked in Santiago. The specific epithet, *candelaris*, speaks for itself—the top branches of the cactus resemble a candelabra. The bottom of the cactus is covered with spines up to 22 centimeters in length—that's about 8 and ½ inches. The spines cover the bottom six feet of the main stem, and at that length would deter even the hungriest guanaco, which is a native camelid herbivore in South America. The cactus drops out altogether at 2800 meters, where more shrubs and other desert plants begin.

Deserts are located just above tropic zones usually on the south and west of large continents. Deserts are classified by rainfall, not by temperature, so there are a number of cold deserts, such as the Gobi Desert in China and Mongolia or the Great Basin Desert in the U.S. In fact, the largest desert in the world is a cold desert—it's the Antarctic desert, and this desert is larger than the Sahara, and receives only slightly more annual precipitation than the Sahara. These deserts definitely go below freezing in winter time. Usually, deserts are areas that receive less than around 10 inches of precipitation a year. For comparison sake, New York City receives over 40 inches of precipitation a year.

Desert plants have an amazing number of adaptations to help them survive the driest places on our planet. For this reason, I think botanists are naturally drawn to deserts.

I need to take a minute here and explain something. You see, there is a big difference between botanists and horticulturalists. This is a difference I've struggled with all my life as a botanist. People think because I'm a botanist, I must be really good at growing plants. I know many botanists who are exceptionally good gardeners as well, but I'm not one of them.

My secret confession is that I like to look at plants in their native habitats, but I'm not so good at remembering to water, or repot, or fertilize my houseplants. And, I'm even less good at watering and pruning and planting outdoor plants. As a good botanist, though, I've come to love gardening with native plants and plants that can deal with the weather of the climate where I live. I've ripped up most of the grass in the front yard to save water. This is called xeriscaping, from the Greek *xeric* meaning dry and scape for landscape. For the longest time, my Dad thought I was saying zero-scaping with a z instead of *xeric*, with an x. Xerophytes are plants adapted to live in dry places, such as deserts.

One of the most interesting adaptations to the desert is another pathway to photosynthesis. When we talked about grasses, we learned about the C4 pathway of photosynthesis, where the outside cells have a different enzyme called PEP carboxylase that only picks up carbon dioxide. Then, that CO_2 is moved into the regular C3 cycle that most plants have.

The new pathway that some desert plants have is called CAM, which stands for crassulacean acid metabolism. The word crassulacean refers to the plant family in which this pathway was first discovered. Like all plant families, the family name comes from the type genus, followed by the suffix -aceae. In this case, the genus *Crassula* is the type genus, which comes from the Latin word *crassus*, for thick. It was the name Carl Linnaeus gave to this group of plants because of their thick leaves. A common plant in this family is the jade plant or *Crassula ovata*. A common trait amongst CAM plants is that they have succulent leaves because their leaves are full of water.

It does seem a bit counterintuitive that a plant growing in a desert would have leaves full of water. If the plant were to have open stomata, the pores in their leaves that take in CO_2, the water loss would be extreme. CAM plants deal with this by closing their stomata during the day and opening them at night. Although photosynthesis can't take place at night, CAM plants convert the carbon dioxide they pick up at night into malic acid and store it in the vacuole until daylight. The vacuole is the large liquid filled sac that takes up most of the plant cell volume.

When it's sunny again, the plant goes through the light reactions just as a C3 plant would, only the carbon comes from the malic acid, not directly from the stomata. All plants can use C3 photosynthesis, and some are able to use all three types of photosynthesis, C3, C4, and CAM at different times.

The other advantage here is that the main enzyme of C3 photosynthesis, RuBisCO, cannot come into contact with oxygen. This is good because as temperatures increase, RuBisCO's affinity for oxygen will also increase. Since the plant doesn't use oxygen, this affinity for oxygen reduces the plant's efficiency.

It's the malic acid that the plant is storing at night that makes the leaves of a CAM plant acidic, with a lower pH, at night, which is why it's called crassulacean acid metabolism. As this malic acid is used during the day in the light reactions, the leaves become basic, with a higher pH.

The CAM pathway has been detected in more than 1000 flowering plants in about 17 different families—about 0.3% of all flowering plants. Plants that are able to do CAM exist in a range of CAM activity. Plants can be obligate, meaning they are CAM all the time. Other plants might be facultative CAM, which means that when conditions are good, the plant would just perform normal C3 photosynthesis. But, when conditions get dry or hot, the plant can switch to CAM photosynthesis. Plants using CAM lose about 1/10 as much water per unit of sugars made compared to those using the standard C3 photosynthesis. But there is a trade-off, the overall rate of photosynthesis is slower, so CAM plants grow more slowly than most C3 plants

The CAM photosynthetic pathway is also found in members of the bromeliads, family Bromeliaceae, and in some orchids, Orchidaceae. Both of these families have a large number of epiphytes or species that live in tree canopies. This correlation of epiphytes and CAM is strong in the orchids, but not so much in the bromeliads. What does cause CAM in a bromeliad? It's uncertain, but it is interesting that not all members of these groups have the CAM pathway. This provides evidence that the CAM pathway evolved multiple times in the history of these families.

The CAM photosynthetic pathway is also found in the cactus family, the Cactaceae. The cactus family has about 1600 species, mostly distributed in the New World, or Western Hemisphere. There is one genus, that has dispersed to tropical Africa, including Madagascar, Sri Lanka, and Southern India. This genus is epiphytic, which means that the cactus lives on other plants; it was most likely distributed by birds. But, mainly, cacti are a Western Hemisphere group. All the cacti can perform CAM, though most seedlings use the C3 pathway. The only adult cacti to use C3 photosynthesis is the primitive Pereskia.

Many people think that the Cactaceae are unmistakable. For the large part, this is true. Cacti are recognized by their green, photosynthetic stems and what some people might think of as their needles. These are actually modified leaves called spines. No doubt the function of the spines is to deter animals from eating them. Many cacti are not too palatable because of the acids in their tissues, but those tissues do contain a lot of water. Despite the protection these plants have, they are important water sources to animals in deserts. I've seen desert bighorn sheep knock over a barrel cactus with their hooves and then the whole herd will nibble at it for a few days, not just for the carbohydrates, but for its water as well. Certainly, the fruits of many cacti are important food sources, and not only for animals.

In Mexico, nopalitos are a common vegetable. These are the pads of the Opuntia, or beavertail, cactus. In botany-speak, the pads are called cladodes. The needles are removed, and the cladodes are sliced thin and sautéed with eggs or meat or other vegetables. They can be pickled, too. To some people, the mucilage in the pad is reminiscent of okra. This mucilage is actually a complex carbohydrate that's really good at storing water. So, it's just another way for the plant to store water in the cladodes. Sheep, goats, and cows will also eat the pads as fodder. Because of the water content in the cladodes, sheep will essentially stop drinking water when they are fed cactus.

It's not just the vegetative parts of cactus that are eaten. Cactus fruits are also important for a variety of desert animals, and humans eat the fruits of the Opuntia as well. Colloquially known as tunas, the cactus pear, which is not a pear at all, is enjoying some fame amongst foodies for its high vitamin c content and its unique taste, which is a bit tart like kiwi.

Cacti, especially Opuntia can grow very efficiently in the arid lands. Within two years, a cultivated Opuntia can begin to bear fruit, but not all cactus grow so quickly. When you think of the giant cactus in Arizona, the one with arms on the license plate of the state, that's a Saguaro cactus, and they are very slow growing. They can take 75 years just to produce one arm. Saguaro, and other arborescent cacti, can also produce a very interesting looking form called cristate or crested, and botanists still debate the cause of this form. Some suggest it's a genetic mutation, but others think it's caused by lightning strike or a freeze during a particular time of development, and recent evidence suggests it's caused by a viral infection.

I did say that cacti were relatively easy to identify, and that's because there is another group of plants that can look a lot like cacti, and these are the euphorbs, or the family Euphorbiaceae. They, too, can have succulent stems, they can also have spines, and they can also use CAM photosynthesis, though not every member of the Euphorbiaceae does this. So, how would you tell these two families apart if you saw them in a desert garden? Euphorbs typically have milky sap, so if you make a tiny scratch in the leaf, it will usually produce a white latex. This latex is poisonous and probably developed to deter insect herbivores.

Although members of the Euphorbiaceae are all over the world, in the deserts of Africa and Madagascar, they look for all the world like cacti. This is the classic example of convergent evolution. This occurs where two, totally unrelated groups, like cacti and euphorbs, both converge on a particular morphology, in this case, succulent stems for water storage and CAM photosynthesis and leaves reduced to spines to prevent, or at least slow down herbivory. This convergence is also interesting because some euphorbias don't look anything like a cactus. For example, the most familiar euphorb in the U.S. is probably the poinsettia plant of Christmastime fame.

Poinsettias are short-day plants, and you may recall from our lecture on circadian rhythms, that this means the plant needs a long night to flower. Long nights in the Northern Hemisphere, such as Mexico, part of the native range of poinsettias, occur around the winter solstice, around Christmas time. Even though the flowers of this plant are actually very tiny, the bright red leaves attract pollinators, and as there isn't much else flowering this time of year, they get noticed.

Another Christmastime plant is the Christmas cactus, and this is an actual cactus. The leaves are again the modified small spines on the margins of the flattened stems. Christmas cactus is a plant that flowers around Christmastime too, just like the poinsettia because it, too, is cued by the short day, or rather, the long night.

There are other plants in deserts that are commonly called cactus that aren't cactus at all. Examples of such plants might be agave, aloe vera, and Joshua tree. All of these plants are actually monocots, so they are more closely related to lilies than they are to cacti. The Cactaceae are eudicots, and they are more closely related to carnations.

When people think of deserts, they don't usually think of trees, but trees do live in the deserts. The cactus we met in the beginning of this lecture, *Browningia candelaris*, is considered an arborescent cactus, and there are a number of arborescent forms growing in the desert, though they are not all cactus. Another desert arborescent form is the Joshua tree. Apparently, this tree was named by the early Mormon settlers who believed it resembled the biblical figure of Joshua holding his hands up in prayer. The scientific name is *Yucca brevifolia*, and yes, Joshua trees are related to non-tree forming yuccas. Joshua trees are actually in the Asparagaceae, and if that sounds like asparagus, that's right. They are related to asparagus. The agaves are in this family as well. Agaves may not look like asparagus, but the flowering stalk of an agave does resemble a giant asparagus.

Another arborescent genus in the Asparagaceae is the Dragon Tree. The genus name for Dragon Trees is *Dracaena*, which comes from a Greek word meaning female dragon. There are about 120 species in this genus, and there is one species, *Dracaena cinnabari*, or Dragon's Blood Tree that only grows on the island of Socotra, off the coast of Yemen.

This tree has leaves that grow very close together to reduce the evapotranspiration. By being really close together, the tree's canopy can actually maintain a small amount of humidity because there is less air flowing around each leaf. This air around each leaf that has reduced mixing with other air is called a boundary layer.

The trees also tend to grow in clumps because the seedlings are shaded by an adult plant—a phenomenon called nurse plants. When I showed a picture of the Dragon's Blood tree to my son, he said, no, that is called root tree, and the branches do look like roots in the air.

Because leaves can lose so much water, many desert plants will flower during the dry season and wait for periodic rains before they produce leaves. The elephant tree of the Baja peninsula is a good example. It produces these small yellow flowers during the dry season, which are an important source of nectar for pollinators.

This delay in leaf development also enables insects to see the very small flowers without a covering of leaves. Because flowers typically don't photosynthesize, they don't transpire nearly as much, so there's little water loss for the plant to have flowers out. As you might guess, this tree, which is really more of a shrub, is called the elephant tree because of its thickened tree trunk, which stores water. A thickened, water-storing stem is called a caudiciform trunk.

A tree that actually grows where elephants live and has a similar sort of water-storing mechanism is the famous Baobab tree of the African plains, though they also live in India and Madagascar, and the island of Socotra. One species even lives in Australia, where individuals up to 1500 years old are purported to be the oldest living organisms in Australia. Also in Australia, the baobab is grown commercially. When it's 6 or 8 weeks old, the small plants are harvested for their roots, which are something like a cross between a carrot and a water chestnut. Apparently, all parts of the baobab are edible.

All the regions where baobabs grow are savanna, which might be desert-like half the year and grassland-like the other half of the year when the rains come. Since the adaptations of the baobab tree are mostly to save water, we'll consider them here. Also known as bottle trees, these trees can grow up to 15m in circumference and can hold up to 4500 liters of water, which is almost 1200 gallons. In addition to all the water these trees store, they also produce fruits that are eaten to cure a myriad of ailments in Africa, plus the fibers of the bark are also used to make cloth. It's no wonder this tree is sometimes called the tree of life.

Like other plants in arid regions, the baobabs drop their leaves during the dry season. They will then put out leaves and flowers before the rains arrive again. In studies, the dominant cue seems to be photoperiod, as it is for many plants. In short, baobabs will synchronously flower and put out leaves when the critical day length is slightly longer than 12 hours.

There's another really interesting desert tree that is often called the Dr. Seuss tree, and it's called the Boojum tree. This tree, *Fouquieria columnaris* is in the Foqueriaceae, which is just fun to say. This also called the ocotillo family. The Boojum tree is endemic to the Baja California Peninsula and the surrounding areas. The name Boojum Tree was introduced by Godfrey Sykes of the Desert Laboratory in Tucson, Arizona. He took it from Lewis Carroll's poem "The Hunting of the Snark."

Like the other member of this family, the ocotillo, both of these plants will flush with leaves in response to rainfall. The ocotillo, *Fouquieria splendens*, grows throughout the Sonoran desert, and if you see it in July before the monsoons, the plant looks like dead sticks coming right up out of the ground. The ocotillo stems also have spines, like a cactus, but they have an unusual formation. The spines are made by leaves. First, leaves grow out from the main stem on an oversized petiole, which is the leaf stem. When those leaves die, the petiole remains and stiffens into the sharp spines. No other plant produces spines like this.

After the spines are established, small buds grow in the space between the stem and the spine, and these are called axillary buds. From these buds, a different type of leaf grows that is small with short petioles. These leaves appear within 24–48 hours after a rain, but fall off after a few weeks. Amazingly, ocotillo may grow new leaves seven or eight times a year, which is astonishing given the arid environments in which they live and the energetic cost and water loss potential of having leaves. This ability to produce leaves will even occur in rootless segments of ocotillo stems if they've been watered.

There are some plants that don't tolerate dryness at all. We've already learned that perennial plants, like the ocotillo and the baobab, will time certain events, like flowering, based on when rainfall occurs. Even most deserts will have a time of year when it rains. Annual plants will time their entire lives around these rainfalls. Every spring in the Mojave Desert, for a few weeks, the desert blooms with many desert annuals. These small wildflowers will begin to germinate hours after a rainfall, and then grow quickly and produce flowers, fruits, and seeds before the really hot and dry weather of the summer begins. The seeds are then dispersed into the soil by summer and wait through the winter until the next rain. The California desert grassland of Antelope Valley is awash in California poppies every March and April.

A different way to avoid the stress of the dry season is the become dormant altogether. A striking example of a plant like this is the resurrection plant. This plant is a lycopod, which makes it a relative of ferns.

For a bit of review, do you remember the major separation between ferns and the rest of the vascular plants? Ferns reproduce by spores. So, the resurrection fern is actually native to the southwestern United States. This plant can look dead for all the world during the dry season, and then when the rains come, it can completely revive. It's like plant hypersleep for the desert.

If we knew how this process worked exactly, or which specific genes turned metabolism on and off, even in a plant, it might be a link to controlling animal metabolism.

So, the resurrection plant prepares to have a slow down in metabolism. Like a bear preparing to go in the den for winter, the plant fattens up. It starts to store extra sugars, like sucrose, and amino acids, which make up proteins. Next, it will start to produce antioxidants. Last of all it produces proteins that help it survive called dehydrins and expansins. These proteins help the internal organelles and the cell walls toughen up for the dry spell, quite literally.

Of course, another desert adaptation is to avoid leaves altogether. Some plants don't even photosynthesize. Although these plants had the ability to make chlorophyll, over time they were selected for, to lose this ability, and instead use sugars from a host plant.

These are parasitic plants, and dodder is a good example. Dodder is a bright orange plant that can often grow on creosote shrubs in the desert.

Another very interesting growth form isn't even a plant at all, but it's so important to the desert and to the survival of plants that we have to consider it. The famed desert crust sits on top of desert soils and looks like the top of burnt toast. It sort of looks like someone has come along with a torch and lit the top layer of the soil on fire. However, this layer isn't dead at all—in fact, it's vital to the conservation of the soil because it helps to keep the top layers of soil from blowing away. This crust also helps to absorb water quickly, which helps reduce flash flooding. And, this crust also provides much-needed nitrogen to desert soils, which are low in nutrients.

So, what is this crust you may ask? It's another symbiosis. Like the lichens we learned about earlier, this crust is a whole community, but this one consists of cyanobacteria, which are blue-green algae even though they're bacteria and not algae. The cyanobacteria are the main site of photosynthesis, but they join with other organisms, like mosses, green algae, lichen, microfungi, and other bacteria. That's a symbiosis of three different kingdoms of life. These organisms all live together in a thin layer that has hardened a bit like a crust on top of the soil. Botanically, the layer is called cryptobiotic crust. Crypto comes from the Greek word *kruptos* for hidden because many times, this layer isn't even visible to the naked eye.

Desert botanists are very keen to avoid walking on cryptobiotic crust because despite growing in the desert, the crust is very fragile. The crust usually gets started by a cyanobacteria that grows in filaments. These filaments are encrusted in sheaths made of minerals, which provide protection in the desert climate. The cyanobacteria can be fairly dry most of the year, but with rainfall, it will absorb a lot of water and grow. Because they're so small, the cyanobacteria don't grow very fast, but as they do, they leave their sheaths behind in the soil.

These sheaths are sticky, and as they are left behind, they adhere to soil particles and begin to form the crust. Later, other organisms grow in the crust, like fungi. This is a good place to grow now because the cyanobacteria can also fix nitrogen, so they can provide this important mineral in a usable form for the other organisms living there. Lastly, mosses and then vascular plants will begin to colonize areas where a good crust has built up. Sometimes, these crusts can be up to 15 centimeters deep.

So, I hope after this lecture, you now see deserts, not as a wasteland, but as a botanic wonderland. Much of the vast variety of desert plant forms are adapted to use water quickly when it's available, and then to withstand the certain periods of drought. It's this necessary adaptation that gives desert plants so many unique forms.

So, while deserts may not be the most diverse plant community in terms of species, their exceptional diversity of morphology means that deserts should be near the top of any list of plants worth watching.

LECTURE 21

HOW TEMPERATE TREES CHANGE COLOR AND GROW

Trees are the largest, oldest, and tallest organisms on the planet. Temperate forests are dominated by trees. A tree differs from a shrub by having one main stem. Trees have wood, which distinguishes them from herbaceous plants. Unlike grasses, which grow from their base, trees grow from their tips. The total number of plant families varies with taxonomic opinion, but it's somewhere around 300 to 600 families—and many of them have trees.

HOW LEAVES CHANGE COLOR

- Deciduous means to lose the leaves, and in a temperate forest, that means to lose the leaves for the winter. Temperate trees do this because the trade-off of keeping a leaf from freezing during wintertime doesn't offset the photosynthetic gain. The sunlight is too weak, the days are too short, and the temperatures are too cold to make a living. So, the tree hibernates for winter. Before the winter comes, though, the leaves change colors.

- There are 2 main pigments that are seen when leaves change colors. Yellow and orange colors occur when the leaf has carotenoid pigments. Carotenoids have long hydrocarbon chains, which are made up of a carbon skeleton surrounded by hydrogens. Because of the hydrocarbon chains, carotenoids tend to repel water and are thus hydrophobic.

- The other major pigment class responsible for color in leaves are the anthocyanins, which are generally red and made up of carbon rings. Anthocyanins are made in the cytoplasm of the cell, where carotenes are made in plastids, usually chromoplasts, which are organelles like chloroplasts in the cell.

- These 2 pigments, carotenoids and anthocyanins, are primarily responsible for the autumn colors that light up the forest. The primary reason we see these colors is because the plant needs to break down the major green pigment, chlorophyll, which is responsible for photosynthesis. The plant needs the nitrogen that is in the chlorophyll molecule.

- Most ecosystems are nitrogen limited. If the tree can store nitrogen, it will be ready to go in the spring when it's time to put out the leaves and flower buds. So, as the plant breaks down the green chlorophyll to reabsorb the nitrogen, the other pigments, the carotenoids and anthocyanins, become visible.

- The mechanisms for their appearance, however, are quite different. Carotenoids are present in the leaf through the summer. Their purpose is to act as an antioxidant. As an accessory pigment, they can absorb extra energy if the chlorophyll molecules are at capacity. When the fall comes, these carotenoids just become more visible.

- Anthocyanins, the red pigments, are different. While present in small amounts during the summer in the leaf, they are actually synthesized in the fall as the plant begins to break down chlorophyll. Many very young leaves also have high concentrations of anthocyanins.

- This pigment helps protect a developing or dying leaf in 2 ways. First, because anthocyanins absorb the highest visible light, blue, it can protect the chloroplasts. This protection occurs because high-energy light can cause the production of free radicals, which can damage the leaf. Anthocyanins can also help neutralize chemicals in the leaf that might slow down photosynthesis.

- There are trees that just don't turn red in the fall; instead, they turn yellow. The reasons for why some trees produce anthocyanins and why others don't are not well understood. It could be that the ability to produce anthocyanins is not in the genetic history of some trees, so they don't have that ability. There could be a cost to producing anthocyanins that we're unable to detect. There could also be other environmental factors that might make anthocyanin production less efficient.

- After the leaves change color, they fall off. It's easy to understand that leaves need to go away for the winter, or for a dry spell, but leaf drop is also a good way to curb herbivore populations. If the leaves are gone, the insects and other herbivores will also go away, or at least their populations will drop, or they will switch to eating something else, which gives the tree time to recover.

- The process of leaf drop, called leaf abscission, is the same process for flowers dropping petals and ripe fruits falling off of plants. When the leaf drops, it sometimes leaves a leaf scar on the branch. By counting the leaf scars, the age of the tree can be determined. If you look closely, you can usually see the leaf scars on the twigs of most deciduous trees.

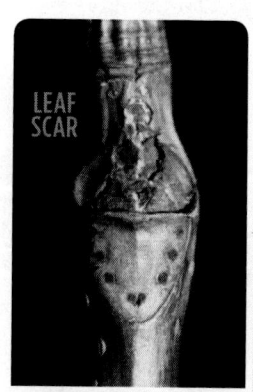

COMMON HORSE CHESTNUT

Aesculus hippocastanum

LEAF AND TREE GROWTH

- Broadleaf trees have many shapes. Early leaves of broadleaf trees, such as magnolias, had no teeth along the margins and no lobes. Lobing and teeth are more common in temperate trees. This is easily seen in the genus *Quercus*, the oaks. Tropical oaks have more trees with unlobed leaves, whereas lobes are the norm of the north. Lobing might be popular in the north because it expands the surface area without increasing the weight of the leaf.

NORTHERN RED OAK
Quercus rubra

- Even on the same individual tree, the size of the leaf will probably vary. Leaves that are lower in the canopy tend to be thinner and have a larger surface area to capture more light. These leaves are called shade leaves, and they are usually a darker color because chlorophyll *b* is very abundant in these leaves. The purpose of chlorophyll *b* is to improve light-capturing capability of the chloroplast.

- Leaves higher in the canopy are called sun leaves, and these leaves are thicker with multiple layers of photosynthesizing cells in the leaf because of the high light levels.

- Even the way the leaves hang on the trees will differ. Trees that are adapted to hot climates will have leaves that hang almost vertically, so as to avoid direct sunlight, which can cause too much water loss.

- Moving down from the canopy, we come to the stem, which is the trunk in trees. The wood is created by cells dividing in the vascular cambium. Whether this growth is fast or slow will generally determine the hardness of the wood. Trees that are slow growing have cells that are closer together, with less water, so the wood is harder. Fast-growing wood has cells with more water, so the wood is lighter, and usually softer.

- In addition, conifer trees lack the wood fiber cells in their wood that flowering trees have. So, the wood of conifer trees is generally referred to as softwood and that of deciduous trees is called hardwood.

- Even among the flowering trees, the hardness of the wood can vary dramatically. The weight of wood is due to cellulose and lignin in the cell walls. Cellulose is a polymer of glucose that is similar to starch. Lignin is a chemical made up of many carbon rings, called benzene rings, attached together. Lignin's ring structures give hardness and strength to the wood.

- The density of wood, and thus whether it is buoyant in water, is due to components of the cell walls. The presence of many air cavities within the cell walls of the wood make the wood become water logged. Also, the thickness of the cell walls and the amount of lignin they contain determine weight and density of the wood. Some wood is so dense that when it's dry, it sinks in water.

- Have you ever seen a tree that is leaning over, and then it straightens itself? Conifers and deciduous trees do this in 2 separate ways. As a result, there are 2 separate reactions for how trees can grow in response to a stimulus that would cause leaning. In both cases, the cells in the trunk will "react" to gravity to right the tree, so all such wood is called reaction wood. There are 2 ways to react: Conifers push themselves right, and deciduous trees pull themselves into line.

COAST LIVE OAK

Quercus agrifolia

- The pushing or pulling has to do with how the cells are growing on which side of the trunk. In conifers, the wood on the inside angle of the lean will thicken and grow with many spaces between the tracheids, which will also have a higher lignin content. This will make the wood denser and harder, but also weaker—mostly because the wood cells are not as orderly as they are in normal wood. The wood is in compression because it's pushing the rest of the tree, so it's called compression wood.

- Deciduous trees, by contrast, will pull themselves straight and produce tension wood. The cells opposite the angle of the lean will contract at the top of the stem, which will eventually pull the tree to the correct side. These cells will have gelatinous fibers but little lignin—practically the opposite of compression wood in conifers.

- The reaction wood isn't only present when trees lean; it's always in the branches so that they remain at the correct angle. This kind of wood growth can also help keep tree canopies in the light. The reaction wood growth is controlled generally by plant hormones.

TREE ROOTS

SAGUARO

Carnegiea gigantea

◇ Many trees have pretty shallow roots. Typically, most tree roots are in the top 6 to 18 inches of soil. This is usually where there will also be the greatest concentration of water, nutrients, and oxygen.

◇ The tree cacti, such as saguaro and Joshua trees, don't have very deep roots. There's no water table in the desert, so there's no need to have deep roots.

◇ Some trees have a taproot, and these are primarily found where there are water tables, such as the Midwest. Trees with a taproot include hickory, walnut, white oak, and European hornbeam.

◇ Regardless of how they grow, the tree root systems extend outward really far—usually about 2 to 3 times larger than the canopy area. Despite the extra area under the ground, the actual biomass distribution is different. The root-to-shoot ratio describes how much biomass the tree has put into belowground, or root, growth, versus aboveground, or shoot, growth. For most trees, the root-to-shoot ratio is around 1 to 4, so the tree aboveground is about 4 times heavier than the roots.

◇ Tree roots get a lot of help from mycorrhizal associations, the fungal partners that extend the surface area, and thus absorptive power, of the roots. Yet tree roots can forage through the soil. According to David Eissenstat, a professor of woody plant physiology at Penn State, thinner tree roots will send out exploratory roots or use the fungal partners to locate areas in the soil that have a high mineral content that are then exploited. Trees with thin roots that do this are oaks and maples.

◊ Other trees with thicker roots, such as pines and tulip poplars, don't use this strategy. These are the slow-and-steady types that produce thick roots, which will last a longer time and pay out just as many nutrients, but over a longer time frame.

TULIP POPLAR
Liriodendron tulipifera

READINGS

Coombes, *The Book of Leaves*.

Lee, *Nature's Palette*.

Thomas, *Trees*.

QUESTIONS

1 We still don't really understand how different soil types might cause different colors to appear in autumn leaves. Why do you think this is still an unknown area of botany?

2 How do you feel about the genetic modification of the American chestnut? Is it more acceptable to modify an organism to help it survive? Would it make a difference if that organism were on the brink of extinction?

LECTURE 21 TRANSCRIPT
HOW TEMPERATE TREES CHANGE COLOR AND GROW

I did my Ph.D. at the University of California, Los Angeles. Nestled at the base of the Santa Monica mountains, UCLA is remarkably close to some beautiful natural areas despite its namesake of the second largest city in the U.S. Running along the fire roads through the brush, it was always a treat to come into the shade of the Coast Live Oak. This is an oak, but it's also an evergreen. And because they're evergreen, these trees offer shade year round, with small, somewhat leathery leaves. Sometimes, an oak tree would mast, meaning it would drop a massive number of fruits. And since it's an oak tree, the fruit is an acorn.

It always amazed me to think of the early Chumash Indians of California eating these acorns. How did anyone decide that something as hard as a rock would be edible? Clearly, raw acorns are not edible, at least not by us. Numerous squirrels and birds use them as a food source, and many animals will cache the acorns away for the winter. When they forgot where they planted an acorn, a new tree is born. In order to eat the acorns, the Chumash would soak them and then roast them. So, all acorns are edible and some don't even need to be soaked, like the Swamp White Oak. I've never eaten an acorn, a fact I'd like to remedy, and this makes me wonder why they've never been domesticated.

I'm not the first botanist to wonder about this, and certainly many have tried. There seem to be two good reasons the oak has never been domesticated. First, oaks are slow-growing, and many would-be farmers want profit sooner than an oak tree would allow. For comparison, an almond planted into the ground can grow into a nut-bearing tree in as little as four years. However, a planted acorn may not produce acorns for two or three decades. Still, pecan trees can take that long to produce fruit, so that can't be the only reason.

A more compelling reason for the lack of a domesticated acorn is in the genes, so to speak. Again, if we look at the almond for comparison, we'll find that some almonds are indeed very bitter. But, if an ancient farmer planted almonds from an occasional non-bitter mutant almond tree, about half the nuts from the resulting tree would be equally non-bitter. This is because bitterness is controlled by one dominant gene in the almond tree. If the gene is dominant, that means that a non-bitter tree would have to have two recessive alleles for it to be non-bitter.

When this non-bitter tree mates, i.e. when the farmer picks an almond from the non-bitter tree and plants it, chances are 50% that the new tree will also be non-bitter if it was mated with a heterozygote, that is a tree with one dominant bitter allele and one recessive non-bitter allele.

However, bitterness in oaks is controlled by more than one gene. So, even if a farmer were lucky enough to find a non-bitter mutant oak, almost all the resulting acorns would still be bitter. That, combined with the slow growth, probably kicked the acorn out of the running to be America's top nut.

What are oaks so bitter about? Certainly, there are many tannins in oak, as any wine drinker understands from drinking wine that's been aged in wood oak barrels, but why the acorns have tannins, too, is a bit of a mystery.

Still, the oaks are an impressive group. All the oaks belong to one genus, *Quercus*, which is Latin for oak. I know that seems a bit circular, but that's how it is. Almost every U.S. state has at least one native oak. There are three exceptions. I wonder if you can easily guess two of them? If you thought Alaska because it's too cold and Hawaii because it's too isolated, you're right. Surprisingly, Idaho has no native oaks. It's cold and dry, but so is Montana, and the hearty bur oak is able to make it there. Oaks are the National Tree, official or unofficial, in European countries, too, including *Quercus alnifolia* in Cyprus, *Quercus suber* in Portugal, remember the cork tree, and *Quercus robur* in England, Poland, and all three of the Baltic States. Oak was declared the national tree of the United States in 2004.

This leads to the interesting question of why the temperate forests aren't nearly as diverse as tropical forests? One reason might be that the temperate forests have had to move northward and southward with advances and retreats of glaciers. As such, most temperate forests are less than two million years old, often much less, for New York City's Central Park, it was less than 20,000 years ago. Another reason might have to do with lower energy inputs. There's less sunlight in the temperate forests, especially in winter.

The temperate forest is, as its name suggests, dominated by trees. A tree differs from a shrub by having one main stem. And, of course, trees have wood, which distinguishes them from herbaceous plants. Unlike grasses, which grow from their base, trees grow from their tips. A branch that is 5 feet off the ground in one year will still be 5 feet off the ground in 20 years, as long as the tree is still alive.

Trees are another example of convergent evolution. The tree form probably came about numerous times. We see it in hundreds of unrelated plant families. The total number of plant families varies with taxonomic opinion, but it's somewhere around 300–600 families, and many of them have trees.

Trees are the largest, oldest, and tallest organisms on the planet, and we have a relationship with trees. We name trees. Hyperion is the name of the tallest Redwood. Methuselah is the name of the oldest Bristlecone. Trees have been used as markers for directions and property lines. Each state or jurisdiction records its tallest trees, and in the U.S., these are called Champion trees or Venerable Trees. There's even a National Big Tree Program in the U.S., a National Register of Big Trees in Australia, and so on. So much is online, of course.

We've considered coniferous trees, which are most common in the higher latitudes of boreal forests, and in the higher elevations of temperature forests. Then, there are evergreen flowering trees, like the coast live oak that grows in the American South. But another forest type that we're not so familiar with is the eucalyptus forest. Although most eucalyptus are evergreen, a few are drought deciduous, meaning they lose their leaves for the dry season. The genus *Eucalyptus* has about 700 species in it, only nine of which exclusively live outside Australia. So, in Australia, these trees live in every conceivable habitat—one even lives close to treeline—the snow eucalyptus.

The common name for eucalyptus is the gum tree because the bark will exude a plant gum when broken. This gum has many tannins in it, which serve to protect the tree when it's injured. The eucalyptus are also the tallest flowering trees. Not only that, but they are amazingly fast growing. Their growth rates can be up to a meter a year. That means in 10 years, a eucalyptus could be 10 meters tall, or 33 feet. However, because temperatures in most of Australia are tropical or subtropical, most botanists wouldn't consider any part of Australia to be dominated by a temperate forest by a North American or European standard.

So, with that detour let's focus on the temperate deciduous forests. Deciduous means to lose the leaves, and in a temperate forest, that means to lose the leaves for the winter. Temperate trees do this because the trade off of keeping a leaf from freezing during winter time doesn't offset the photosynthetic gain. The sunlight is too weak, the days too short, and the temperatures are too cold to make a living. So, the trees hibernate for winter.

Before the winter comes, though, the leaves change colors. Why do the leaves do this and why do trees have different colors? The why leaves change colors is easier to answer than why there are different colors for different trees. There are two main pigments that are seen when leaves change colors, sort of a yellow-orange color are when the leaf has carotenoid pigments. Beta-carotene is the main pigment that gives carrots their orange color, and this is a type of carotenoid.

Carotenoids have long hydrocarbon chains. These chains are made up of a carbon skeleton surrounded by hydrogens. Because of the hydrocarbon chains, carotenoids tend to repel water and are thus hydrophobic. The other major pigment class responsible for color in leaves are the anthocyanins. These pigments are generally red and made up of carbon rings. Anthocyanins are made in the cytoplasm of the cell where carotenes are made in plastids, usually, chromoplasts which are organelles like chloroplasts in the cell.

These two pigments—carotenoids and anthocyanins—are primarily responsible for the autumn colors that light up the forest. The primary reason we see these colors is because the plant needs to break down the major green pigment, chlorophyll, which is responsible for photosynthesis. The plant needs the nitrogen that is in the chlorophyll molecule. Most ecosystems are nitrogen-limited. If the tree can store nitrogen, it will be ready to go in the spring when it's time to put out the leaves and flower buds. So, as the plant breaks down the green chlorophyll to reabsorb the nitrogen, the other pigments—the carotenoids and anthocyanins—become visible.

The mechanisms for their appearance, however, are quite different. Carotenoids are present in the leaf through the summer. Their purpose is to act as an antioxidant. As an accessory pigment, they can absorb extra energy if the chlorophyll molecules are at capacity. When the fall comes, these carotenoids just become more visible.

Anthocyanins, the red pigments, are different. While present in small amounts during the summer in the leaf, they are actually synthesized in the fall as the plant begins to break down chlorophyll. It turns out that many very young leaves also have high concentrations of anthocyanins. This pigment helps protect a developing or dying leaf in two ways. First, since anthocyanins absorb the highest visible light, the blue, it can protect the chloroplasts. This protection occurs because high-energy light can cause the production of free radicals, which can damage the leaf. These are the same sort of free radicals that cause aging in humans. Anthocyanins can also help to neutralize chemicals in the leaf that might slow down photosynthesis.

A great experiment to test this advantage of anthocyanin was conducted at Harvard Forest by David Lee, a professor at Florida International University. If anthocyanins protect the photosynthetic apparatus as the leaf is winding down, leaves with this pigment should be able to reclaim more nitrogen than leaves that didn't have anthocyanins. By collecting red leaves and yellow leaves, Dr. Lee and his team found that the red leaves did have less nitrogen when they fell, which means that these leaves were able to reclaim more of the nitrogen. This hypothesis of anthocyanins increasing nutrient reabsorption was supported.

If anthocyanins allow more reabsorption of nitrogen, why don't all plants make anthocyanins when their leaves are senescing? There are trees, however, that just don't turn red in the fall, they turn yellow, witch hazel is a good example. The reasons for why some trees produce anthocyanins and why others don't aren't well understood. It could be that the ability to produce anthocyanins is not in the genetic history of some trees, so they just don't have that ability.

There could also be a cost to producing anthocyanins that we're unable to detect. Even more, there could be other environmental factors that might make anthocyanin production less efficient.

What is very clear, however, is that after the leaves change color, they will fall off. It's not easy saying goodbye. Certainly, it's easy to understand that the leaves need to go away for the winter, or for a dry spell, but leaf drop is also a good way to curb herbivore populations, too. If the leaves are gone, the insects and other herbivores will also go away, or at least their populations will drop, or they will switch to eating something else, which gives the tree time to recover.

The process of leaf drop, called leaf abscission, is the same process for flowers dropping petals, and ripe fruits falling off of plants. The process always begins with the formation of an abscission zone. This zone will typically be at the base of the leaf, flower, fruit, or whatever plant part is about to fall off. Next, sometimes even months later, the cells in this zone become activated by a plant hormone, usually ethylene, which is the same gaseous hormone that causes fruits to ripen. However, it's not just the amount of ethylene that does the regulation. Another plant hormone, auxin, is also involved. The balance of these two hormones seems to trigger the actual abscission, which involves digestion of the cell walls in the abscission zone. When auxin levels in the plant decline, the cells in the abscission zone become more receptive to ethylene.

For the leaf, flower, or fruit to drop, the walls of the cells in the abscission zone must be digested by enzymes, such as cellulase and polygalacturonase. These enzymes are synthesized by the abscission zone cells and then released into the cell wall. The cell walls then weaken and break, and the plant part is released.

When the leaf drops, it will sometimes leave a leaf scar on the branch. By counting the leaf scars, the age of the tree can be determined. If you look closely, you can usually the leaf scars on the twigs of most deciduous trees. Those in the ash genus, or *Fraxinus*, are pretty noticeable. And, you don't have to explain why you're looking at leaf scars. You can just smile and point out, it's an ash.

When you're walking through a forest in autumn, seeing the leaves on the ground, it's hard not to notice the variety of leaf shapes. Needles in conifers are all basically one shape—like needles.

Why do broadleaf trees have so many leaf shapes? Early leaves of broadleaf trees, like magnolias, had no teeth along the margins and no lobes, the margins of magnolias are in botany-speak, entire. Lobing and teeth are more common in temperate trees. This is easily seen in our friendly genus, *Quercus*, the oaks. Tropical oaks have more trees with unlobed leaves; whereas, lobes are the norm in the north. Lobing might be popular in the north because it expands the surface without increasing the weight of the leaf.

Even on the same individual tree, the size of the leaves will probably vary. Leaves that are lower in the canopy tend to be thinner and have a larger surface area to capture more light. These leaves are called shade leaves, and they are usually a darker color because chlorophyll b is very abundant in these leaves. The purpose of the chlorophyll b is to improve the light-capturing capability of the chloroplast. Leaves higher in the canopy are called sun leaves, and these leaves will be thicker with multiple layers of photosynthesizing cells in the leaf because of the high light levels.

Even the way the leaves hang on the trees will differ. Trees that are adapted to hot climates will have leaves that hang almost vertically, so as to avoid direct sunlight, which can cause too much water loss.

Moving down from the canopy, we come to the stem, which is the trunk in trees. The wood is created by cells dividing in the vascular cambium. Whether this growth is fast or slow will generally determine the hardness of the wood. Trees that are slow growing have cells that are closer together, with less water, so the wood is harder. Fast growing wood has cells with more water, so the wood is lighter and usually softer.

In addition, conifer trees lack the wood fiber cells in their wood that flowering trees have. So the wood of conifer trees is generally referred to as softwood and that of deciduous trees is called hardwood. You've heard the phrase hardwood floors, which means that the wood came from something like an oak or a maple as opposed to a pine. The problem with these descriptors are that some conifers have hard and dense wood, like the yew tree, which is a conifer. Other so-called hardwoods, or flowering trees, have very light wood, like the balsa tree. So, botanists don't usually refer to conifers as softwoods. We call them conifers because they have cones.

Even among the flowering trees, the hardness of the wood can vary dramatically. The weight of the wood is due to cellulose and lignin in the cell walls. Cellulose is a polymer of glucose that is similar to starch. Lignin is a chemical made up of many carbon rings, called benzene rings, all attached together. Lignin's ring structures give hardness and strength to the wood.

The density of wood, and thus whether it is buoyant in water, is due components of the cell walls. First, the presence of many air cavities within the cell walls of the wood will actually make the wood become water logged. Also, the thickness of the cell walls and the amount of lignin they contain will determine weight and density of the wood. Some wood is so dense that when it's dry, it will actually sink in water.

Also, thinking about tree trunks, have you ever seen a tree that's leaning over, and then it sort of straightens itself? Conifers and deciduous trees do this in two separate ways. As a result, there are two separate reactions for how trees can grow in response to a stimulus that would cause leaning. In both cases, the cells in the trunk will react to gravity to right the tree, and so all such wood is called reaction wood. But, there are two ways to react—conifers push themselves right and deciduous trees pull themselves into line. This is a sort of reaction that occurs.

The pushing or pulling has to do with how cells are growing on which side of the trunk. In conifers, the wood on the inside angle of the tree will thicken and grow with many spaces between the tracheids, which will also have a higher lignin content. This will make the wood denser and harder, but also weaker—mostly because the wood cells are not as orderly as they are in normal wood. The wood is in compression because it's pushing against the rest of the tree, so it's called compression wood.

Deciduous trees, by contrast, will pull themselves straight and produce tension wood. The cells opposite the angle of the lean will contract at the top of the stem, which will eventually pull the tree to the correct side. These cells will have gelatinous fibers, but little lignin—practically the opposite of compression wood in conifers.

The reaction wood isn't only present when trees lean, it's always in the branches so that they remain at the correct angle. This sort of wood growth can help keep tree canopies in the light. The reaction wood growth is controlled generally by plant hormones with auxin and ethylene regulating push of compression growth in conifers and auxin and gibberellin controlling the pull of tension growth in deciduous trees.

Most people probably envision a tree's roots as going deep into the soil. In fact, many trees have pretty shallow roots. Typically, most tree roots are in the top 6–18 inches of soil. This is usually where there will also be the greatest concentration of water, nutrients, and oxygen—remember roots need oxygen because they don't photosynthesize.

Ironically, the tree cacti, like Saguaro and Joshua Trees, don't have very deep roots at all. There's no water table in the desert, so there's no need to have deep roots. So, it's a little counterintuitive that trees growing in dry habitats will have fairly shallow roots.

Some trees do have a taproot, and these are primarily found where there are water tables, such as the midwest. Trees with a taproot include hickory, walnut, white oak, and European hornbeam.

Regardless of how they grow, the tree root systems extend outwards really far—usually about 2–3 times larger than the canopy. Despite the extra area under the ground, the actual biomass distribution is different. The root to shoot ratio is just as it might sound—it describes how much biomass the tree has put into belowground, root growth, versus above ground, or shoot growth. For most trees, the root to shoot ratio is around 1:4, so that the tree above ground is about four times heavier than the roots.

As we learned, tree roots get a lot of help from mycorrhizal associations, the fungal partners that extend the surface area, and thus the absorptive power of the roots. Yet, tree roots can actually forage through the soil. According to David Eissenstat, a professor of woody plant physiology at Penn State, thinner roots will send out exploratory roots or use the fungal partners to locate areas in the soil that have a high mineral content. These so-called hot spots of minerals are then exploited. Trees with thin roots that do this are oaks and maples.

Other trees with thicker roots, like pines and tulip poplars, don't use this strategy. These are more the slow-and-steady types that produce thick roots, which will last a longer time and pay out just as many nutrients, but over a longer time frame.

And, now, dispensing with roots, a tale of woe. A tale of fruits of a tree. Here, and then gone, and perhaps to come again. In his book *Walden*, Thoreau writes, "When chestnuts were ripe I laid up half a bushel for winter. It was very exciting at that season to roam the then boundless chestnut woods of Lincoln." *Walden* was written in 1854. Now, there are no boundless chestnuts anymore.

The American Chestnut Tree was almost wiped out by chestnut blight. A blight is any sort of disease that causes rapid chlorosis, which is a yellowing of the leaves. A blight will then quickly cause browning of the leaves, and then death of the plant. The chestnut blight is caused by the fungus, *Cryphonectria parasitica*. This fungus was accidentally introduced into the United States over a century ago with the import of Asian chestnuts, which are resistant to the blight. Now, some chestnuts remain, but there are few compared with their former numbers.

William Powell is a Professor at the State University of New York, College of Environmental Science and Forestry. He and his team have been searching for a way to help the chestnut regain its population by looking for genes that are resistant to the chestnut blight. At first thought, using the genes from an Asian species that has resistance might seem like the way to go—just interbreed these two species. That kind of breeding, though, can have unintended consequences.

Usually, we think of a hybrid as having more vigor, and this is true of hybrids that make it to market, such as hybrid seeds. But, in nature, a hybrid can actually show decreases in fitness. In the case of the chestnut, sometimes it can lead to internal kernel breakdown or IKB. This causes the chestnut to rot on the tree, and it's a rot caused by incompatible genetics, not by a pathogen.

Also, breeding American Chestnuts with other chestnuts will dilute the American Chestnut genome, such that in the future, it may not have the genetic machinery to fight off other homegrown pathogens. For these reasons, Dr. Powell and his colleagues are searching the genes of other plants, and they found one in bread wheat.

This gene from bread wheat gene an enzyme called oxalate oxidase. This enzyme detoxifies oxalate, which is a naturally-occurring substance found in a wide variety of foods. Oxalates are used in the metabolism of many plants and animals, including human metabolism The enzyme oxalate oxidase breaks down oxalate that the fungus uses to destroy the chestnut trees. This enzyme, oxalate oxidase, is also found in all of the grain crops and many other foods.

Dr. Powell added the gene to make oxalate oxidase to the chestnut genome. Since the chestnut genome contains about 40,000 other genes, this is a very small alteration compared to the products of many traditional breeding methods. In typical cross breeding to create hybrids, tens of thousands of genes are added. This method of inserting this one oxalate oxidase gene produces an American chestnut that's genetically over 99.999% identical to wild-type American chestnuts, according to Dr. Powell. More importantly, it's able to withstand chestnut blight.

According to several food blogs, the American Chestnut is supposedly even more delicious than the Asian or European varieties. The American chestnut also grows quickly and produces a rot-resistant wood—a valuable timber product. The chestnut is an interesting case of genetic modification used for conservation.

Lastly, did you know that trees sleep? In 2016, botanists from Austria, Finland, and Hungary used laser scanners to study the circadian patterns of trees. As it turns out, trees sleep at night, too The laser scanners showed that the trees droop their leaves and branches during night. Although the changes were only about a 10 cm droop for trees about 5 meters tall, it was measurable and consistent. The next step in the research is to do a whole-plant water-balance measurement since the researchers believe the changes occur due to water status of the tree.

So, if you're sleepy when you wake up tomorrow morning, check the tree branches nearby and see if the trees are sleepy, too, or if their branches are perky and ready to go.

LECTURE 22

ALPINE COLD MAKES PLANTS DO FUNNY THINGS

Alpine generally means above the trees. Moving above the trees, we reach the land of alpine plants, where most plants have a short stature to deal with conditions that are too harsh for trees. When we come to mountains within the tropics—between the Tropic of Cancer and the Tropic of Capricorn—then we are arriving in what are called tropical alpine zones. Tropical alpine systems provide some of the most unique plants on the planet. But there are also alpine plants in temperate regions.

ALPINE PLANTS

- Alpine plants face a number of problems. They typically have a short growing season. In temperate regions, it can be as short as 6 weeks. If they're growing on rocky soil, it can be very dry. It's usually very windy above the tree line, and it can freeze pretty much any night of the year. In addition, light levels can be extraordinarily high on cloudless days. To meet these challenges, many temperate alpine plants are small and low to the ground.

- There are a number of different adaptations that alpine plants have to cope with these stresses. One adaptation is their small stature. Being small is good for a number of reasons. First, plants can overwinter under the snow, so as not to freeze. Although they are dormant, their tissues still need protection from the extreme cold that can occur in the alpine. Being under the snow for the winter provides a blanket of temperatures that won't go much below freezing, unlike the air temperatures above the snow.

- Small stature is also great because the ground absorbs solar radiation. Alpine soil surface temperatures above 100° Fahrenheit have been measured in Rocky Mountain National Park. At night, the soil will reradiate the heat to the air above it. This reradiation can help keep plants from freezing.

- The plant morphologies of alpine plants are also related to helping keep warm. There are 2 unique growth forms among alpine plants. One is the rosette, which is classically defined as having a set of leaves that radiates out from a single location on a stem, with very small distances between where these leaves insert on the stem. Rosettes typically have a circular outline or shape. Some rosettes hug the ground while others tower on a long stalk or even a trunk. The leaves in a rosette are arranged so

TOWER OF JEWELS
Echium wildpretii

that there is very little overlap, and this is an efficient way for the leaves to gather sunlight.

◇ Unlike the tree rosettes of the tropical alpine, the temperate alpine rosettes grow almost flush with the ground. In addition to absorbing heat from the ground, this growth form also prevents the leaves from losing too much moisture.

◇ Because the leaves are against the ground, there is little wind blowing over the stomata, which are pores for the exchange of carbon dioxide and oxygen. When these stomata are open, as they must be for photosynthesis to occur, evaporation of water from the leaves will occur, and minimizing this loss is paramount for the plant.

SILVERSWORD

Argyroxiphium sp.

◇ A rosette also has a sort of parabolic shape in the leaves that can concentrate heat to the center of the plant, which is often where the meristem, or growing part of the plant, is located.

◇ The other unique growth form is that of the cushion plant. These unique plants are called cushions because they have a mounded, hemispherical appearance, like that of a pincushion.

◇ The cushion growth form solves several alpine growing problems. First, the form creates a boundary layer within its small canopy. Just like the rosette leaves growing against the ground, cushion plant leaves lose less water because there is not much wind movement through the cushion.

◇ Cushion plants can also store a fair amount of heat in their canopies. Just as the soil will warm up during the day, so will a cushion.

- Cushions also collect dead material at the base of their canopies, which basically makes a compost heap at the base of the cushion. As more debris collects, the cushion grows over it as it gets larger, creating a bigger and more mound-shaped plant. And all of this decomposing debris also creates some heat and provides more nutrients to the plant.

AVOIDING FREEZING

- Alpine plants have interesting physiological traits to avoid freezing. The problem for alpine plants is not so much the cold temperatures in winter. By the time winter hits, plants are already dormant for the year. They look dead above ground, but their roots are still living, and they are hardened for the winter.

- The major stress for alpine plants involves a late freeze—that is, a freeze that would happen in the very late spring after plants have already started their comeback from winter, or after they've started to flush, which is when they start to produce their summer leaves and flowers. Early fall freezes can also damage a plant if it hasn't already become frost hardened for the winter.

- Plants have essentially 2 ways to avoid freezing: supercooling and supersaturation. In supercooling, water can go down to temperatures below freezing, but plants avoid the formation of ice crystals in their cells. When an ice crystals form, it's called nucleation, and when that happens, ice will begin to form rapidly. But if there isn't anything for the water to form a nucleation event around, then water won't freeze—even below the normal freezing point.

- That's what the plant is doing when it uses supercooling. The water is below 0° Celsius, but the water has been moved into vacuoles and into cell walls. For an herbaceous plant, this mechanism is only good for a few hours and can usually only avoid damage at temperatures down to −10° Celsius.

- For woody tissues, dense, thick walls can prevent water from nucleating, so these tissues can survive freezing in temperatures down to −50° Celsius, which is the freezing point for supercooled water. But even the most winter-hardy plants cannot survive formation of ice crystals in the cytoplasm, which is not within the contents of the cell.

- When plants use supersaturation to avoid freezing, they are basically using the same principle behind antifreeze and road salts. By adding solutes to the water in their cells, plants can reduce the freezing point, but usually only by up to −5° Celsius.

- Aside from growth form and leaf physiology, many plants avoid freezing by insulating their leaves with hairs, also called trichomes. Many alpine plants look white because the leaves are covered in a very dense, thick mat of woolly-like hairs, and these are for insulation. These hairs can also prevent the leaf from getting too wet, and they can shield the leaf from fungal spores. These hairs also act as sunscreen.

EDELWEISS
Leontopodium alpinum

- The alpine sun can be very strong because there is less filtering by the atmosphere at high elevations. Just as alpine plants can avoid freezing by being close to the ground, they can also risk heat damage on warm days.

- In a paper from 1984, a botanist measured the surface temperature of a cushion plant at 45° Celsius, or 113° Fahrenheit. He then reported that this same part of the plant had brown leaves the next day. One way to avoid this overheating is by keeping the leaves shaded with many hairs. Interestingly, ultraviolet radiation doesn't seem to have that great of an effect on plant growth, not even at alpine elevations.

- It's not just heat damage that plants at high elevations have to worry about. The strong sunlight can also cause photo-oxidation, in which too much light overexcites the electron transport chain in the light reactions, and electrons can run amok.

- When this happens, a plant can undergo photorespiration as a way to deal with the excess energy. This process, photorespiration, is photosynthesis happening in the presence of light. Although this process protects the plants from damage, it costs the plant in terms of lost carbon fixed into sugars. The plant can avoid this with some hairs on its leaf.

ALPINE FLOWERS

- For an alpine plant, there are 3 ways to flower: early, midseason, and late. Early-flowering plants flower as the snow is melting. An example is the glacier lily. Midseason plants flower at the peak of the growing season. Many cushion plants fall into this category. The last category is late flowering, which flower after the main part of the growing season is over. The genus *Sedum*, also called the stonecrops, tend to flower a bit later.

GLACIER LILY
Erythronium grandiflorum

ENGLISH STONECROP
Sedum anglicum

- All the early-flowering and many midseason-flowering alpine plants preform their flowers during the end of the previous growing season, which ensures that they are ready to go when the temperatures get warm enough. Preformed flower buds are more common in alpine areas than in other habitats.

- Some plants that flower won't reproduce sexually. This can happen through apomixis, which is seed development without pollination. Although seeds are produced, they are clones of the parent plant. But even clones can benefit from the advantages of seed dispersal; that is, even though that seed may be a clone, it can still travel farther than vegetative reproduction through a runner or a rhizome.

- Alpine plants with a wide distribution are able to do this facultatively. When conditions are very harsh and the elevation is very high, these species will tend to produce more clonal seeds. As the elevation decreases and growing conditions become more favorable, plants will produce more seeds through sexual reproduction. Apomixis can also be a successful strategy because of the volume of seeds produced. Dandelions are a great example of this strategy.

- Alpine areas, at least those in North America, are known for their variations in color. Alpine plants come in a wide variety of pinks, reds, whites, yellows, and purples, including some blue flowers. This is not the case with all alpine flowers.

- A curious thing about the alpine flora of New Zealand is that it is mostly white. According to Sir Alan Mark, the premiere alpine botanist in New Zealand, the lack of color is due to a lack of pollinators that prefer such flowers. Colorful flowers tend to be visited by insects that can see color and are attracted to color. In other alpine regions, this would include long-tongued bees, which are absent in the alpine of New Zealand, and butterflies, which are scarce.

⬥ Another curious, but very effective, attribute of virtually all alpine flowers is heliotropism, or sun tracking, in which the flower tracks the Sun as it moves across the sky. The alpine snow buttercup is a tiny yellow flower that does this. The prevailing thought is that flowers track the Sun to keep the center of the flower warm, which would be attractive to pollinators in the cool alpine air.

READINGS

Körner, *Alpine Plant Life*.

Rundel, Smith, and Meinzer, *Tropical Alpine Environments*.

Zwinger and Willard, *Land above the Trees*.

QUESTIONS

1. Why do you think giant rosettes flourish in tropical alpine regions but fail to grow in temperate alpine regions?

2. What advantages does the cushion plant growth form offer to another developing plant?

LECTURE 22 TRANSCRIPT
ALPINE COLD MAKES PLANTS DO FUNNY THINGS

The tropics, let's remember, are anywhere on the globe between the Tropic of Cancer and the Tropic of Capricorn. When we come to these mountains within these regions, then we are arriving in what is called the tropical alpine zone. This is near and dear to my heart because one of my study species for my dissertation was a giant cushion plant in the Andes mountains, a plant called *Azorella compacta*. Believe it or not, this is in the carrot family, the Apiaceae. Of course, the plant, called llareta by the local Aymaran people, doesn't look anything like a carrot, but it does have a taproot, and this is found in many of the plants with a mound shape morphology, like the Azorella.

Llareta grows at elevations between 13,000 and 16,000 feet in the high northern Andes. While there, I found the highest plant recorded, at 17,224 feet. They grow where no trees grow, and so they were exploited for the railroad and road building in this area. They are still occasionally harvested and dried and used as fuel since they have half the caloric value of coal. This plant has been tested for many medicinal compounds since the local people use it to treat glaucoma and diabetes.

My work with this plant looked at questions of physiology and general ecology. How did it grow? What are its photosynthetic levels like? How does it survive the incredible diurnal temperature variation of tropical alpine systems, which can be summer every day and winter every night? What sort of habitat does it prefer? Are there young plants? Is the population still viable? These were the questions I sought to answer. I will give you the great reveal at the end of this lecture.

For now, let us continue our tour of tropical alpine systems since they provide some of the most unique plants on the planet. Though giant cushions, like llareta, aren't limited to tropical alpine regions, there are other plants that are found in this kind of habitat.

The giant rosette plant growth form is only found in tropical alpine regions, and it's not even found in all of those regions. A rosette is classically defined as having a set of leaves that radiates out from a single location on a stem, with very small distances between where these leaves insert on the stem. The green leaves of a dandelion plant are an ordinary, non-giant example of a rosette. Rosettes typically have a circular outline or form. Some rosettes will hug the ground while others tower on a long stalk or even a trunk. The leaves in a rosette are arranged so that there is very little overlap, and this is an efficient way for the leaves to gather sunlight.

Here are some examples of giant rosettes, giant lobelias in Africa, Espeletias and Puyas in the Andes, Cyathea in New Guinea, Silverswords in Hawaii, and Teide's bugloss in the Canary Islands, though they are just outside of the tropics.

The giant Espeletias in Venezuela keep their dead leaves wrapped around their stems, which has been shown to insulate the trunk. Just a bit of insulation can keep the outer stem tissues from freezing, which is important when temperatures can drop below freezing any night of the year.

Although Teide's bugloss doesn't quite live in the tropics, it's definitely an alpine rosette, and it's so unique that I have to mention it here. During its first year, it forms a rosette of leaves, then it forms a giant flowering stalk, up to nine feet high its second year, and then it dies. Talk about going out with a bang. This sort of lifestyle is called monocarpic, which translates to one fruit. It's a fairly common lifestyle among rosettes. The century plant is another common name given to many desert agaves that show the same lifestyle of grow for awhile, reproduce, and then die.

The Hawaiian Silversword is another monocarpic, tropical alpine rosette. If forms a giant flowering stalk as well, though the Silverswords are not biennial, and they can live up to 50 years before flowering and dying.

I love tropical alpine plants, especially, the giant rosettes because they are so unique, but we should think about temperate alpine plants too. Alpine generally means above the trees. We get the word alpine from the Latin *alpinus*, which refers to the Alps mountains in Europe.

To a botanist, alpine is anything above the trees. Using the word above usually refers to elevation, since trees drop out at certain elevations around the world. But, alpine tundra flora is a lot like arctic tundra, which is too far North to support trees, so it's above the trees in higher latitudes.

Thinking about where treeline occurs around the world is interesting. Christian Korner is a botanist at the University of Basel in Switzerland. He literally wrote the book on Alpine Treelines. His study of what forms treeline take around the world is exhaustive in nature. He says that high elevation tree limit is basically determined by a growing season that is at least 3 months long, where the mean air temperature for the year reaches at least 6.4 ° C. This is called an isotherm, which translates to same temperature.

No matter where you go in the world, wherever the yearly mean temperature is at least 6.4 ° C, there will be trees. That's 43.5 ° F. For example, treeline in Colorado is around 11,500 feet. The treeline on Cotopaxi in Ecuador, close to the equator is much higher, around 15,700 feet. The average yearly temperature in Ecuador, even above 15,000 feet is warmer than the average yearly temperature in Colorado. The elevation difference for tree lines is about 3500 feet, and the difference in latitude is roughly 39 ° North, so a very crude calculation suggests that the tree line might be about 111 feet higher for each ° of latitude away from the Equator

But it's not that simple, You have to look at the isotherms. Some of you may be familiar with Mount Washington in New Hampshire. Its treeline is only about 4,500 feet. Why so low—less than half that of Colorado. Using what we know about isotherms, we have to think about the average yearly temperature. Colorado has a continental climate, so while winters are cold, summers are very warm. Even in the mountains, summer temperatures can be 80 ° F, if just for an hour or so. The White Mountains of New Hampshire, on the other hand, have a coastal climate, with cooler and cloudier summers. There has actually never been a temperature recorded higher than 70 ° on Mt. Washington, even in summer.

Treeline is an interesting concept because it typically means where the last individual trees drop out. By contrast, usually, the word timberline refers to where the bulk of the forest abruptly ends. In between, above the timberline, but below the true treeline, there will be a few tree islands. The member trees of these islands are called krummholz trees.

The word krummholz is German for crooked wood, and it refers to the gnarled shape that the trunks take on with response to extreme winds. These are shorter evergreen trees, usually spruces and firs, which are stunted in growth but will actually move around the landscape, as growth will typically occur on the leeward side of the tree island and the windward side will gradually die off. According to research done on tree islands of Engelmann spruce and Subalpine Fir, the leeward parts of the tree islands expand at average rates of 1.5–2.6 centimeters per year because of the rooting of the low horizontal branches. By contrast, the windward ends of the tree islands recede by average rates of 0.9–1.9 centimeters per year due to needle death by freezing or desiccation.

With a Northern Hemisphere bias, we might think that all trees at treeline are evergreen conifers, but globally, this isn't so. In Australia, there are eucalyptus at treeline, where they have small, waxy leaves to endure the cold. In New Zealand, treeline is made up of Nothofagus trees, which are a type of beech tree, but these trees are still evergreen, even though they are flowering trees. One thing I like to show people is how abrupt treeline is in New Zealand as compared with Colorado. Again, this probably has to do with the abruptness of the isotherms.

Moving above the trees, we reach the land of alpine plants. Here, most plants will have a short stature to deal with the conditions that are too harsh for trees. Alpine plants face a number of problems. They typically have a short growing season. In temperate regions, it can be as short as 6 weeks. If they're growing on rocky soil, it can be very dry. It's usually very windy above the treeline, and it can freeze pretty much any night of the year. Along with this, light levels can be extraordinarily high on cloudless days. To meet these challenges, many temperate alpine plants are small and low to the ground. In fact, I like to think of alpine botany as belly botany because I'll often lie on the ground to look at plants close up with my hand lens.

There are a number of different adaptations alpine plants have to cope with these stresses. Let's think first about their small stature. Being small is good for a number of reasons. First off, plants can overwinter under the snow, so as not to freeze. Although they are dormant, their tissues still need protection from the extreme cold that can occur in the alpine. Being under the snow for the winter actually provides a blanket of temperatures that won't go much below freezing, unlike the air temperatures above the snow. This is one reason why a climate-related loss of winter snow could be damaging, not just for skiers, but also for alpine plants.

Small stature is also great because the ground absorbs solar radiation. Alpine soil surface temperatures above 100 °F have been measured in Rocky Mountain National Park. At night, the soil will reradiate the heat to the air above it. This re-radiation can help keep plants from freezing.

The plant morphologies of alpine plants are also related to helping keep warm. There are two unique growth forms among alpine plants. One is the rosette form that we already met earlier. Unlike the tree rosettes of the tropical alpine, the temperate alpine rosettes grow almost flush with the ground. In addition to absorbing heat from the ground, this growth form also prevents the leaves from losing too much moisture. Since the leaves are against the ground, there is little wind blowing over the stomata, or those pores for the exchange of carbon dioxide and oxygen. When these stomata are open, as they must be for photosynthesis to occur, evaporation of water from the leaves will occur, and minimizing this loss is paramount for the plant. A rosette also has a sort of parabolic shape in that the leaves that can concentrate heat to the center of the plant, which is often where the meristem, or growing part of the plant, is located.

The other unique form that we haven't yet met is that of the cushion plant. Anyone who knows me or has heard me talk will know of my love for these unique plants. They are called cushions because they have a mounded, hemispherical appearance, like that of a pincushion. I'm so drawn to these plants because of their beauty and their hardiness.

The cushion growth form solves several alpine growing problems. First, the form creates a boundary layer within its small canopy. Just like the rosette leaves growing against the ground, cushion plant leaves will lose less water because there is not much wind movement through the cushion. Cushion plants can also store a fair amount of heat in their canopies. Just as the soil will warm up during the day, so will a cushion. I've measured cushion plants being 12 ° C warmer than the outside temperature, especially first thing in the morning when the air temperature is usually coldest. Twelve ° can often mean the difference between freezing tissues or not freezing.

Lastly, cushions will also collect dead material at the base of their canopies. This basically makes a compost heap at the base of the cushion. As more debris collects, the cushion grows over it as it gets larger, creating a bigger and more mound shaped plant. And, all of this decomposing debris will also create some heat and provide more nutrients to the plant.

Some of my recent research has involved looking at the role these plants might play in the restoration of alpine systems. In Colorado, we have 53 mountains that are over 14,000 feet tall. These mountains are known as the 14ers. Many people come from all over to climb the 14ers and while many trails in Colorado are deserted, there will always be a few people on a 14er trail in summer, if not an outright parade.

The trails up these 14ers were often first created by early mountaineers, and may be very direct, but are oftentimes not the best paths for avoiding soil erosion. Many of these early, heavily-used, social trails up the mountains are prone to widening over time, and this increases trampling of the vegetation.

Although alpine vegetation is tough against environmental hazards, it doesn't hold up so well to repeated trampling. Sure, there are some mountain goats and mountain sheep that find their way to these high areas, but there aren't steady herds of grazers, so the plants are not well adapted to repeated trampling. As a result, a non-profit called the Colorado 14ers Initiative builds better trails up the 14ers and restores old trails that are eroded. When I first started volunteering with the 14ers Initiative in 2010, we were tasked to dig up pieces of vegetation and then transplant them into the old trail. This would help the old trail recover. In our orientation to begin work, they asked our group not to transplant the cushion plants. They said that the cushions wouldn't survive if transplanted.

Now, being a lover of cushion plants, I obeyed that summer, but later that year, I started searching the literature. There was only anecdotal evidence that this might be true, but nothing definitive. So, the following summer, I set about creating a research project to determine what happens if cushions are transplanted. As it turns out, cushions are just as good at surviving transplanting as grasses.

Not only will cushions survive, but they will help other plants survive as well. Because their canopies are less dry, less windy, and overall warmer, they are great places for other plants to germinate. Because many alpine plants will send out vegetative stems, they'll spread out from the cushion. And in this way, cushions are helping to facilitate the restoration of these high alpine trails.

Alpine plants also have some pretty cool physiological traits to avoid freezing as well. The problem for alpine plants is not so much the cold temperatures in winter. By the time winter hits, plants are already dormant for the year. They look dead above ground, but their roots are still living, and they are hardened for the winter.

The major stress for alpine plants involves a late freeze. This is a freeze that would happen in the very late spring after plants have already started their comeback from winter, or as botanists would say after they've started to flush—when they start to produce their summer leaves and flowers. Early fall freezes can also damage a plant that hasn't already become frost-hardened for winter.

How can leaves that are exposed to cold temperatures not freeze? Plants have essentially two ways of avoiding freezing, supercooling and supersaturation. In supercooling, water can go down to temperatures below freezing, but plants can avoid the formation of ice crystals in their cells. When an ice crystal forms, it's called nucleation, and when that happens, ice will begin to form rapidly. But, if there isn't anything for the water to form a nucleation event around, then water won't freeze—even below the normal freezing point.

That's what the plant is doing when it uses supercooling. The water is below 0 °C, but the water has been moved into the vacuoles and into cell walls. For a herbaceous plant, this mechanism is only good for a few hours and can usually only avoid damage at temperatures down to -10 °C. For woody tissues, dense thick walls can prevent water from nucleating, so these tissues can survive freezing in temperatures down to -50 °C, which is the freezing point for supercooled water. But, even the most winter-hardy plants cannot survive formation of ice crystals in the cytoplasm, that is not within the contents of the cell.

When plants use supersaturation to avoid freezing, they are basically using the same principle behind antifreeze and road salts. By adding solutes to the water in their cells, plants can greatly reduce the freezing point, but usually by up to −5°C.

Aside from growth form and leaf physiology, many plants avoid these freezing by insulating their leaves with hairs, also called trichomes. Many alpine plants look white because the leaves are covered in a very dense, thick, mat of wooly like hairs, and these are for insulation. These hairs can also prevent the leaf from getting too wet, and they can shield the leaf from fungal spores. And, these hairs also act as a sunscreen.

The alpine sun can be very strong because there is less filtering by the atmosphere at such high elevations. Just as alpine plants can avoid freezing by being close to the ground, they can also risk heat damage on warm days. In a paper from 1984, a botanist measured the surface temperature of a cushion plant at 45°C, this is 113°F. He then reported this same part of the plant had brown leaves the next day. I see this sort of damage on cushion plants all the time. One way to avoid overheating is by keeping your leaves shaded with many hairs. Interestingly, ultraviolet radiation doesn't seem to have that great of an effect on plant growth, not even at alpine elevations.

It's not just the heat damage that plants at high elevations have to worry about. The strong sunlight can also cause photooxidation. This is a process where too much light overexcites the electron transport chain in the light reactions, and electrons can run amok. When this happens, a plant can undergo photorespiration as a way to deal with the excess energy. This process, photorespiration, is just the way it sounds—it's respiration happening in the presence of light. Although this process protects the plants from damage, it costs the plant in terms of lost carbon fixed into sugars. Best to avoid it with some hairs on the leaf.

So, alpine plants have a lot to deal with, and we haven't even talked about flowers. For an alpine plant, there are three ways to flower and they are as described, early, midseason, and late. Early flowering plants will flower as the snow is melting. An example here is the glacier lily. Midseason plants will be at the peak of the growing season, many cushion plants fall into this category, like moss campion, which is, of course, not a moss, since it has flowers. The last category is late flowering, which will occur after the main part of the growing season is over. The genus *Sedum*, also called the stonecrops tend to flower a bit later.

All the early flowering and many mid-season flowering alpine plants will preform their flowers during the end of the previous growing season, and this ensures that they are ready to go when the temperatures get warm enough. Preformed flower buds are more common in alpine areas that in other habitats.

Some plants that flower won't actually reproduce sexually. This can happen through apomixis, which is seed development without pollination. As you may have guessed, although seeds are produced, they are clones of the parent plant. But even clones can benefit from the advantages of seed dispersal. That is, even though that seed may be a clone, it can still travel farther than vegetative reproduction through a runner or a rhizome. Alpine plants with a wide distribution are able to do this facultatively, which means when conditions are very harsh and the elevation is very high, these species will tend to produce more clonal seeds. As the elevation decreases and growing conditions become more favorable, plants will produce more seeds through sexual reproduction. Apomixis can also be a successful strategy because of the volume of seeds produced. Dandelions are a great example of this strategy. It's also one reason why they're so successful.

Alpine areas, at least those in North America, are known for their variations in color. Alpine plants come in a wide variety of pinks, reds, whites, yellows, and purples, including some blue flowers. This is not the case with all alpine flowers. A curious thing about the alpine flora of New Zealand is that it is mostly white. According to research, about 77% of the alpine flora in New Zealand is white.

The Rocky Mountains might be about 20% white. So, this is curious—why the lack of color in New Zealand's alpine flora? According to Sir Alan Mark, the premiere alpine botanist in New Zealand, it's because of a lack of pollinators that prefer such flowers. Colorful flowers tend to be visited by insects that can see color and are attracted to color. In other alpine regions, this would include long-tongued bees, which are absent in the alpine of New Zealand, and butterflies, which are scarce.

Another curious, but very effective, attribute of virtually all alpine flowers is heliotropism or sun-tracking. This is just the way it sounds, the flower will track the sun as it moves across the sky. The alpine snow buttercup is a tiny yellow flower that does this. Two botanists, Maureen Stanton and Candace Galen, wanted to know how much this ability influenced the reproduction success of this flower. The prevailing thought is that flowers track the sun to keep the center of the flower warm, which would be attractive to pollinators in the cool air.

They found that, sure enough, flowers that tracked the sun the best had temperatures that were several degrees warmer than the ambient air temperatures. In particular, flies were found more often on flowers that were good trackers, mainly because the flies stayed on these flowers a longer time. They also tethered the flowers at random angles that prevented sun tracking, and they found that tethered flowers at all angles set fewer seeds than untethered flowers.

I said at the beginning of the lecture, I would tell you the secrets of my study species, the llareta, a giant cushion plant in the Andes mountains. Its population has rebounded nicely. Although there are very few huge plants, it's still unclear whether the largest plants, which can be up to 12 meters across are actually one plant or several plants that have grown together. The scientific name of this plant is *Azorella compacta*. The *compacta* part refers to how dense the canopy is. I can stand on the plant and not fall through. It looks and feels like a moss covered boulder. These cushions are not at all like our temperate cushions—they're much hardier. Trying to separate out individuals would require a chainsaw or genetic analysis.

These plants, like many of the alpine plants we've discussed, have particular habitat preferences. They grow on equator-facing slopes, and they prefer to grow next to large rocks. This is called a nurse rock effect. Big rocks heat up a lot during the day and then reradiate heat back out at night, so it's common to see plants growing right up and over the rocks. And water will often collect in the shade of a rock, and that's another advantage.

The photosynthetic rates of these plants, like many alpine plants, is not particularly high. In fact, I found that they are very slow growing. In 14 years, some plants hadn't grown at all, and others had actually shrunk. This can happen when the plant absorbs water from its tissues for growth in times of prolonged drought. Given this growth rate and the size of these plants, they could be thousands of years old. Only carbon dating would give accurate ages because of the variability in their growth rates.

I also found these plants at a higher elevation than they had ever been recorded, just over 17,000 feet, and this is happening with plants around the world. In 2012, botanists found six vascular plants, growing in a single patch, at an elevation of 6150 meters above sea level, almost to the summit of Mount Shukule II in the Western Himalayas of India. This is 20,177 feet, which is much higher than has been reported previously.

Though this conquest of new elevations is likely a retreat from warming temperatures, it also shows the amazing adaptability of alpine plants.

LECTURE 23

BAD PLANTS AREN'T SO BAD

Poisonous plants, invasive plants, and carnivorous plants are not similar in terms of their habitats, chemistry, ancestry, or even the way they make a living. It's interesting that not all plants are poisonous, given the wide variety of those that are, and even a plant that is harmful at one level might be beneficial in a different dose or context. Invasive species can be feared, but some research says that they increase diversity, and, at least for some invasive plants, animals in their new habitats sometimes learn to eat them, and other plants will eventually compete with them. Carnivorous plants are revered by humans because they go against the norm of what we expect plants to do.

POISONOUS PLANTS

- There are many poisonous plants, such as poison ivy and poison oak. The poison in poison oak and poison ivy is urushiol, which causes a red, itchy rash and maybe a puss-filled blister.

- If you rub against the ortiga plant, expect a rash and even hives. This plant is a type of stinging nettle, called nettle tree. Both of these plants come from the Urticaceae family, which is basically a whole family of stinging plants. Many of these plants have very hairy leaves that are arranged opposite of each other on the stem.

- Botanically, the stinging hair is just a trichome, which is what all plant hairs are called. In the case of the nettle, the trichome is made of brittle silica, the primary component of glass, and it's equipped with a small spherical bulb on the tip. When the trichome is disturbed—for example, by your hand accidentally brushing up against it—the trichome will break and stick into your skin, and the toxin from the bulbous tip will move into your skin.

ORTIGA BRAVA
Urera baccifera

- The nettle toxin is composed of several chemical compounds. It contains histamine, which triggers an inflammatory response, and this is why you take an antihistamine to treat the symptoms of nettle. The toxin also contains the neurotransmitters acetylcholine and serotonin, which is supposed to be a happy neurotransmitter, but in this case, it causes pain. It also has a few acids, including formic acid, the stinging component of ant venom. There is some evidence that stinging nettle has antimicrobial properties and could be useful for ulcers.

- There are many other poisonous plants, such as oleander, castor bean, deadly nightshade, and poison hemlock. The cecropia tree is a fast-growing pioneer tree in Central America. A pioneer is a species that grows best in a disturbed area, such as a landslide or an area that has a gap in the forest canopy, perhaps from a tree falling down. Most pioneers are relatively fast growing.

CASTOR BEAN
Ricinus communis

- Although the cecropia contain latex, this doesn't fend off all of the herbivores that would feast on them. Their ultimate protection comes from their symbiosis with ants. A whole genus of ants, *Azteca*, is found with different species of cecropia. The tree provides a home and snacks for the ants in exchange for protection. The ants will attack any animal that tries to eat the cecropia, and they will nibble off any plants that try to grow on it.

RED CECROPIA
Cecropia glaziovii

INVASIVE PLANTS

◇ Another type of plant that is to be feared is an invasive plant, which is a plant that is not only a nonnative plant in that it has been introduced from elsewhere, but also that it's invaded into natural ecosystems and is taking over the habitat.

◇ One reason why invasive plants are able to take over so well is that, typically, none of their enemies or competitors get introduced too, so there are no other organisms to keep it in check, so it grows out of control.

◇ Cheatgrass is a good example. Cheatgrass came to the United States in grain feed accidentally and was well distributed across the West by 1930. Cheatgrass gets its name from the way it would cheat ranchers out of good forage for their livestock. Because cheatgrass flourishes in disturbed habitats, it's difficult to eradicate.

◇ Another invader found in the West is the diffuse knapweed, *Centaurea diffusa*. A knapweed is a shrub in the sunflower family, and it's sort of a cross between a thistle and a tumbleweed. Not good as forage and an excluder of native plants, diffuse knapweed is a pretty unwelcome visitor from the Mediterranean region.

DIFFUSE KNAPWEED
Centaurea diffusa

◇ Unlike some plants that are successful in new habitats because their competitors and herbivores didn't come with them, knapweed outcompetes native North American grasses more so than grass species from its old neighborhood.

- Knapweed's offense is a cocktail of exudates from its roots that prevent the growth of grasses from North America. Apparently, grasses from the Mediterranean are immune to these exudates, or at least not as repelled by them. This type of chemical defense against competitors is called allelopathy, and native shrubs, such as creosote, do it as well.

- A more famous invasive plant, at least in the Southern United States, is the infamous kudzu vine, which was brought to the United States from Japan to control soil erosion around the time of the Great Dust Bowl in the 1930s. It's rumored that this plant covers more than 7 million acres and that it will take over the South, but a 2015 issue of *Smithsonian* reports that the U.S. Forest Service assessed that the actual area consumed by kudzu was only about 227,000 acres of forestland.

KUDZU

Pueraria montana var. *lobata*

- Not just small herbaceous plants are invasive. In the Eastern Unites States, the Norway maple can displace the sugar maple. They are pretty similar, but their leaves are different: The sugar maples lobes are deeper to the midrib, or the hinge between the 2 parts of the leaves, and the serrations, or teeth, on the margins are different.

JEQUIRITY
Abrus precatorius

◇ As if it weren't bad enough to be deadly or invasive, some plants are both. The *Abrus precatorius*, or jequirity, is both terribly invasive and extremely toxic. A native of India, it is a member of the pea family, Fabaceae. This plant has become naturalized throughout the tropics and subtropics, including Florida. The entire plant contains the toxin abrin, which is similar in chemical makeup to the toxin ricin, which is the poison found in castor bean. If the plant's seeds are chewed, the toxin can cause death.

CARNIVOROUS PLANTS

◇ There are about 600 species of carnivorous plants, and they can live in a variety of habitats. The most well-known carnivorous plant, the Venus flytrap, is native to North and South Carolina, where it lives in bogs and swamps. However, carnivorous plants can also live in sandy or rocky areas, in forests, and even in bodies of water.

- Carnivorous plants still photosynthesize, but one thing that unites most of the habitats where they live is the paucity of minerals in the soil. These plants are often located in environments that are low in nitrogen. A soil that is low in nitrogen would likely be a wet soil that has leached much of its nitrogen away. As rain constantly bathes the soil in water, the minerals, including nitrogen, dissolve into the water and are carried off by gravity and erosion. This process is called leaching.

- At the other extremes, desert soils and soils that are very sandy or rocky also lack nitrogen because there is little decayed organic material in them.

- Because carnivorous plants live in regions without a lot of nitrogen, they have found a way to augment their diet with insects. Generally, plants will secrete an enzyme that will speed up the decomposition of the insect, and then the plant will absorb the minerals that were in the insect's body.

VENUS FLYTRAP
Dionaea muscipula

- Carnivory probably evolved several times among the flowering plants. There are 2 main ways to be a carnivorous plant, both of which involve liquids: A plant can trap insects in water, such as the carnivorous tank plants, or a plant can trap insects on sticky leaves. The wide variety of ways leaves have evolved to do this trapping shows the flexibility of leaf development.

- Water traps are found in plants like pitcher plants. Even though the most famous carnivorous plant, the Venus flytrap, is in one of these families, the Droseraceae family, it doesn't have a water trap or a sticky leaf. Other members of the Droseraceae family are the sundews, which are so called because they look like they are glistening with dew in the sun.

- These plants have flattened leaves that are covered with trichomes, or leaf hairs, and these hairs have glands at that base that produce digestive enzymes that decompose the trapped prey. These digestive enzymes increase in production once a prey has been captured, peaking about 4 days after an insect has been captured.

SUNDEW
Drosera sp.

◊ The Venus flytrap is only one species in one genus, *Dionaea*. There are trichomes that line the inside of 2 modified leaves that form the trap. Two trichomes have to be triggered within a short time for the physical stimulus on the leaves to turn into an electrical stimulus inside the leaves. The electrical stimulus goes to the midrib, or the hinge between the 2 parts of the leaves.

◊ What happens next is basically an educated guess because there isn't strong evidence to support the exact mechanism. In the big picture, changes will happen in the cells of the midrib that will cause a snap-buckle response. In their open state, the flytrap's leaves are spring loaded. Then, when an insect lands on a flytrap's leaf, the midrib cells swell with water, snapping the leaves closed to a position that has no stored energy. Apparently, it's fairly difficult to reset the trap. A Venus flytrap can only reset about 7 times before it dies.

READINGS

Largo, *The Big, Bad Book of Botany*.

Laws, *Fifty Plants That Changed the Course of History*.

Stewart, *Wicked Plants*.

Wilhelm, Porembski, Seine, and Theisen, *The Curious World of Carnivorous Plants*.

QUESTIONS

1   Would you plant a nonnative plant in your garden? Why or why not?

2   Why do you think people are so fascinated by carnivorous plants?

LECTURE 23 TRANSCRIPT
BAD PLANTS AREN'T SO BAD

When we really think about it, most plants should be feared—at least a little. Even the healthiest thing on the planet can become toxic if we eat too much of it. So it is with most plants, even those that we use as medications. A little knowledge about botany can go a long way in determining how much fear is necessary.

One example where a little knowledge about plants being extraordinarily useful comes from a flower called foxglove. This small herb—and herb just means a non-woody plant—called *Digitalis purpurea*, provides a major heart medication. This plant grows wild around Europe but it's also been in cultivation for at least the past century because of its pretty flowers. Doctors have known that foxglove could stimulate the heart since the late 1700s.

An 18th-century botanist, named William Withering discovered that a patient suffering from congestive heart failure improved after using a traditional herbal remedy. He discovered this while he was a doctor at Birmingham General Hospital in England. In 1775, Withering determined that foxglove tea could slow down the heart rate while increasing the strength of the heart's contractions, which could help the kidneys filter fluid out of the lungs. He then characterized the active ingredient in foxglove as digitalis, which is the primary ingredient in the drug digoxin, which is still used for patients with congestive heart failure.

Like many things in biology, there is a dose-response curve to digitalis. I like to think about this as a bell-shaped curve. In science, we call the bell-shaped curve a normal distribution. Basically, this is the goldilocks approach, meaning not enough is no good and too much is no good. Thus, not enough of a medication, and there is no effect. Too much of a medication, and there is death, as in the case of foxglove. The entire plant is toxic, and it took many botanists over the centuries to get the dose right. It took a doctor with an interest in plants to bring foxglove into mainstream medication.

Some plants, on the other hand, really are out to get us, like poison ivy or poison oak. I've never seen so much poison oak in my whole life as I did at the Jasper Ridge Biological Station, which is near the Stanford University campus. Now, poison oak is a misnomer. It's not an oak at all, but a member of the Anacardiaceae, which is the cashew family. Mangoes are also in this family. Now although poison oak isn't an oak, it is definitely a shrub, and it is definitely poisonous.

The poison in poison oak and poison ivy and poison sumac is urushiol. It gets on your skin and then moves through the pores in your skin to your bloodstream where cells pick it up and present it to your helper T cells, who then basically call out the SWAT team of your immune system, which kills all the urushiol-presenting cells—those were the cells that picked up the chemical in the first place, but some healthy cells get killed off too. This response generally causes a big red itchy rash and maybe a puss filled blister. The itch is from proteins called cytokines, which are inflammatory proteins produced by the helper T cells. There are some people whose helper T cells don't remember urushiol, and they never have a reaction.

Poisonous plants have been part of human culture for millennia. Socrates committed suicide with water hemlock. This plant, hemlock, causes a neuromuscular blockage. This plant looks deceptively close to many edible plants, particularly wild carrot and wild celery—all are in the carrot family, Apiaceae. The alkalide that gives hemlock its toxicity, coniine, has a close structure to nicotine.

And, there are so many poisonous plants. A colleague of mine does fieldwork in Costa Rica, where she told me tales of the evil ortiga. Rub against this plant and expect a rash and even hives. This plant turns out to be a type of stinging nettle, called nettle tree. What's interesting is that both of these plants come from the Urticaceae family, which is basically a whole family of stinging plants. Many of these plants have very hairy leaves that are arranged opposite of each other on the stem. Not every plant in this family stings, but so many of them do have these stinging hairs that the type genus for the family, *Urtica*, is Latin for sting, or burn. The medical term for hives, urticaria, also comes from the family name Urticaceae.

Instead of stinging nettle, it should more likely be called injecting nettle, since this is what it does. Basically, the stinging hair is just a trichome, which is what all plant hairs are called. In the case of the nettle, the trichome is made of brittle silica, the primary component of glass, and it's equipped with a small spherical bulb on the very tip. When the trichome is disturbed, say by your hand accidentally brushing up against it, the trichome will break and stick into your skin and the toxin from the bulbous tip will move into your skin. Now, if that isn't a botanical hypodermic needle, I don't know what is.

The nettle toxin is composed of several chemical compounds. It contains histamine, which triggers an inflammatory response, and this is why you take an antihistamine to treat the symptoms of nettle. The toxin also contains the neurotransmitters acetylcholine and serotonin, which is supposed to be a happy neurotransmitter, but in this case, it causes pain. Lastly, it has a few acids, including formic acid, the stinging component of ant venom.

You may have heard that stinging nettle is supposed to have medicinal properties. There is some evidence that it has anti-microbial properties and could be useful for ulcers. I didn't know, however, that there was a stinging nettle eating contest, but there is. Apparently, if you fold the leaf over, all the stinging parts will press together. Apparently, the flavor is something between arugula and spinach.

Of course, there are many other well known poisonous plants out there like oleander, castor bean, deadly nightshade, poison hemlock and many others. When I think about a plant to truly fear, I choose Cecropia. Most North American residents won't know the Cecropia tree. It's a fast-growing pioneer tree in Central America. A pioneer is a species that grows best in a disturbed area, like a landslide, or an area that has a gap in the forest canopy, perhaps for a tree falling down. Most pioneers are relatively fast growing.

Although the Cecropia contain latex, this doesn't fend off all of the herbivores that would feast on them. Their ultimate protection comes from their symbiosis with ants. A whole genus of ants, *Azteca*, are found with different species of Cecropia. The trees actually produce hollow stems to house these ants. A queen Azteca ant looking for a new home need only chew through a thin membrane that covers the opening to the hollow stem, and she has a new palace for her brood. Lucky for the queen ant, the area around the membrane doesn't have latex ducts that might otherwise prevent her from moving in.

The tree also encourages the ants to stay for awhile by offering them special oil-rich droplets called Müllerian bodies. These tasty treats are harvested by the ants, but apparently, the treat doesn't contain any sugar until it's ready to be harvested, so the ants don't take them before they're ready.

So, this all sounds very nice, what's to be feared? The tree provides a home and snacks for the ants as well as protection. The ants will attack any animal that tries to eat the Cecropia, and they will nibble off any plants that try to grow it. I learned the hard way about how effective these Azteca ants can be when I was innocently admiring the leaves of a Cecropia—talk about ants in my pants.

Another type of plant to be feared is an invasive plant. This term describes a plant that is not only a non-native plant in that it has been introduced from elsewhere, but also that it's invaded into natural ecosystems and is taking over the habitat. One reason why invasive plants are able to take over so well is that typically, none of their enemies or competitors get introduced too, so there are no other organisms to keep it in check, so it grows out of control. Cheatgrass is a good example.

At least in the Western U.S., cheatgrass might be the most hated of all the introduced grasses. Cheatgrass came to the U.S. in grain feed accidentally that was well distributed across the west by 1930. Cheatgrass gets its name from the way it would cheat ranchers out of good forage for their livestock. Since cheatgrass flourishes in disturbed habitats, it's difficult to eradicate.

Ironically, cheatgrass turns out to be decent forage for cattle in the early spring. That's because cheatgrass is a winter annual, so it germinates in the fall and grows until the temperatures are too low. But the roots can survive the winter, so it can live through the winter and be the first grass to green up in the spring, and that's when it needs to be grazed. It is actually very good early spring forage for livestock, but once it starts to die in the late spring, it dies off very fast, leaving the ground parched for the native grasses and encouraging range fires.

Another invader in the west is the diffuse knapweed, *Centaurea diffusa*. A knapweed is a shrub in the sunflower family, and it's sort of a cross between a thistle and a tumbleweed. Not good as forage and an excluder of native plants, diffuse knapweed is a pretty unwelcome visitor from the Mediterranean region. Unlike some plants that are successful in a new habitat because their competitors and herbivores didn't come with them, knapweed outcompetes native North American grasses more so than grass species from its old neighborhood. Knapweed's offense to is a cocktail of exudates from its roots that prevent the growth of grasses from North America. Apparently, grasses from the Mediterranean are immune to these exudates, or at least not has repelled by them. This type of chemical defense against competitors is called allelopathy and native shrubs, like creosote, do it as well.

A more famous invasive plant, at least in the southern U.S. is the infamous kudzu vine. According to Amy Stewart, author of *Wicked Plants: The Weed that killed Lincoln's Mother and Other Botanical Atrocities*, writes that, to the Japanese, the word kudzu means, rubbish, waste, or useless scraps. Not the best impression for this plant. Kudzu was brought to the US to control soil erosion around the time of the Great Dust Bowl in the 1930s. It's rumored that this plant covers over 7 million acres and that it will take over the South. It's a fast grower, but it's no bamboo.

A 2015 issue of *Smithsonian* reports that the U.S. Forest Service assessed that the actual area consumed by kudzu was only about 227,000 acres of forestland, which for comparison is about one-sixth the size of Atlanta. The same report estimated that Asian privet, another invasive plant, covered about 3.2 million acres.

People often ask me if I have a favorite plant, and I do, it's the alpine forget-me-not that I shared with you in the very first lecture. No one ever asks me if I have a least favorite plant, but I have one of those, too, it Kochia. As far as weeds go, it's not terrible. It's not poisonous, and livestock can eat it, though it can make them sick because of occasional high nitrate levels. Kochia is native to Asia and was introduced to the U.S. from Europe as an ornamental. Looking at the plant, it's hard to understand why. Like most members of the family Chenopodiaceae, to which Kochia belongs, it has small inconspicuous green flowers. How could tiny green flowers be ornamental? It has what I would consider a weedy habit, that is sort of tall and spindly looking. The word weed is a social construct. A weed is any plant growing where you don't want it growing.

I suppose I dislike the Kochia so much because it is flowering if you can call it that, all over the neighborhoods of my fair city. Every roadside ditch, every empty weed lot, every unmowed section along a bike trail, is rampant with Kochia. There doesn't seem to be a good way to control Kochia except to use a lot of herbicide on it or to plant other plants that will keep it at bay. I suppose it's a reminder to me that there are patches of ground around the city that no one pays attention to—that no one cares about, and I feel kind of bad for these patches. If only someone cared enough to plant some nice native shrubs to keep the Kochia out, this patch would have a chance. Then I feel guilty that it's not me planting those native shrubs, and that is why I hate Kochia.

Not just small herbaceous plants are invasive, either. In the Eastern U.S., the Norway Maple can displace the Sugar Maple. They are pretty similar, but if you look closely at the leaves, the Sugar Maple lobes are deeper to the midrib and the serrations, or teeth, on the margins are different.

As if it weren't bad enough to be deadly or invasive, some plants are both. The *Abrus precartorius*, or jequirity is both terribly invasive and extremely toxic. A native of India, it is a member of the pea family, the Fabaceae, with long, pinnate, or feather-like, leaves. The seeds of this plant are red with black on the tips, and they are used as beads for percussion instruments and jewelry. This plant has become naturalized throughout the tropics and subtropics, including Florida. The entire plant contains the toxin abrin, which is similar in chemical makeup to the toxin ricin, which is the poison found in Castor Bean. Usually, ingestion of the whole seeds won't cause death because of a tough seed coat. However, if the seeds are chewed, the toxin can cause death. In 2015, bracelets made from the seeds were removed from a gift shop in the U.K. because abrin is a controlled substance under the U.K. Terrorism Act.

This next group of plants isn't feared by humans. Perhaps they are more like revered by humans because they go against the norm of what we expect plants to do. This brings us to the full-on, real, and larger than life carnivorous plants.

Most people who couldn't even identify a dandelion are still pretty excited about carnivorous plants. These plants are are the botanical exception rather than the botanical rule, and as we know—there are very few rules in botany.

There are about 600 species of carnivorous plants, and they can live in a variety of habitats. The most well-known carnivorous plant, the venus flytrap, is native to North and South Carolina where it lives in bogs and swamps. But, carnivorous plants can also live in sandy or rocky areas, in forests, and even in bodies of water.

Carnivorous plants still photosynthesize, but one thing that unites most of the habitats where they live is the paucity of minerals in the soil. These plants are often located in environments that are low in nitrogen. A soil that is low in nitrogen would likely be a wet soil that has leached much of its nitrogen away. Leaching is sort of like pouring water over the coffee. When you first pour the water over the coffee, it's nice and strong. If you have to reuse those grounds, your coffee is going to be weak. By the third pour over, it's hard to see how that liquid resembles coffee at all. The same is true for soils. As rain constantly bathes the soil in water, the minerals, including nitrogen, dissolve into the water and are carried off by gravity and erosion. This process is called leaching. At the other extreme, desert soils, and soils that are very sandy or rocky, also lack nitrogen because there is little decayed organic material in them.

Because carnivorous plants live in regions without a lot of nitrogen, they have found a way to augment their diet with insects. Generally, plants will secrete an enzyme that will speed up the decomposition of the insect, and then the plant will absorb the minerals that were in the insect's body.

Carnivory probably evolved several times among the flowering plants. It occurs in about a dozen genera and around 600 species of plants. There is sort of two main ways to be a carnivorous plant, both of which involve liquids. A plant can trap insects in water, like the carnivorous tank plants, or a plant can trap insects on sticky leaves. The wide variety of leaves that have evolved to do this trapping shows the flexibility of leaf development, a trait biologists would call plastic. A trait that can have a lot of different forms is said to be a plastic trait. So, carnivorous plants display the plasticness of the leaf.

Water traps are found in plants like pitcher plants. Even though the most famous carnivorous plant, the venus fly trap, is in one of these families, the Droseraceae, it doesn't have a water trap or a sticky leaf. Other members of the Droseraceae family, are the sundews, so-called because they look like they are glistening with dew in the sun, but these plants look more like fly paper.

These plants have flattened leaves that are covered with trichomes, which is what botanists call leaf hairs, and these hairs have glands at the base that produce digestive enzymes that decompose the trapped prey. Again, the trichomes glistening with the enzymes earn the plants the title of sundew. These digestive enzymes, including protease and phosphatase, increase in production once a prey has been captured, peaking about four days after an insect has been captured. Plants take their time—it's not like they have to go somewhere. There are about 150 different species of sundews, both temperate and tropical.

OK, what you've really been waiting for. How does the Venus flytrap work? The Venus flytrap is only one species in the whole genus *Dionaea*, which is Greek for daughter of Dione, where Dione is the Greek mother goddess whose daughter was Aphrodite. So what we're talking about is the Aphrodite Genus. The specific epithet, *muscipula*, is Latin for mousetrap. Though I've never seen one so big as to actually eat a mouse, unlike "The Little Shop of Horrors."

First, there are trichomes that line the inside of two modified leaves that form the trap. Two trichomes have to be triggered within a short time for the physical stimulus on the leaves to turn into an electrical stimulus inside the leaves. This electrical stimulus is much like a human action potential that allows our nerves to fire. The electrical stimulus goes to the midrib of the leaf, which is the hinge between the two parts of the leaf.

What happens next is basically an educated guess because there isn't strong evidence to support the exact mechanism. In the big picture, changes will happen in the the cells of the midrib that will cause a snap buckle response. In their open state, the flytrap's leaves are spring-loaded, like an inside-out contact lens. Then, when an insect lands on a flytrap's leaf, the midrib cells swell with water. The swelling of the cells in the midrib snaps the leaves closed to a position that has no stored energy. Apparently, it's fairly difficult to reset the trap. A venus flytrap can only reset about seven times before it dies.

In 2016, Ranier Hedrich from the University of Würzburg also discovered that the plant will then start to secrete more digestive enzymes the more the trapped insect inside moves around. If two touched trichomes close the trap, three touched trichomes start the flow of enzymes that will digest the insect.

But flytraps need energy to power the movements. In a study from 2015, botanists at the University of Bristol discovered that the pitcher plant, *Nepenthes gracilis*, actually uses the force of falling rain to flick ants into the pitcher trap, where they drown and are digested by enzymes. The nitrogen from these decomposed ants is then absorbed by the plant.

The top of each pitcher is enclosed by a hood-like lid, which is attached to a flexible hinge. When the rain drops fall on this lid, it springs down and then up, flinging the ants into the pitcher. This movement of this trap is an order of magnitude faster than the snapping flytrap—too fast to be observed with the naked eye.

There are even underwater carnivorous plants like the bladderwort. The bladderwort genus, *Utricularia*, actually contains about 225 species. It is the largest genus of all the carnivorous plants, so something about underwater carnivory is pretty successful. Although the genus name sounds a bit like Urtica or Urticaceae, this genus isn't in that family or even related to the nettles. The bladderworts are in the family Lentibulariaceae, which is entirely comprised of carnivorous plants. Utricularia, the genus name for bladderworts, comes from the Latin word *utriculus*, which means small skin. This refers to the small bladders that trap insects. This is how the plants get their common name bladderwort. The small sacs that grow on the plants resemble tiny wine skins or wine bladders that were used when this name came about.

The word wort is an Old English word for plant, as we saw with the non-vascular liverwort and hornwort. It's W-O-R-T, not wart, not needing Compound W.

Bladderworts are flowering plants distributed around the world, and the genus has a wide variety of color types and flower types. Although they are famed for having underwater traps, most of the bladderworts are actually terrestrial, living in soils that are constantly moist or waterlogged for part of the year. And although they are terrestrial flowering plants, they lack true roots and instead have small root-like structures called rhizoids. Because of this, they generally don't grow more than a foot tall. They can be up to a meter when growing in water, however.

The traps, or bladders, in Utricularia are on the shoots, or the on the leaves. The bladders are generally small, about a centimeter in length and form a hollow space inside. The hollow space, the inside of the bladder, is covered by a thin membrane that has trichomes on it. This hollow space is covered by a velum, a membranous plant material that curves over the hollow space and acts as a door. Cells at the bottom of the door are especially flexible to form a hinge. The outer cells of the bladder excrete a sugary mucus, which attracts insect prey.

The basic principle that the traps use is suction. Inside the bladder, water has been removed, so there is a negative pressure. When a visiting insect brushes against these trichomes, the pressure is just enough to distort the membranous door, which swings the bladder in and sucks water and the insect inside. The door immediately closes, and glands surrounding the door secrete mucus to seal the trap.

When the prey has been digested, the trap has to be set again. This is done with the help of glands located inside the bladder. First, a concentration gradient of chloride ions is set up with the help of active transport. That means, the plant has to use cellular energy, which is the molecule ATP, to drive chloride ions into the outer cell layer of the bladder.

Once the concentration of the chloride ions is sufficiently high in this outer layer, water molecules will follow because of osmosis. Remember, osmosis is the diffusion of water, which will move from a high concentration of water, inside the bladder, to a low concentration of water, in the outer cell layer where all the chloride ions are located.

So, these plants are not similar in terms of their habitats, chemistry, ancestry, or even the way they make a living. It's interesting to me that not all plants are poisonous, given the wide variety of those that are. And dose-response curves for plants like foxglove suggest that even a plant harmful at one level might sometimes be beneficial in a different dose or context. Invasive species can be very feared, but there's also some research that says they increase diversity, and, at least for some invasive plants, animals in their new habitats sometimes learn to eat them, and other plants will eventually compete with them.

Whether apparently poisonous, invasive, or carnivorous, these plants should at least be respected, if not feared—and possibly admired. Even bad plants aren't so bad.

LECTURE 24

MODIFYING THE GENES OF PLANTS

Genetically modified organisms (GMOs) are typically created by transferring a piece of DNA from one organism into another organism. The genome of an organism is the combination of all the DNA. If you add a snippet of DNA from some other organism and insert it into the original organism, you have modified it. Humans can manipulate genes within an organism to provide all kinds of benefits, both economic and ecologic.

GENETICALLY MODIFIED ORGANISMS

- People are skeptical that we understand all the repercussions of genetically modifying organisms—particularly those that we eat. But there are compelling reasons to want to genetically modify an organism.

- One of the most popular GMOs that virtually everyone in the United States has eaten is Bt corn. The "Bt" stands for *Bacillus thuringiensis*, which is a type of bacteria that lives naturally in the soil. Organic farmers have known since the 1960s that it produces a protein that that can ward off a major corn pest, the European corn borer. A borer is a type of moth that has a larval stage that eats the leaves and stems of the corn.

- If this type of bacteria was harmful to the European corn borer, what was it specifically that harmed the moth? Because most bacteria cause harm by creating proteins called endotoxins, the Bt endotoxin was the likely candidate, and Bt was already sprayed on plants to kill pests.

- To genetically modify the corn, there needs to be a piece of DNA from another organism, and the donor organism in this case is *Bacillus thuringiensis*. This bacterium has a gene that produces the protein that is the Bt endotoxin.

- So, if you can get this one gene from the *Bacillus thuringiensis* into the corn, then the corn plant can make its own pesticide against the European corn borer. If the corn plant has the gene from the *Bacillus thuriengiensis*, then it can create Bt endotoxin and destroy its own pests, with no help from pesticides needed.

- In this way, one might think that everyone would love GMOs, as they could bring about the end to the use of pesticides. Since the invention of Bt corn, there is now Bt cotton, Bt potatoes, Bt eggplant, and Bt soybeans. The idea is that if the plants are making their own defense, then we are not using pesticides on them, and workers who would normally have to treat these plants are also not being exposed.

- Although there is speculation about many different negative consequences of Bt crops, the strongest and most potentially damning evidence thus far comes from natural selection.

- These corn borers and other lepidopterans, such as butterflies and moths, are just trying to make a living. If any of them just happen to have a bit of resistance to this Bt toxin, they are going to do well, while all of their friends die off. The ones that have resistance find mates that also have resistance, because they're the only ones left, and they mate.

- Chances are that many of their offspring will also have resistance to the Bt endotoxin, and suddenly Bt corn doesn't work as well. There is evidence of instances where this has happened. Just because the toxin is produced by the plant doesn't mean that insects won't evolve to resist the toxin. Pesticide resistance develops no matter where the pesticide comes from.

INSERTING GENES

⬥ Interestingly, you insert a gene from a bacterium into the genome of a corn plant by using another bacterium. Bacteria are so useful in this process because they contain something that most eukaryotic cells, such as our cells, don't have. They have plasmids, which are small circular pieces of DNA that are separate from the regular chromosomes of the bacteria. These plasmids are one of the reasons that bacteria can develop antibacterial drug resistance so quickly—because these plasmids can be shared between bacteria.

⬥ The DNA within the plasmid can be sliced up using restriction enzymes. An enzyme is just a protein that speeds up a reaction. A restriction enzyme will search for particular sequences of DNA and cut the DNA in the same location. When this cut occurs, the ends of the DNA are now sticky, which means that they will glom onto another piece of DNA, if they happen to meet up with one. The DNA they're going to meet up with is the donor gene from the bacteria.

⬥ Now there are gene guns, which use nanometer-sized gold or tungsten bullets that are coated with the DNA of interest to insert into the plant's DNA. Gene gun technology was invented in the mid-1980s.

THE POTENTIAL HARM OF GMOS

⬥ There are several concerns about the potential harm of GMOs, but most of these concerns don't have strong evidence—yet.

⬥ One concern about GMOs is that you're putting in a different gene, which will make a new protein, that wasn't found in the original organism. For example, if you put a gene to make a bacterial protein into a soybean, people who have never been allergic to soybeans might now be allergic to a GMO soybean because it has a new protein from the bacteria. The concern seems to be one of knowledge to the consumer and potential for allergies.

- Another concern with GMOs is that the gene that is put in the crop will find its way into other plants, causing them to be resistant to herbicides or be able to make their own pesticides. The concern is that the gene of interest—for example, the Bt gene that was bred into corn—would find its way into a relative of corn.

- Corn is a grass, and there are many grasses that are considered weeds, and many plants cross-pollinate to form new varieties. Because plants hybridize, the concern is that the pollen from Bt corn may fertilize a non-corn grass and give it the Bt gene, which would prevent it from being harmed by the corn borer.

- So far, there's not strong evidence of superweeds, but there has been a problem with pollen contamination. Corn pollen is very light and can easily travel miles away and pollinate other cornfields. Pollen from Bt corn has made its way onto fields that are not Bt corn, thus conferring the Bt gene to that corn, even though the farmer hasn't paid for it and may not want it. Because corn is wind pollinated, it would be very difficult to keep Bt pollen off a nearby cornfield if you were trying to maintain a non-GMO crop.

- The concern is protecting farmers who don't want their crops cross-pollinated with GMOs, as well as consumer confidence that a small grower who claims that his or her corn is non-GMO should actually be able to protect that crop from GMO pollen.

- Another concern with GMOs is that friendly insects, or at least insects that aren't pests, will eat the plant that makes pesticide and get killed, or pollen from a plant that has been genetically modified to make its own pesticide will get eaten by a friendly insect, which will die.

- Other arguments against GMOs pertain to sustainable agriculture and profit for smaller farmers versus larger corporations. Depending on the type of GMO used, the seed might have to be purchased every year. If left to their own reproductive strategy, genetically modified genes may or may not be passed onto offspring. To ensure that every plant has the gene, a farmer must buy all new seeds. If farmers have to buy these seeds every year, then they are making less money.

- The contrary argument is that companies that did the research to develop the seed deserve the money to pay for that research and development and that the farmers are saving money by buying fewer pesticides. But pests such as the corn borer will continue to develop a resistance to the Bt toxin, even if the corn produces it in its own Bt toxins. The problem of resistance still remains.

- Another concern about GMOs is that biotechnology will not solve world hunger problems. In most parts of the world, hunger is an economic or political problem, or a problem of distribution—not one of production. When countries have a drought or a severe food shortage, the main issues become distribution of donated food.

- But GMOs can provide new varieties of crops that would be more drought tolerant or that would produce more nutritious foods. Scientists in Egypt have developed a drought-resistant form of wheat by inserting a gene from barley. Experiments indicated that the new wheat strain outperformed the non-GMO variety under normal rainfall conditions. Currently, Egypt has to import wheat because it can't meet its need, but if the wheat range could expand, then perhaps Egypt could meet its demand.

- Because many of the objections to GMOs can, in principle, be met through further research into GMOs, the eventual promise of GMOs can seem limitless. For example, plants can be genetically engineered to make vaccines.

- Because of the contention with GMOs, some botanists wondered if there was an alternative pathway to changing aspects of plants without changing their genes. In a 2016 report, scientists in Germany showed that just spraying RNA on leaves of barley reduced the amount of fungal infection of the sprayed leaves. Apparently, the RNA they sprayed on the barley leaves was able to silence, or turn off, some major genes in the fungal pathogen. Techniques like this demonstrate that there are different ways to give plants resistance to disease without modifying their genes.

- There is an even newer and greater technique on the plant biotechnology horizon. Scientists have observed certain patterns in the genomes of some bacteria. The pattern is that some base pairs, which make up the DNA, would be repeated over and over again. In between these repeated segments would be normal areas of DNA. These patterns are called clustered regularly interspaced short palindromic repeats (CRISPR).

- Why the bacteria had these repeats was a mystery until scientists figured out that these repeats matched the DNA of viruses that prey on bacteria. CRISPR holds onto pieces of viral DNA so that the bacteria will recognize future attacks from viruses, much like our own immune system. The bacteria with CRISPR also have a group of enzymes called CRISPR-associated proteins that will cut DNA in precise locations. Using these 2 tools, scientists can add and delete genes from genomes with unprecedented accuracy and speed.

READINGS

Endersby, *A Guinea Pig's History of Biology*.

Slater, Scott, and Fowler, *Plant Biotechnology*.

Wood and Habgood, *Why People Need Plants*.

QUESTIONS

1. How do you feel about genetically modified organisms? Does it make any difference if the modified piece of DNA is from another organism or if the primary modified organism is created in the lab using CRISPR technology?

2. How has this course changed the way you think about botany?

LECTURE 24 TRANSCRIPT
MODIFYING THE GENES OF PLANTS

My hope throughout this course has been to open your eyes to the science of botany that is all around us. For this concluding lecture, we're going to do that with genetics, which has some of the most surprising aspects of all.

Looking backward, what would you say? Which plant was the most important to introducing the world to genetics? Maybe the pea plant, because of Gregor Mendel? Personally, I think corn could be a contender—a corn-tender. While it is true, that Mendel's work was the basis for understanding patterns of plant inheritance of physical traits, the genetic basis for inheritance was far from being understood in Mendel's time.

But by the time botanists started working with corn in the 1920s, they had a pretty good idea that the genes were responsible for the physical characteristics that showed up in organisms. But, corn continued to corn-fuse the early geneticists because it did some unusual things.

First of all, corn is very promiscuous in that its pollen will travel very far and different varieties of corn will hybridize easily. More than that, a single cob would only have one mother, but it could have dozens of fathers from pollen all over the place. It's really a-maize-ing.

As you may recall from our lecture on grasses, the grass seed and fruit are fused into a single caryopsis. The silks coming out of the corn on the cob you buy at the store are the leftover stigmas where the different pollen grains landed to fertilize each fruit, also known as a corn kernel.

This is why one cob can be variegated, that is, have several different colors of kernels. OK, not too corn-fusing yet, but corn didn't seem to follow a Mendelian pattern. That might be understandable because scientists knew that some genes that were on the same chromosome wouldn't always be inherited together because of crossing over. This occurs when pieces of chromosome break off and reattach to other chromosomes that are the same size and shape. These same size and shape chromoosomes are called homologous chromosomes, where one comes from mom, and one came from dad. This process happens during the creation of the sex cells—the eggs and sperm.

This got geneticists thinking about how these genes really worked. It was one thing to understand inheritance, but the entire suite of genes, the genome, is in every single cell in an organism. How does a root cell know how to turn on the genes that are important for roots? How does a leaf cell know how to make the proteins that will ultimately be responsible for photosynthesis?

A young woman studying at Cornell University made this her life's work. She was the first to use a technique called squashing with corn cells. By slicing sections of corn into very thin sections and then putting these sections on a slide and then literally squashing the top of the coverslip with one's thumb, the chromosomes in the cell became visible. Using this technique, she was the first to identify chromosomes in corn and the first to identify all 10 of them. She also noted what she called knobs on the end of these chromosomes so that she could identify each chromosome uniquely. She was the first to actually see the process of crossing over occur when looking at the chromosomes in the microscope.

Barbara McClintock was interested in the variegation of corn. The kernels themselves would be striped or multicolored, and that was not Mendelian in the least. It was more like pre-Mendelian blending, but corn was more complicated. Even when corn was specially bred, the patterns of variegation couldn't be predicted. She wanted to know what was turning genes off and on causing variegation. She tried using x-rays on corn plants, and this did cause mutations in the genes. But the mutations were random, and it would take too long to figure out what mutation caused what change in the physical structure.

She then noticed that some corn plants had a fair number of broken chromosomes. She also discovered that these broken chromosomes were heritable. By carefully looking at where the chromosome had broken and what physical characteristic went with it, she could determine what a particular gene should have been doing. She thought that broken chromosomes might be linked with a pattern in the variegation, but she didn't know how.

Her most famous discovery occurred as she observed that when her chromosomes broke, pieces of DNA would move and reinsert themselves in another part of the chromosome. She thought these broken pieces of DNA were turning genes off and on, and so she called them controllable elements. Other scientists believed her observations but questioned her interpretation. Although her controllable elements were not turning other genes off and on, she eventually won a 1983 Nobel Prize for for her discovery of mobile genetic elements. The name of her controllable elements is now transposons. Sometimes they're called jumping genes because, in a sense, this is what they do.

Botanists are just now discovering the usefulness of jumping genes. It seems that jumping genes can create new combinations in the genome, which can lead to new adaptations that might be beneficial for the plant. For example, a 2015 paper indicates that transposons provide gene variation that helps the corn plant cope with environmental stress.

The study of genes has come a long way since Barbara McClintock's time. We now understand that nature uses gene jumping. In fact, it's known that bacteria and viruses can and do facilitate gene jumping across species. This has been the biological background inspiring human efforts to manipulate genes within the organism to provide all sorts of benefits, both economic and ecologic.

So humans have been creating genetically modified organisms or GMOs. In practice, these are typically created by transferring a piece of DNA from one organism into another organism. Remember that DNA is deoxyribonucleic acid or the blueprint of the organism. The genome of an organism is the combination of all the DNA. So, if you add a snippet of DNA from some other organism and insert into the original organism, you've modified it.

GMOs have been labeled Frankenfoods after Victor Frankenstein's famous creation. This isn't entirely accurate since, in the book, Mary Shelley never talks about using parts of any other organism that isn't human to bring the monster to life, but still, I think you see the point. Like bringing the dead back to life, people are skeptical that we understand all the repercussions of genetically modifying organisms, particularly, those organisms that we eat.

There are actually compelling reasons to want to genetically modify an organism. One of the most popular GMOs that virtually everyone in the U.S. has eaten is Bt corn. The Bt stands for *Bacillus thuringiensis*, which is a type of bacteria. These bacteria live naturally in the soil, minding their own business, and organic farmers have known since the 1960s that this bacterium, *Bacillus thuringiensis*, produces a protein that that can ward off a major corn pest, the European Corn Borer. A borer is a type of moth that has a larval stage, a caterpillar if you will, that eats the leaves and stems of the corn. So, if this bacteria was harmful to the European Corn Borer, what was it specifically that harmed the moth? Since most bacteria cause harm by creating proteins called endotoxins, the Bt endotoxin was the likely candidate, and Bt was already sprayed on plants to kill these pests.

Now, in order to genetically modify the corn, there needs to be a piece of DNA from another organism, and the donor organism, in this case, is the *Bacillus thuringiensis*. It turns out that this bacterium has a gene, one gene, which produces the protein that is the Bt endotoxin. So, if you can get this one gene from the *Bacillus thuringiensis* into corn, then the corn plant can make its own pesticide against the European Corn Borer. If the corn plant has the gene from the *Bacillus thuringiensis*, then it can create Bt endotoxin and destroy its own pests, no help from pesticides needed.

In this way, one might think that everyone would love GMOs as they could bring about the end to the use of pesticides. Since the invention of Bt corn, there is now Bt cotton, Bt potatoes, Bt eggplant, and Bt soybeans. The idea is that if the plants are making their own defense, then we are not using pesticides on them, and workers who would normally have to treat these plants are also not being exposed.

So, what's the concern? Why would countries like Switzerland ban the use of Bt crops? Although there is speculation about many negative consequences, the strongest and most potentially damning evidence thus far, comes from natural selection.

Remember, these corn borers and other lepidopterans—that's a fancy word for butterflies and moths—are just trying to make a living. If any of them just happen to have a bit of resistance to this Bt toxin, they are going to do well, while their friends all go kaputt. So, the ones that have resistance find mates and they also have resistance since they're the only ones left, and they mate. Chances are, many of their offspring will also have resistance to the Bt endotoxin and suddenly Bt corn doesn't work as well. There is evidence of these instances where this has just happened. Just because the toxin is produced by the plant, doesn't mean insects won't evolve to resist the toxin. Pesticide resistance develops no matter where the pesticide comes from.

I think it's interesting to see just how this research is possible in the lab. How do you actually insert a gene from a bacteria into the genome of a corn plant? Interestingly, you do it by using another bacteria. Bacteria are so useful in this process because they contain something that most eukaryotic cells, like our cells, don't have. They have plasmids. Plasmids are small circular pieces of DNA that are separate from the regular chromosomes of the bacteria. These plasmids are one of the reasons that bacteria can develop anti-bacterial drug resistance so quickly—because these plasmids can also be shared between bacteria.

The DNA within the plasmid can be sliced up using restriction enzymes. An enzyme is just a protein that speeds up a reaction. A restriction enzyme will cut DNA at precise places. For example, if you were to use the Control Find function to search this lecture in a text document, you could search for the term GMO.

In much the same way, restriction enzymes will search for particular sequences of DNA, and they will cut them in the same location. When this cut occurs, the ends of the DNA are now what scientists call sticky. This basically means that they will glom onto another piece of DNA if they happen to meet up with one. Well, you can guess what kind of DNA they're going to meet up with. Yes—our donor gene from the bacteria.

The most common way to create GMOs is with the use of a bacterium called *Agrobacterium tumefaciens*. This bacteria is pretty interesting because, in nature, it infects plants and causes Crown Gall Disease, which is a sort of tumor-like growth in the plant. It gets the plant to make this tumor-like growth by infecting it with its own bit of DNA, which is called transferred DNA, or tDNA, for short.

This tDNA is on a plasmid, and because the plasmid causes tumors, it's called a tumor-inducing plasmid, a Ti plasmid. So, the plasmids of the *Agrobacterium tumefaciens* are called Ti plasmids because they are normally tumor-inducing. But, remember what we said—that a plasmid was extra. The bacteria can function normally without it. So, one removes the Ti plasmid, cuts it with the restriction enzyme, and inserts the donor gene into the plasmid. Then, one reinfects the plant host with the modified plasmid, which will insert itself into the plant's DNA, just like it would in nature. Then, instead of the plant producing the tumor-like growth, the plant will produce whatever gene you inserted into the Ti plasmid. In the case of Bt corn, the plant will now make the Bt endotoxin.

wThis was how genetic engineering got started, and it can still be done this way. But now there are actual gene guns. These use nanometer-sized small gold or tungsten bullets that are coated with the DNA of interest to insert it into the plant's DNA. First invented in the mid-1980s, the gene gun technology was the way Roundup Ready soybeans were created.

Roundup is trademarked name for a powerful herbicide called glyphosphate. Remember, an herbicide kills plants. A pesticide kills pests, which are usually insects. Roundup and any generic product containing glyphosphate will kill pretty much any plant. This chemical works by shutting down an enzyme that plants need to synthesize at least three amino acids, which are building blocks of proteins. The enzyme is called 5-enolpyruvylshikimate3-phosphate or, EPSP synthase. When this EPSP synthase enzyme is shut down, the reaction to make the amino acids also stops, and the plant can't survive.

Generally, glyphosphate is virtually non-toxic to vertebrates and insects. Since only plants and microbes produce the EPSP synthase enzyme, only plants and microbes will be harmed by the glyphosphate, or Roundup, which knocks out the enzyme.

Now, because this glyphosphate is so damaging to plants, the way Roundup Ready crops work is a bit different than the way Bt crops worked. With Bt crops, the plants manufactured their own pesticide. With Roundup, plants can't produce their own glyphosphate because they would inhibit their own EPSP synthase and kill themselves. So, back in the 1980s, there was a search for a gene that would enable plants to use a different enzyme other than EPSP.

Enter our old friend, the crown gall-forming bacteria, *Agrobacterium tumefaciens*. Turns out this bacteria wasn't killed by glyphosphate. Scientists working for the Monsanto company located a gene that was responsible for producing a different enzyme, not EPSP, to make the essential amino acids that the plant needs. So, with the finding of this new bacterial gene, all that had to be done was find a way to get that bacterial gene into soybeans, and then voila—one could spray all the Roundup desired around the soybeans to kill off any weeds, while the soybeans themselves would not be harmed.

Turns out that getting the Roundup Ready gene into soybeans wasn't that easy. Monsanto had to partner with some other companies, and this was where the technology called the gene gun, allowed scientists to get the Roundup Ready gene into soybean plants. Once those plants produced seeds with the same gene, Roundup Ready soybeans went to market.

What's interesting in the history of this is that Monsanto essentially gave away this technology. They received a one-time payment of $450,000 in 1992 for the gene from the company DuPont. Now, DuPont could sell Roundup Ready seeds. Monsanto's profit would come from greater sales of Roundup.

Now, before we get too excited about Roundup Ready crops, we have to remember that natural selection is always present. The axiom nature finds a way has some truth to it. What do you suppose will happen when a lot of undesired plants are exposed to a lot of glyphosphate? Remember, some of those plants will naturally have some resistance to the glyphosphate. Just like when a certain cold virus goes around, some people won't get it—they're resistant. So, beginning in 2000, the first glyphosphate-resistant horseweed, *Conyza Canadensis*, is found. Where is it found? You guessed it—right smack in the middle of a soybean field. Later, other resistant weeds were discovered. Of course, good old scientific ingenuity doesn't sit still either, so Monsanto developed a second generation Roundup Ready group of crops. And so it goes, the arms race of trying to outpace nature.

Now, glyphosate harms an enzyme humans don't even produce. We can't even make the amino acids that the enzyme catalyzes. Bt corn makes a pesticide that doesn't harm us—only caterpillars. So, why all the fuss about GMOs? There are several concerns about the potential harm of GMOs, but most of these concerns don't have strong evidence—yet. Let's look at those concerns and determine the latest evidence that confirms or refutes these concerns.

Allergies. One concern about GMOs is that you're putting a different gene, which will make a new protein, that wasn't found in the original organism. For example, if you put a gene to make a bacterial protein into a soybean, people who have never been allergic to soybeans might now be allergic to a GMO soybean because it has a new protein from the bacteria. The concern here seems to be one of knowledge to the consumer and potential for allergies. That is, think about if you have a peanut allergy. You go to buy some corn chips and suppose the chips were made with GMO corn. What happens if the GMO corn contains one protein from a peanut? Maybe this protein will cause you no harm, but maybe it will. Moreover, maybe you, as a consumer, just feel that you have the right to know if your corn chips were made with GMO corn.

Superweeds. Another concern with GMOs is that the gene that is put in the crop will find its way into other plants causing them to be resistant to herbicides or be able to make their own pesticides. Now, remember I said earlier that there is good evidence that some weeds are becoming resistant to RoundUp? Certainly, that is not good news, but this is a different concern. The concern here is that the gene of interest, say the Bt gene that was bred into corn would find its way into a relative of corn. Corn is a grass, and there are lots of grasses that are considered weeds, and many plants cross-pollinate to form new varieties. Since plants hybridize, the concern is that the pollen from Bt corn may indeed fertilize a non-corn grass and give it the Bt gene, which would prevent it from being harmed by the corn borer.

So far, there's not strong evidence of superweeds, but there has been a problem with pollen contamination. We said at the beginning of this lecture that corn pollen is very light and can travel miles away and pollinate other cornfields. As might be expected, pollen from Bt corn has made its way onto fields that are not Bt corn, thus conferring the Bt gene to that corn, even though the farmer hasn't paid for it, and may not even want it. Because corn is wind pollinated, it would be very difficult to keep Bt pollen off a nearby corn field if you were trying to maintain a non-GMO crop. The concern here is protecting farmers who don't want their crops cross-pollinated with GMOs, as well as consumer confidence that buying corn from a small grower who claims it is non-GMO, then that farmer should be able to protect his crop from GMO pollen. Easier said than done.

Another concern with GMOs is that friendly insects, or at least insects that aren't pests, will eat the plant that makes pesticide and get killed. Or, pollen from a plant that has been genetically modified to make its own pesticide will get eaten by a friendly insect and that friendly insect will die. What if a monarch butterfly is eating its preferred food source, the milkweed, and that milkweed is covered with corn pollen containing the corn toxin? This is exactly the sort of research that was taking place when scientists first noticed a decline in monarch butterfly populations. After much research, it was concluded that monarchs were not eating enough Bt pollen to be affected and that habitat loss in Mexico was a more likely reason for the decline.

Other arguments against GMOs pertain to sustainable agriculture and profit for smaller farms versus larger corporations. Depending on the type of GMO used, the seed might have to be purchased every year. If left to their own reproductive strategy, genetically modified genes may or may not be passed onto offspring. To ensure that every plant has the gene, a farmer must buy new seeds. If farmers have to buy these seeds every year, then they are making less money. The contrary argument is that companies that did the research to develop the seed deserve the money to pay for that research and development and the farmers are saving money by buying fewer pesticides. But, as we heard earlier, pests such as the corn borer, will continue to develop a resistance to the Bt toxin, even if the corn produces it in its own Bt toxins. The problem of resistance still remains.

Another concern about GMOs is that biotechnology will not solve world hunger problems. In most parts of the world, hunger is an economic or political problem, or a problem of distribution, not one of production. When countries have a drought or a severe food shortage, the main issue becomes the distribution of donated food.

But GMOs can provide new varieties of crops that would be more drought-tolerant or new varieties that would produce more nutritious foods. Scientists in Egypt have developed a drought-resistant form of wheat by inserting a gene from barley. Experiments indicated that the new wheat strain outperformed the non-GMO variety under normal rainfall conditions. Currently, Egypt has to import wheat because it can't meet its need, but if the wheat range could expand, then perhaps Egypt could meet its demand.

In a very famous case of GMO technology, a research group in Switzerland worked to incorporate a gene from a daffodil into rice. The idea was that children in some areas where rice is a central crop were not getting sufficient vitamin A, which can cause blindness and other symptoms.

The GMO rice, called Golden Rice, incorporated a gene for the plant pigment beta-carotene into the rice, so that it now contained vitamin A. Beta-carotene is the orange pigment that gives carrots their color, and it's rich in vitamin A. The researchers developed this rice for purely humanitarian reasons and were going to give the rice away, but they've been stymied by GMO opponents who argue that a monocrop of rice is not healthy for sustainable agriculture and that vitamin A can be found in many different crops. However, in many areas where vitamin A deficiency is a problem, rice is already the main crop and may be the only food a person eats on a daily basis.

Because many of the objections to GMOs can, in principle, be met through further research into GMOs, the eventual promise of GMOs seems limitless. Plants can be genetically engineered to make vaccines. In 2007, Japanese scientists developed a new vaccine against cholera that could be administered by eating rice. These scientists genetically modified two rice strains to carry the CTB gene, which codes for a protein found in cholera. When people eat the GMO rice, they get exposed to a protein normally found in cholera. This exposure would then cause the people to develop antibodies against this protein, which would then improve their immunity to the disease. Interestingly, the researchers picked rice as the plant to modify because it mostly isn't digested in the stomach. That means that roughly three-quarters of the protein of interest can make it through the stomach, to be absorbed through the mucus lining of the small intestine.

In just one more example of the wonders of GMOs, Henry Daniell and his colleagues at the University of Central Florida, have genetically engineered plants to deliver insulin to patients with juvenile onset diabetes. Currently, patients with this sort of diabetes have to give themselves injections multiple times a day or they have to wear a pump that injects insulin under their skin. It's a constant roller coaster of measuring and correcting blood sugar. Dr. Daniell's method uses plants specifically because the genetically engineered lettuce plants produce insulin inside the cell walls. The cell walls are made up of cellulose, and they don't break down initially until good bacteria in the gut break them down. These bacteria then absorb the insulin and release it into the bloodstream.

Still, because of the contention with GMOs, some botanists wondered if there wasn't an alternative pathway to changing aspects of plants without actually changing their genes. In a 2016 report, scientists in Germany showed that just spraying RNA on leaves of barley actually reduced the amount of fungal infection of the sprayed leaves. Apparently, the RNA they sprayed on the barley leaves was able to silence, or turn off, some major genes in the fungal pathogen. Techniques like this demonstrate that there are different ways to give plants resistance to disease without modifying their genes.

And, there is an even newer and greater technique on the plant biotechnology horizon. Scientists have observed certain patterns in the genomes of some bacteria. The pattern is that some base pairs, which make up the DNA, would be repeated over and over again. In between these repeated segments would be normal areas of DNA. They called these patterns, clustered regularly interspaced short palindromic repeats or CRISPR for short.

Why the bacteria had these repeats was a mystery until scientists figured out that these repeats matched the DNA of viruses that prey on bacteria. CRISPR holds onto pieces of viral DNA so that the bacteria will recognize future attacks from viruses, much like our own immune system. The bacteria with CRISPR also have a group of enzymes called CRISPR-associated proteins, or CAS for short, that will cut the DNA in precise locations. Using these two tools, scientists can add or delete genes from genomes with unprecedented accuracy and speed.

The study of plants has now gone far beyond the early botanists like Linnaeus and Mendel. More universities have chosen to address these changes by setting up a Department of Plant Sciences, which is just a way of saying that more than one science is contributing.

Although there have been some reports that botany is a declining field, I think that's just an image problem. To me, botany is a beautiful word that incorporates all the study of plants, from microbes to mountain ecosystems, all with a strong grounding in knowledge of plants as whole organisms. I think botany has just morphed into a combination of molecular science, cell biology, and plant ecology.

I do think about the decline of the traditional field guide, but when I see all of the new incredible apps that are out there to identify plants, I'm not worried. I know the next generation of botanists will do great things.

As we wrap up our course, I also want you to consider an idea I've been thinking about for a few years now. I call this idea Natura Revelata. Though the concept of appreciating nature has been around a long time, Natura Revelata is the idea that nature can reveal things to us. These things could be physical or spiritual, emotional or intellectual. And revelata also suggests the word revel, which means to celebrate.

So, continue your explorations in botany, You might learn something more about conifers in the winter, wildflowers in the spring, deciduous trees in summer, and fruits in the fall. Take a field class or field hike to learn more about local plants—there's a plant identification walk in Denver I've taken three times because of those darn yellow composites—the DYC's. There's always more to learn.

And as we learn about the natural world, we might also learn more about ourselves, and, I hope, we also learn to celebrate the intricacies of nature. This celebration is, in my mind, the antidote to our consumer culture. A walk in nature, learning the names of plants, is uniquely calming and satisfying. I hope that the experience of botany all around you becomes a constant source of satisfaction.

So, give yourself some time to learn the names and stories of plants, or at least see them, and look a little more closely. There is a joy in botany, captured in the expression Natura Revelata, that will be yours.

BIBLIOGRAPHY

Attenborough, D., and N. Graham. *The Private Life of Plants: A Natural History of Plant Behaviour*. Princeton University Press, 1995.

Bernhardt, P. *Wily Violets and Underground Orchids: Revelations of a Botanist*. University of Chicago Press, 2003.

Blackburne-Maze, P. *Fruit: An Illustrated History*. Firefly Books, 2003.

Bock, Jane, et al. *Identifying Plant Food Cells in Gastric Contents for Use in Forensic Investigations: A Laboratory Manual*. U.S. Department of Justice, National Institute of Justice Research Report, January 1988.

> https://www.ncjrs.gov/pdffiles1/Digitization/110008NCJRS.pdf.

Bone, M. *Steppes: The Plants and Ecology of the World's Semi-Arid Regions*. Timber Press, 2015.

Buchmann, S. *The Reason for Flowers: Their History, Culture, Biology, and How They Change Our Lives*. Scribner, 2016.

Capon, B. *Botany for Gardeners: An Introduction and Guide*. Timber Press, 1990.

Chamovitz, D. *What a Plant Knows*. Scientific American/Farrar, Straus and Giroux, 2013.

Chase, T. D., and R. Llewellyn. *Seeing Seeds: A Journey into the World of Seedheads, Pods, and Fruit*. Timber Press, 2015.

Cooke, I. *Grasses and Bamboos: A Practical Guide*. Ball Publishing, 2006.

Coombes, A. J. *The Book of Leaves: A Leaf-by-Leaf Guide to Six Hundred of the World's Great Trees*. Edited by Zsolt Debreczy. University of Chicago Press, 2010.

Crane, P., and P. von Knorring. *Ginkgo: The Tree That Time Forgot*. Yale University Press, 2015.

Crawford, R. M. *Plants at the Margin: Ecological Limits and Climate Change*. Cambridge University Press, 2008.

Darke, R. *The Encyclopedia of Grasses for Livable Landscapes*. Timber Press, 2007.

Endersby, J. *A Guinea Pig's History of Biology*. Harvard University Press, 2008.

Essig, F. B. *Plant Life: A Brief History*. Oxford University Press, 2015.

Farjon, A. *A Natural History of Conifers*. Timber Press, 2008.

Farmer, E. E. *Leaf Defence*. Oxford University Press, 2014.

Fenner, M. *Seeds: The Ecology and Regeneration in Plant Communities*. CABI Publishing, 2000.

Fenner, M., and K. Thompson. *The Ecology of Seeds*. Cambridge University Press, 2005.

Gibson, D. J. *Grasses and Grassland Ecology*. Oxford University Press, 2009.

Gilbert, E. *The Signature of All Things: A Novel*. Penguin, 2013.

Glover, B. *Understanding Flowers and Flowering: An Integrated Approach*. Oxford University Press, 2007.

Halle, F. *In Praise of Plants*. Timber Press, 2002.

Hanson, T. *The Triumph of Seeds: How Grains, Nuts, Kernels, Pulses, and Pips Conquered the Plant Kingdom and Shaped Human History*. Basic Books, 2016.

Harder, L. D., and S. C. Barrett. *Ecology and Evolution of Flowers*. Oxford University Press, 2006.

Harkness, P. *The Rose: An Illustrated History*. Firefly Books, 2003.

Hodge, G. *Practical Botany for Gardeners*. University of Chicago Press, 2013.

Hogarth, P. J. *The Biology of Mangroves and Seagrasses*. Oxford University Press, 2007.

Howell, C. E. *Flora Mirabilis: How Plants Have Shaped World Knowledge, Health, Wealth, and Beauty*. National Geographic, 2009.

Karban, R. *Plant Sensing and Communication*. University of Chicago Press, 2015.

Kesseler, R., and W. Stuppy. *Seeds: Time Capsules of Life*. Firefly Books, 2006.

Kimmerer, R. W. *Gathering Moss*. Oregon State University Press, 2003.

King, J. *Reaching for the Sun: How Plants Work*. Cambridge University Press, 2011.

Körner, C. *Alpine Plant Life: Functional Plant Ecology of High Mountain Ecosystems*. Springer Press, 2003.

Kourik, R. *Understanding Roots: Discover How to Make Your Garden Flourish*. Metamorphic Press, 2015.

Large, M. F., and J. E. Braggins. *Tree Ferns*. Timber Press, 2004.

Largo, M. *The Big, Bad Book of Botany: The World's Most Fascinating Flora*. William Morrow Paperbacks, 2014.

Laws, B. *Fifty Plants That Changed the Course of History*. Firefly Books, 2011.

Lee, D. *Nature's Palette: The Science of Plant Color*. University of Chicago Press, 2007.

Mabey, R. *Cabaret of Plants*. W. W. Norton Press, 2015.

MacAdam, J. W. *Structure and Function of Plants*. Wiley-Blackwell, 2009.

McKell, C. M. *The Biology and Utilization of Shrubs*. Academic Press, 1988.

Mehltreter, K., L. R. Walker, and J. M. Sharpe, eds. *Fern Ecology*. Cambridge University Press, 2010.

Moore, M. *Medicinal Plants of the Mountain West*. Museum of New Mexico Press, 1979.

Moran, R. C. *A Natural History of Ferns*. Timber Press, 2004.

Nobel, P. S. *Environmental Biology of Agaves and Cacti*. Cambridge University Press, 2003.

Richardson, D. M. *Ecology and Biogeography of* Pinus. Cambridge University Press, 1998.

Royte, E. *The Tapir's Morning Bath*. Houghton Mifflin Company, 2001.

Rundel, P. W., A. P. Smith, and F. C. Meinzer. *Tropical Alpine Environments: Plant Form and Function*. Cambridge University Press, 1994.

Slater, A., N. W. Scott, and M. R. Fowler. *Plant Biotechnology*. Oxford University Press, 2003.

Speer, J. H. *Fundamentals of Tree-Ring Research*. University of Arizona Press, 2010.

Stewart, A. *Wicked Plants: The Weed That Killed Lincoln's Mother and Other Botanical Atrocities*. Algonquin Books, 2009.

Stuppy, W. *Fruit: Edible, Inedible, Incredible*. Earth Aware Editions, 2013.

Thomas, P. A. *Trees: Their Natural History*. Cambridge University Press, 2014.

Trewavas, A. *Plant Behaviour and Intelligence*. Oxford University Press, 2014.

Trimble, S. *The Sagebrush Ocean: A Natural History of the Great Basin.* University of Nevada Press, 1999.

Tudge, C. *The Tree: A Natural History of What Trees Are, How They Live, and Why They Matter.* Crown Publishers, 2007.

Vogel, S. *The Life of a Leaf.* University of Chicago Press, 2012.

Whitelock, L. M. *The Cycads.* Timber Press, 2002.

Wilhelm, B., S. Porembski, R. Seine, and I. Theisen. *The Curious World of Carnivorous Plants: A Comprehensive Guide to Their Biology and Cultivation.* Timber Press, 2007.

Wood, C., and N. Habgood. *Why People Need Plants.* Kew Publishing; Open University, 2010.

Zwinger, A. H. and B. E. Willard. *Land above the Trees: A Guide to American Alpine Tundra.* Johnson Books, 1996.

IMAGE CREDITS

title page	© Ohmega1982/iStock/Thinkstock.
lecture titles	© andreakaulitzki/iStock/Thinkstock.
5	© dhobern/flickr/CC BY 2.0.
7	© Jannoon028/iStock/Thinkstock.
8	© USDA.
25	© Jirawat Jerdjamrat/iStock/Thinkstock.
26	The Teaching Company Collection.
28	The Teaching Company Collection.
43	© PFMphotostock/iStock/Thinkstock; © pellaea/flickr/CC BY 2.0.
44	© treegrow/flickr/CC BY 2.0.
45	© jacilluch/flickr/CC BY-SA 2.0.
46	© Martin Cooper/flickr/CC BY 2.0.
48	© jph9362/iStock/Thinkstock.
66	© pichaitun/iStock/Thinkstock.
67	© Holcy/iStock/Thinkstock.
68	© Ian_Redding/iStock/Thinkstock.
83	© GaryKavanagh/iStock/Thinkstock.
85	© Dorling Kindersley/Thinkstock.
86	The Teaching Company Collection.
87	© Gerald Holmes, California Polytechnic State University at San Luis Obispo, Bugwood.org/CC BY-NC 3.0.
89	© tonrulkens/flickr/CC BY-SA 2.0.
103	The Teaching Company Collection.
104	© Starr Environmental/flickr/CC BY 2.0.
106	The Teaching Company Collection.
107	© pawpaw67/flickr/CC BY-SA 2.0.
108	© jhfearless/flickr/CC BY 2.0; The Teaching Company Collection.
123	© Starr Environmental/flickr/CC BY 2.0.

Page	Credit
124	© Alexander Rozhenyuk/Hemera/Thinkstock; © Elena Elisseeva/Hemera/Thinkstock; © Gratysanna/iStock/Thinkstock.
125	© BSANI/iStock/Thinkstock; © Rob Routledge, Sault College, Bugwood.org/CC BY 3.0; © born1945/flickr/CC BY 2.0.
126	© Bruce Ackley, The Ohio State University, Bugwood.org/CC BY 3.0.
142	The Teaching Company Collection.
143	The Teaching Company Collection.
146	© Designua/Shutterstock.
162	The Teaching Company Collection; © Nadezhda_Nesterova/iStock/Thinkstock.
164	© inra.dist/flickr/CC BY 2.0.
181	© MWCPhoto/iStock/Thinkstock.
183	© v_apl/iStock/Thinkstock.
185	© t_y_l/flickr/CC BY-SA 2.0; © Psumuseum/Wikimedia Commons/CC BY-SA 3.0; © alexlomas/flickr/CC BY 2.0.
200	© Dave Powell, USDA Forest Service (retired), Bugwood.org/CC BY 3.0.
202	© Arturo Reina Sánchez/Wikimedia Commons/CC BY-SA 3.0.
203	© Paul Wray, Iowa State University, Bugwood.org/CC BY 3.0; © Mark-Poley/iStock/Thinkstock; © Sten Porse//CC BY-SA 3.0.
204	© Bill Cook, Michigan State University, Bugwood.org/CC BY 3.0.
205	© tamara_kulikova/iStock/Thinkstock; © picture_istock/iStock/Thinkstock.
221	© Scott Zona/Wikimedia Commons/CC BY 2.0.
222	The Teaching Company Collection.
226	© AlessandroZocc/iStock/Thinkstock; © John Tann/flickr/CC BY-SA 2.0; © Björn S.../flickr/CC BY-SA 2.0; © John D. Byrd, Mississippi State University, Bugwood.org/CC BY 3.0; © Jan Samanek, Phytosanitary Administration, Bugwood.org/CC BY 3.0.

Page	Credits
227	© Dave Powell, USDA Forest Service (retired), Bugwood.org/CC BY 3.0; © emmor/iStock/Thinkstock; © jacilluch/flickr/CC BY-SA 2.0; © Daniel J. Layton/Wikimedia Commons/CC BY-SA 4.0; © ArturoYee/flickr/CC BY-SA 2.0.
228	© John D. Byrd, Mississippi State University, Bugwood.org/CC BY 3.0; © Deanster1983 who's mostly off for a while/flickr/CC BY-SA 2.0; © Aries Tottle/flickr/CC BY-SA 2.0; © Muffet/flickr/CC BY 2.0; © Aries Tottle/flickr/CC BY-SA 2.0; © RPFerreira/iStock/Thinkstock.
229	© asavliuk/iStock/Thinkstock; © wplynn/flickr/CC BY-ND 2.0; © siur/iStock/Thinkstock; © DeaPeaJay/flickr/CC BY-SA 2.0; © Michael Wunderli/flickr/CC BY-SA 2.0.
246	© AJC1/flickr/CC BY-SA 2.0.
247	© mb-fotos/iStock/Thinkstock.
248	© DanielaC173/flickr/CC BY-SA 2.0.
263	© johan63/iStock/Thinkstock.
264	The Teaching Company Collection.
265	© Jonas Janner Hamann, Universidade Federal de Santa Maria (UFSM), Bugwood.org/CC BY 3.0.
266	© Starr Environmental/flickr/CC BY 2.0.
267	© Amawasri/iStock/Thinkstock.
282	© TANAKA Juuyoh (田中十洋)/flickr/CC BY 2.0.
283	© HildaWeges/iStock/Thinkstock.
284	© standret/iStock/Thinkstock.
285	© Purestock/Thinkstock.
300	© richcarey/iStock/Thinkstock.
304	© hakoar/iStock/Thinkstock.
319	© odmeyer/iStock/Thinkstock.
322	© cassinga/iStock/Thinkstock.
323	© Starr Environmental/flickr/CC BY 2.0.
328	© Starr Environmental/flickr/CC BY 2.0.
339	© pjmalsbury/iStock/Thinkstock.

Page	Credit
340	The Teaching Company Collection.
341	© Lomvi2/Wikimedia Commons/CC BY-SA 3.0.
344	© Forest and Kim Starr, Starr Environmental, Bugwood.org/CC BY 3.0; © Dun.can/flickr/CC BY 2.0.
358	© RobIre/iStock/Thinkstock.
360	© Joanna Wnuk/Hemera/Thinkstock.
362	© andrey_zharkikh/flickr/CC BY 2.0.
363	© Morn the Gorn/Wikimedia Commons/CC BY-SA 3.0.
378	© joedecruyenaere/flickr/CC BY-SA 2.0.
379	© Patrick_Lienin/iStock/Thinkstock.
380	© Tomas Castelazo/Wikimedia Commons/CC BY 3.0.
396	© Bill Cook, Michigan State University, Bugwood.org/CC BY 3.0.
397	© Maria_Ermolova/iStock/Thinkstock.
399	© Franco Folini/flickr/CC BY-SA 2.0.
400	© Lacy Mayberry/iStock/Thinkstock.
401	© magicflute002/iStock/Thinkstock.
415	© slava296/iStock/Thinkstock.
416	© AlexTerrill/iStock/Thinkstock.
418	© adrianciurea69/iStock/Thinkstock.
419	© billmiky/flickr/CC BY-ND 2.0; © randihausken/flickr/CC BY-SA 2.0.
434	© Dick Culbert/flickr/CC BY 2.0.
435	© Forest and Kim Starr, Starr Environmental, Bugwood.org/CC BY 3.0; © Eurico Zimbres/Wikimedia Commons/Public Domain.
436	© Matt Lavin/flickr/CC BY-SA 2.0.
437	© Kerry Britton, USDA Forest Service, Bugwood.org/CC BY 3.0.
438	© Pthltl/iStock/Thinkstock.
439	© monica-photo/iStock/Thinkstock.
440	© Koldunov/iStock/Thinkstock.
455	© jcarillet/iStock/Thinkstock.

NOTES